W9-AVC-535

Simulation — Gaming in the Late 1980s

Proceedings of the International Simulation and Gaming Association's 17th International Conference

Other Pergamon Titles of Related Interest

COLMAN
Game Theory and Experimental Games
The Study of Strategic Interaction

CROOKALL
Simulation Applications in L2 Education and Research
(System, Vol. 13, No. 3)

MOONEN & PLOMP
Eurit 86
Development of Educational Software & Courseware

SENDOV & STANCHEV
Children in an Information Age: Tomorrow's Problems Today

STAHL
Operational Gaming: An International Approach

ZETTERSTEN
New Technologies in Language Learning

Related Pergamon Journals

Computers & Education
Computers & Operational Research
International Journal of Educational Development
Journal of the Operational Research Society
Language & Communication
Long Range Planning
System

Simulation — Gaming in the Late 1980s

Proceedings of the International Simulation and Gaming Association's 17th International Conference

Université de Toulon et du Var, France, 1–4 July 1987

Edited by

DAVID CROOKALL
Université de Toulon et du Var, France,

CATHY STEIN GREENBLAT
Rutgers University, USA

ALAN COOTE
Polytechnic of Wales

JAN H G KLABBERS
State University of Utrecht, The Netherlands

D R WATSON
Manchester University, UK

PERGAMON PRESS

OXFORD · NEW YORK · BEIJING · FRANKFURT
SÃO PAULO · SYDNEY · TOKYO · TORONTO

U.K. — Pergamon Press, Headington Hill Hall, Oxford OX3 0BW, England
U.S.A. — Pergamon Press, Maxwell House, Fairview Park, Elmsford, New York 10523, U.S.A.
PEOPLE'S REPUBLIC OF CHINA — Pergamon Press, Room 4037, Qianmen Hotel, Beijing, People's Republic of China
FEDERAL REPUBLIC OF GERMANY — Pergamon Press, Hammerweg 6, D-6242 Kronberg, Federal Republic of Germany
BRAZIL — Pergamon Editora, Rua Eça de Queiros, 346, CEP 04011, Paraiso, São Paulo, Brazil
AUSTRALIA — Pergamon Press Australia, P.O. Box 544, Potts Point, N.S.W. 2011, Australia
JAPAN — Pergamon Press, 8th Floor, Matsuoka Central Building, 1-7-1 Nishishinjuku, Shinjuku-ku, Tokyo 160, Japan
CANADA — Pergamon Press Canada, Suite No. 271, 253 College Street, Toronto, Ontario, Canada M5T 1R5

First edition 1987

Library of Congress Cataloging in Publication Data

International Simulation and Gaming Association.
Conference (17th: 1986: Toulon, France)
Simulation-gaming in the late 1980s.
1. Game theory—Congresses. 2. Simulation methods
Congresses. I. Crookall, David. II. Title.
QA269.I558 1986 519.3 87–11304

British Library Cataloguing in Publication Data

International Simulation and Gaming Association. *International Conference (17th : 1986 : University of Toulon)*
Simulation-gaming in the late 1980's : Proceedings of the International Simulation and Gaming Association's 17th International Conference.
1. Game theory 2. Simulation methods
I. Title II. Crookall, David
001.4'24 QA269
ISBN 0–08–034158–6

Acknowledgement
"Icons illustration on the front cover appears by permission of the National Council on Foreign Language and International Studies, USA"

Printed in Great Britain by Hazell Watson & Viney Limited, Member of the BPCC Group, Aylesbury, Bucks

to the memory of
AYMERIC BAILLEUX

Nous courons à toutes jambes vers l'avenir et nous allons si vite que le présent nous échappe, et la poussière de notre course nous dissimule le passé.

BORIS VIAN

L'homme ne pouvant être appréhendé que situé, le peuple, sans même invoquer sa définition, va désirer se voir reconnaitre des droits concrets, opérationels.

AYMERIC BAILLEUX

Contents

Section One Keynote Papers

Editor: David Crookall

Section Two Language and Communication

Editors: D. R. Watson & David Crookall

Section Four Management and Business

Editor: Alan Coote

Section Five Taxonomies and Methodologies

Editor: Jan H. G. Klabbers

Acknowledgements

We wish to thank a large number of people and organizations involved with the Conference. Our first thanks go to those who delivered keynote addresses, as well as to the members of the Honorary and Organizing Committees.

Keynote Speakers

Bob Anderson	Manchester Polytechnic, England.
Richard D Duke	University of Michigan, USA.
Cathy S Greenblat	Rutgers & Princeton Universities, USA.
D R Watson	Manchester University, England.
Denis Meadows	Dartmouth College, Hanover, USA.

Honorary Committee

Pier Ayala	UNESCO, France.
Klaas Bruin	Teacher Training Institute Ubbo Emmius, The Netherlands; Editor *ISAGA Newsletter*.
Arnaldo Cecchini	Instituto Universitario di Architettura di Venezia, Italy; ISAGA President-elect.
Alan Coote	Polytechnic of Wales; Editor *Simulation/Games for Learning*.
Adriana Frisenna	Instituto Universitario di Architettura di Venezia, Italy.
Cathy S Greenblat	Rutgers & Princeton Universities, USA; Editor *Simulation & Games*.
Jan H G Klabbers	State University of Utrecht, Netherlands; ISAGA Secretary General.
Bob Hart	Advisory Unit for Computer Based Education, England.
Hubert Law-Yone	Technion — Israel Institute of Technology.
Jack Lonergan	Language Training London, England.
Allan Martin	University of Leeds, England.
Jacob Ngwa	Regional Pan-African Institute for Development, Cameroon.
Eric Petterson	School of Architecture in Aarhus, Denmark.
Israel Porat	Hebrew University of Jerusalem, Israel.

WALTER E ROHN	Deutsche Planspiel-Zentrale, Germany.
CHRISTOPHER SAWYER-LAUÇANNO	Massachussetts Institute of Technology, USA.
LEOPOLDO SCHAPIRA	Universidad Nacional de Cordoba, Argentina; ISAGA Latin America Secretary.
ANDREW SCOTT	University of North Carolina, USA.
RICHARD SWITALSKI	University of Warsaw, Poland.
ROD WATSON	Manchester University, England.

Organizing Committee (Université de Toulon, unless otherwise stated)

ANDRÉ BERTRAND	UFR de Sciences Economiques.
DANIELLE DUMAS	Dépt de Techniques de Commercialisation, IUT.
PHILIPPE DUMAS	Dépt de Techniques de Commercialisation, IUT.
VÉRONIQUE FOURESTIER	Promovar, Toulon.
LESLIE GREENBLAT	Princeton, USA.
ALISON LE GAT	Laboratorie de Simulation, Faculté de Droit.
MARTINE HARDY	Dépt de Techniques de Commercialisation, IUT.
AMALIA METZGER	Faculté de Droit.
MICHEL PAILLET	Institut des Collectivités Territoriales, Faculté de Droit.
FRANÇOISE PESSEL	Service Universitaire d'Information.
DOMINIQUE TESSIER	Université d'Eté.

Conference Secretary

ANNE-MICHELLE LAMOTTE	Laboratoire de Simulation, Faculté de Droit.

Programme Chair

DAVID CROOKALL	Laboratoire de Simulation, Faculté de Droit; ISAGA President 1986–87.

Support

We would like to express our grateful thanks to the following organizations for their very generous support of ISAGA' 86:

* The British Council, Paris & London, France & England.
* Le Conseil Régional PACA (Provence, Alpes, Côte d'Azur), Marseille, France.

* La Direction de la Coopération et des Relations Internationales du Ministère de l'Education Nationale de la République Française.
* La Direction de la Coopération Scientifique et Technique du Ministère des Affaires Etrangères de la République Française.
* La Faculté de Droit de l'Université de Toulon et du Var, France.
* ICONS, University of Maryland, College Park, USA.
* La Mairie de la Ville de Toulon, France.
* The UNESCO Man and Biosphere Programme, Paris, France.
* L'Université de Toulon et du Var, France.
* USIS — United States Information Service, Paris, France.

Special Thanks

Very special thanks must be given to a number of people who, though not on the Organizing Committee, did much behind-the-scenes work at the Université de Toulon et du Var that was essential to the Conference. These include:

Daniel Bouillet	Faculté de Droit.
Crystel Delaul	Université d'Eté.
Max Eutizi	Faculté de Droit.
Gérard Février	Président de l'Université.
Hélène Le Rouedec	Faculté de Droit.
Danielle Poisson	Université.
Bernard Rieupouilh	Université.
Jean-Pierre Servel	Doyen de la Faculté de Droit.
René Spénato	IUT.

A number of other people also did much to provide support and encouragement at a distance, and our gratitude should be recorded here.

Pier Ayala	UNESCO, Paris, France.
Henri Bartholin	AS Var 2000, Université de Toulon et du Var.
Max Bausset	Université de Toulon et du Var
Robert C Noel	University of California, Santa Barbara, USA.
Hamish Norbrook	BBC World Service, London, UK.
M A Frankel	British Council, Paris, France.
S K Braun	British Council, London, UK.
Betty Taska	USIS, Paris, France.
Jean-Louis Vernet	IUT, Université de Toulon et du Var.
Jonathan Wilkenfeld	University of Maryland, College Park, USA.

Participants must be thanked for making the conference a memorable event; so must session leaders, as well as contributors to these Proceedings. My affectionate gratitude to five close friends should also be recorded, to

Alan Coote, Cathy Stein Greenblat, Jan Klabbers, and *Rod Watson*, for their invaluable help in co-editing these Proceedings, and to *Danny Saunders* (Polytechnic of Wales) for his help. *Barbara Barrett* and *Michèle Norton* of Pergamon Press provided valuable editorial guidance and have been a real pleasure to work with. *Rebecca Oxford* (Center for Applied Linguistics, USA) gave much affectionate encouragement and helped to see the final stages through.

My warm thanks are expressed to the members of the Organizing Committee for all their support and hard work, and to the University Summer School for taking care of the logistics. The final and warmest thanks must go to *Anne-Michelle Lamotte* for her efficiency and charm. She did most of the secretarial work, but was also closely involved with the overall and detailed organization. Without Anne-Mi the Conference would not have been possible. No doubt I will have forgotten to give thanks to some people, and ask them to forgive this oversight.

ISAGA

The International Simulation And Gaming Association was founded in 1970. Its major objective is to further the use and development of simulation and gaming throughout the world. It publishes a quarterly Newsletter, and its official journal is *Simulation & Games: An International Journal of Theory, Design, and Research* (Sage Publications). Membership applications should be sent to Prof. Dr. Jan H. G. Klabbers, Faculty of Sciences, PO Box 80140, 3508 TC UTRECHT, The Netherlands.

***Aymeric Bailleux** (4th April, 1947–9th October, 1986)*

Finally, the dedication to *Aymeric Bailleux*, who died at the age of 39 shortly after the Conference. Aymeric was a prominent faculty member and was greatly appreciated by students and staff alike, for whom he provided much needed inspiration; he did much to make simulation a legitimate activity at the University. His intellectual vision and inimitable wit debunked humbug and cut through hypocrisy, much to the benefit of all who had the privilege of knowing him well. His sudden premature demise is a tragic and irrevocable loss to all, not least to his wife, Dominique, and his son, Thibault. I am proud to dedicate these Proceedings to his memory.

Le Pradet, France — DAVID CROOKALL
January, 1987

Gaming context, communication, reality and future: An introduction

David Crookall
University of Toulon, France

Rebecca Oxford
Center for Applied Linguistics, USA

This volume records the proceedings of the 17th International Conference of the International Simulation And Gaming Association, held at the Université de Toulon et du Var, France, from 1st to 4th July, 1986. The contributions included here present an extremely diverse set of perspectives, accounts and analyses, but together they can be conceived as providing a glimpse of the state of the art and craft of simulation/gaming in the latter part of the 1980's.

The volume contains five main sections, each of which focuses on a cluster of topics and issues. The first section contains the keynote presentations, offering a variety of perspectives on broad areas of interest. This is followed by four sections containing accounts of, or papers given at, sessions. Section Two deals mainly with language and computers. Section Three covers social issues, research, and game design. Section Four concentrates on business and management, while the last section discusses methodologies and taxonomies.

Computers, being essentially content-free, are treated in all sections. Although Section Two contains a higher proportion of papers dealing with computers, this was not by design, their inclusion here having been determined by substance. This ubiquity of computers is indicative of their growing use in a simulation, and it is becoming increasingly difficult to publish a substantively organized collection, such as this one, which contains a section exclusively devoted to computerized simulation. Indeed, it can easily be argued that computers constitute just one in a whole array of simulation support paraphernalia, along with such items as boards, dice, and forms.

Sections Two to Five are headed by their own introductions, so we shall concern ourselves here with a few brief remarks on the keynote papers. We have four very different keynotes; so different, in fact, that trying to draw a common thread is a risky move. On a general level, however, each of these papers deals substantially and subtantively with four closely interrelated and often recurring themes: <u>context</u>, <u>communication</u>, <u>reality</u>, and <u>future developments</u>. Context and communication form a nexus, while they, in their turn, imply and are implied in a concern with realities and future developments. Context can only be understood in its communicational dimension, and communication necessarily becomes meaningful only within a context. Definitions of reality, both reality within games and the reality of games within society, are formulated and communicated contextually, i.e., socially. Context, communication and reality are mutually and actively productive of each other. Furthermore, they are not merely spatial constructs, but temporal as well; they imply not only a past, but also a present and an awareness of the future.

The first theme, context, can be characterized as both the context of simulation/ gaming and the game context, although the precise nature of these contexts varies considerably, from historical, through intellectual, to institutional. The contextual orientation of the paper by Bob Noel et al. is essentially historical, international and intellectual. The paper traces the intellectual development of today's international network gaming, mainly in the form of ICONS (which was a major feature of ISAGA'86). The successive research and technological contexts of network gaming have shaped today's methodologies. On another level, ICONS itself provides a rich context for intercultural communication, and the impact of ICONS on one specific context, that of Latin America, is mentioned. The paper by Cathy Stein Greenblat is concerned more with the institutional (educational and research), as well as the intellectual, context of gaming. Cathy emphatically points out that gamers cannot continue to carry out their activities in a social or institutional vacuum. Simulation/gaming must be part and parcel of the educational context, both enriching it and being recognized by it. Wes Sharrock and Rod Watson remain within a similar purview, that of the educational context, but rather than dealing with this in relation to gamers or their colleagues, they choose to look at the contextual features of game playing. Wes and Rod stress that we should be careful in assuming that merely because we use simulations, we ipso facto solve all these contextual, organizational problems of teacher control, learning responsibility, authority structures, and the like. The context which Bob Anderson talks about in his discussant keynote paper is the intellectual tools or the conceptual apparatus with which gamers avail themselves in their gaming work, whether this be an analytic pursuit or a practical concern. By questioning certain central presuppositions cherished by many gamers, this paper deals with the vital issue of whether the context from which a simulation is drawn can be conceived as some underlying model of some 'external' reality, or whether a simulation should be seen as forming its own context in its own right.

Closely related to context is our second theme of communication, for they presuppose one another. Bob Noel et al. deal with an essential characteristic of network gaming, this feature by definition being communication. Here one of the main communicational threads relates to international telecomputing, but this merely but vitally constitutes a medium for communication at a higher level, that between people across national boundaries. Cathy Greenblat, in her discussion on game design, highlights three communicational problems that gamers need to address in the context of their work, namely: communication among gamers themselves (and ISAGA'86 was an ideal forum for such exchange); communication about individual games, e.g., in game manuals; and, vitally, communicating to non-gamers what it is that gamers are up to, what is involved in designing games, and why games are much more than mere pedagogical technique. Cathy implies that gamers have responsibilities to their discipline, and they include rendering our work public and intelligible by communicating about this as clearly and as explicitly as possible with colleagues, both gamers and others (especially those in positions of power, and therefore of decision-making). Cathy deals mainly with communicating outwards, towards others, while Wes Sharrock and Rod Watson deal with communication patterns within simulation itself, among players, and how these constitute, and are constituted by, participants' definitions of game reality and their communicative competences. This paper also provides a healthy critical appraisal of a number of unexamined assumptions held by gamers, and suggests how we might re-orient our thinking. Finding a direct communication thread in Bob Anderson's paper is a little more difficult, but we may characterize it has a kind of internal and intellectual communication, that in which gamers need to think about the reality problem (and think it through for themselves) and to talk among themselves in open debate about this and related issues.

A third thread running through the keynote papers may be termed the "reality problem"; that is, in a nutshell, what is the relationship, if any, between a game and what we term (external, i.e., non-game) "reality". ICONS, described by Bob Noel <u>et al</u>., is a prime example of how a simulation may parallel another reality, for participants in this global studies simulation are drawn from a number of countries around the world, and the issues and complexities tackled are precisely those characteristic of today's international political realities. Questions revolving around real as opposed to hypothesized international crises and scenarios are explored in depth throughout the simulation. Cathy Greenblat then provides us with a most perspicuous outline of the design process, that is how reality must be studied, conceptualized and translated into a working model. Her examples of BLOOD MONEY and CAPJEFOS vividly illustrate the huge amount of detailed research and subsequent fine-tuning of alternative models that must be accomplished in order to capture the essentiality and relevancy of that which the simulation purports to model. One of the central themes of the two papers by Wes Sharrock and Rod Watson and by Bob Anderson is that of reality; both papers wish to challenge our current thinking on this matter. They open wide a debate which heretofore has only been glimpsed -- this is, how much can a game or simulation actually mirror 'external' reality, i.e., represent some pre-defined system, at least in the eyes of participants <u>qua</u> players? (a question also raised by Cathy Greenblat and Bob Noel <u>et al</u>.). Wes and Rod emphasize the need to take account just as much of what simulation reality is for participants as of how it is conceived by game designers and organizers. Simulation (whether for teaching or for research purposes) is ultimately built, in most cases, to accommodate participants, and we shall therefore neglect their definitions of game reality at our and, crucially, their peril. Bob's central concern is precisely the reality problem, and he outlines a number of presuppositions in the use of models, and suggests why we should maintain a healthy sceptisism of our taken-for-granted conception of modelling. Bob offers little in the way of a clearly-defined, definitive, unquestioned and unquestionable set of alternative options, for so doing would invalidate those very arguments which he wields to question our original assumptions.

The final theme again raises more questions than it offers answers about the <u>future development</u> of simulation/gaming: What are the prospects? Where do we go from here? What guiding principles should we adopt? Bob Noel <u>et al</u>. paint a historical portrait of network gaming developments over the last fifteen years, and this in itself gives us an idea of where things might go. The general impression is that, although we have come a long way since the pioneering days of Bob Noel and Gerald Shure, there is still much room for further considerable development, especially in the areas of multicultural communication and of the "pluralization of the social and behavioural sciences". Another, related, aspect touched on here is the use of network gaming to explore possible future (i.e., hypothetical) situations. Cathy Greenblat's 'vision' of the future of gaming is also grounded in her long experience of the whole area over the years, and this leads to her very legitimate concern with the necessity of raising the general level of awareness in colleagues and, indeed, in society at large, of simulation/gaming. All disciplines, methodologies and perspectives vie for social, professional and institutional prestige, and so far simulation/gaming has been singularly unsuccessful in achieving the recognition it rightly deserves. Cathy emphasizes that, if simulation/gaming is to develop in the near and long-term future, a vital aspect of our gaming work should be to deploy these methods which other areas use to legitimize their activity. One of those methods is effective communication (which brings us back to previous themes and highlights their interdependence). Cathy's concern will strike a chord in every committed gamer, and her paper is a compelling blueprint for that promotional programme. Wes Sharrock and Rod Watson also touch on the need to communicate with the 'outside' world, in this case to avoid "the fallacy of group soliloquy". In recalling the historical context of education, Wes and Rod also tackle a very different, but related, series of questions about the future. A relatively new analytic, non-positivist, paradigm is outlined, particularly as applied to the simulation/gaming field; a quantum shift is needed in our perspective on games and simulations. This concern is taken up again by Bob Anderson. By questioning a number of central assumptions held unawares by the gaming community, Bob offers a number of directions which we might explore in our future thinking about simulation/gaming. This shift and these new directions, coupled with Cathy's blueprint, may prove to be a powerful way of helping to shape the future of simulation/gaming, and of rendering it a (more?) respectable and respected activity.

Taken together, the four keynote papers offer a provocative and intriguing set of ideas, and bring a variety of perspectives on a host of questions into contradistinction. One major question for the future is: Are these varied perspectives and analytic paradigms incompatible or irreconcilable, each leading to views and even developments which cannot be contemplated together, except during moments of creative leaps? We think not; we believe that the differential perspectives and approaches nicely complement each other and are for the reader essential to a full understanding of the wealth offered and sought by the simulation/gaming endeavour. Certainly, these keynotes will serve to provide rich insights into contextual, communicational, and reality aspects of our simulation/gaming activity, and thereby provide a broader guiding outlook for our formulation of future development programmes. These papers may well turn out to be a watershed.

Network gaming: A vehicle for intercultural communication

Robert C. Noel
University of California, Santa Barbara, USA

David Crookall
University of Toulon, France

Jonathan Wilkenfeld
University of Maryland, USA

Leopoldo Schapira
Universidad Nacional de Cordoba, Argentina

ABSTRACT: Intercultural and cross-national communication by means of computers and international data packet switching networks combine with simulation and foreign languages to form a particularly powerful educational and research methodology. The intellectual and historical context of this methodology is outlined, and recent developments discussed. A scenario (the MUF Crisis), complete with additional organiser's notes and participants' messages, is given in full. This was used during the ICONS workshop at ISAGA'86, and illustrates the intricacies of political gaming. A number of comments of the impact of this type of methodology in Latin America end the paper.

KEYWORDS: network gaming, computerized simulation, telecomputing, global studies, international communication, intercultural communication, political gaming, crisis gaming, simulation, scenarios, foreign language learning, translation, Latin America.

ADDRESSES: RCN: Dept of Political Science, University of California, Santa Barbara, California, USA; DC: Laboratoire de Simulation, University of Toulon, Av de l'Universite, 83130 LA GARDE, France; JW: Dept of Government & Politics, University of Maryland, College Park, Maryland, USA; LS: Estrada 10, 5000 Cordoba, Argentina.

INTRODUCTION

In keeping with the theme of this symposium, this paper is about intercultural communications, simulation and the use of information age technologies. It deals with a research program that has combined these three elements into a new methodology. We call it international network gaming. This research program began under the name "POLIS" at the University of California, Santa Barbara (Noel, 1971; 1979). The main focus of the program is now at the University of Maryland, where it is known as ICONS-"International Communications and Negotiations Simulations" (Wilkenfeld and Brecht, 1985). The program has recently been internationalized with the establishment of a center at the University of Toulon, France (Crookall and Wilkenfeld, 1985) and with the active collaboration of scholars at Waseda University, Central Catholic University of Cordoba, and Hebrew University.

This research program is international in substance, as well as organization. POLIS and ICONS exercises have always dealt with topics in international relations. The two principal investigators are international relations scholars, as have been most of the collaborators. Foreign languages have themselves become substantive focii of the program through the contributions of two major collaborators, both of whom are language scholars (Brecht and Crookall).

In the sections that follow, the POLIS/ICONS research program is located in the context of the literature on simulation in the social sciences. Then some of the main developments in the program are noted. There is also a brief illustration of a network crisis game. The paper concludes with a discussion of some of the issues raised at that conference.

INTELLECTUAL CONTEXT OF THE POLIS/ICONS RESEARCH PROGRAM

It is difficult to imagine a more polyglot area of kindred scholarly activities than simulation. It reflects a coming together of a rather amazing variety of intellectual traditions: operations researchers, diplomatic analysts, economists, military analysts, international relations scholars, small-groups and organizational sociologists, experimental social psychologists, mathematicians, computer scientists, educational psychologists, and still others. That there has been mutual enrichment is as undeniable as is the resulting terminological confusion.

One source of the problem lies with the respective roles of computers, on the one hand, and people, on the other. For many, the word "simulation" is used exclusively in connection with computer models. For others, it is construed quite broadly to encompass almost any laboratory representation of social phenomena.

"Political gaming" is not a term one usually associated with computers. It originated with a group of analysts at the RAND Corporation and at MIT who had practical concerns in the area of security policy (Goldhammer and Spier, 1959; Bloomfield and Padelford, 1959). Their purposes were analytic. They were more interested in concrete situations, actual or hypothetical, than in theoretical problems. Their work was essentially idiographic, not nomothetic. Whereas other scholars worked with abstract models, the political gamers attempted to capture more of the unique complexity and subtlety of foreign policy and strategic issues through the use of elaborate "scenarios" (Bloomfield and Whaley, 1965; Helmer, 1960). The scenarios focused on real or plausible situations, some of them set in contemporaneous time, some of them projected into the future. Country teams were often staffed by foreign policy experts and area specialists; and sometimes even high level officials of the American government took part in the games. Their decision alternatives within a game were constrained, not by rigid rules, but by judgements made by umpires as to their plausibility. The umpires also generated responses from the environment and from country teams that were not represented in a game by actual people.

Early political gaming scenarios tended to concentrate on international crises -- thus the term "crisis gaming". Exercises were of varying length; but the periods of real time they represented were seldom more than a few weeks. Variations on the RAND/MIT theme came to be used rather widely in the research and development community, in government and in universities, especially in the United States. While these activities no longer attract the public attention they once did -- some of it for better, some of it for worse -- they continue to this day in many of the same places. Indeed, some U.S. Government agencies have permanent staffs which devote their full time to developing scenarios and conducting foreign policy gaming exercises.

In contrast to political gaming, so-called "man-machine simulations" involve both people and computer models. These are laboratory exercises in which the people provide decision inputs, some of the consequences of which are calculated by a formal model. In the best known example, it was a model of national political economies (Guetzkow, et al., 1963). Decision making teams representing abstract and ficticious nations interacted with their respective national models in an iterative fashion. They also interacted with each other face-to-face, which gave these exercises a gaming character. Much of the thrust of these efforts with man-machine simulations in international studies was directed toward theory building. While none of the early models was well grounded empirically, the idea was to construct reliable vehicles for quasi-experimental research on international behavior. Replicability was essential; it was a nomothetic enterprise.

It is no exaggeration to say that the early "gamers" and the "simulators" in international studies shared little. Abstract models of hypothetical nations held little interest for gamers who were concerned with actual policy problems in all of their idiosyncratic complexity. On the other hand, the lack of experimental control across political gaming exercises, each of which was unique, made generalization impossible. This was the classic trade-off between external and internal validity (Campbell and Stanley, 1966).

In the late nineteen sixties, Shure and his associates introduced a new use of computers that cut across some old distinctions (Shure and Meeker, 1970). In this project, an attempt was made to reconcile the substative sophistication of professional level political gaming with the methodological requirements of nomothetic science. Political games were viewed as "data-making" devices (Verba, 1964), and the approach was primarily inductive. Powerful new computer-based techniques were developed to capture gaming interaction data and to exploit its richness. Among these was one of the first time-shared programs for terminal communications among subjects in laboratory experiments and a sophisticated program for both syntactic and thematic analysis of interaction text (Greenstein, 1972). Communications among participants were restricted to writing, as they had been in many of the earlier RAND and MIT games. What may have been lost in spontaneity was gained by the production of a machine readable transcript of the substantive content of political gaming exercises. Computer-based political gaming was thus born. With the advent of distributed computer networks, computer-based gaming soon became network gaming. This was the work of the POLIS project at UCSB.

CONTRIBUTIONS OF THE RESEARCH PROGRAME

Through the POLIS/ICONS program, what started in Shure's laboratory at the Systems Development Corporation is now a global network for intercultural collaboration in international studies. Throughout this transformation, the focus on fairly complex gaming pushed the state of the art. The resulting methodologies, while emerging from political gaming, have more general applications.

Computer Conferencing and Data Management

Certain computer developments were prerequisite. The integration of communication with the logic and memory of the computer was a major development in informatics. In principle, it permitted the kind of structured communications that political gaming requires plus the ability to search, retrieve and analyze the resulting record by computer. The problem was that these capabilities were not then available outside of a few advanced laboratories (e.g., at MIT and SDC). A time-sharing system was thus written for a new minicomputer at UCSB (Noel & Jackson, 1972). The development of this "POLNET" software went hand in hand with the evolution of network gaming. POLNET was the first minicomputer-based computertized conferencing system (Hiltz and Turoff, 1978). Its most recent version was written for the ICONS project and runs on two different microcomputers (IBM-AT and MicroVAX). POLNET first operated in a prototype packet-switched network (ARPANET). It now runs in a public, value-added network (TELENET).

An examination of other papers in this symposium will reveal that most of the activities that fall under the rubric of simulation games involve face-to-face interaction over periods ranging from an hour or so to a day -- sometimes a weekend. The POLIS/ICONS exercises employ a new kind of format: asynchronous gaming for several weeks, during which period there may also be synchronous gaming interaction. In this, they are closer to their intellectual forbears, the RAND/MIT political games of the late fifties. In a typical exercise, participants meet face-to-face as members of country teams; once or twice a week is common. They discuss the scenario and recent developments and set and adjust policies which serve as guidelines for team action between meetings. Usually, one or more "duty officers" log onto POLNET once or twice a day to receive and to enter messages, communiques, documents, press releases, etc., from the other teams or from the game control team (the umpires). From time to time throughout an exercise, there are synchronous conferences. Some of these are planned as part of the scenario design; others are convened at the request of the participants. We have rarely used structured time periods in the exercises. The clock runs in real time unless a scenario update from the control team advances it.

In network gaming, the greater the longitudinal distance between teams, the more restricted are the "windows" (or time frames) of opportunity for synchronous interactions. For some combinations of locations, there is no window of common working hours at all. This is illustrated in Table 1. While midnight conferences are not uncommon occurrences, the main flow of routine interaction is usually asynchronous.

TABLE 1 Synchronous Interaction Windows among Selected Sites

Window in GMT		Window in local time
08:00-13:00	France	08:00-13:00
	Mid East	10:00-15:00
	Japan	17:00-22:00
23:00-03:00	Japan	08:00-12:00
	California	15:00-19:00
	Md. and Argentina	11:00-15:00
16:00-20:00	California	08:00-12:00
	Md. and Argentina	11:00-15:00
	France	16:00-20:00
	Mid East	18:00-22:00
No common window	Md. and Argentina	No common window
	France	
	Mid East	
	California	

Multi-Scenarios

Another area in which the POLIS/ICONS research program has contributed to political gaming methodology is in scenario design. Most political gaming is bilateral crisis gaming. However detailed the scenarios and extensive the background materials, the exercises usually focus fairly narrowly on the issues related to a specific hypothesized international crisis, as often as not a crisis which directly involves the USA and the USSR. Other issues are, in a sense, held constant for purposes of analysis. This is quite appropriate for a group of experts which seeks to generate ideas about anticipated foreign policy problems. But a single issue focus tends to reduce the possibilities for the kind of issue linkage that were brought to the forefront in policy circles during the Kissinger era. Relationships between intra-issue and inter-issue bargaining have long been of theoretical interest to researchers on negotiations (Winham, 1977).

We sought to hone a more general research tool. Thus, at about the same time as "multipolarity" came into prominence in the international relations literature we began to experiment with mutli-scenarios, which are structured around a fairly complex set of issues. Some scenarios have combined bilateral Soviet-American problems, such as arms control, with a focus on the trilateral relationship between China, the USA and the USSR. Indeed, the "China card" has been played on more than one occasion in our network games. Linkages to regional issues have also been included. The inclusion of a Middle East section in multi-scenarios has added the interesting dimension of patron-client relations to the game matrix, not to mention the conflict between client states themselves. Multi-scenarios can also interleave "low politics" issues with the more compelling issues of "high politics" (compelling in a behavioral sense). North-South issues associated with the call for a "New International Economic Order" have been included, as have issues related to the nuclear non-proliferation regime. Prearranged synchronous conferences have been found to be a useful device for focusing attention on low politics issues such as these. They may also be given a sense of urgency by hypothesized impending events, such as an expected nuclear detonation (a scenario in 1973 did this with India).

The more issues that were included, the greater the number of country teams in the exercises. A dozen is not unusual. For the more complex games, designing and writing the scenario can be rather a task. Table 2 is illustrative of some of the possibilities.

TABLE 2 Illustrative Structure of a Multi-Scenario

Principal Players	Issue 1 Strategic Arms Limit'ns	Issue 2 Trilat'l Relations	Issue 3 Mid-East Conflict	Issue 4 No-So Econ Talks	Issue 5 Nuclear Prolif'n
USA	X	X	X	X	X
USSR	X	X	X		X
CHINA		X			X
JAPAN				X	X
ARGENTINA				X	X
ISRAEL			X		X
EGYPT			X	X	X
SYRIA			X	X	
S. ARABIA			X	X	
JORDAN			X	X	

nb: Other countries simulated by control team when necessary.

In addition to the overall structure of the game, this table shows how analytic subgames flow from the selection of key issues for a multi-scenario. Each subgame has its own incentive structure (i.e., mix of potential payoffs), and each has its list of key players (read down the columns). Each can also be expected to produce its own coalitions. Some possibilities are suggested in Table 3.

TABLE 3 Illustrative Structure of Criss-Crossing Coalitions

Subgame	Expected Coalitions
Trilateral Rels	USA & PRC v USSR
Mid East Confl	Isl & USA v Egt, Jdn, Syr, SAr & USSR
or	Syr & USSR v Jdn, Egt, SAr & USA
NO-SO Ec Talks	USA & Jpn v Arg, Egt & Jdn
Non-Prolif Reg	USA, USSR & Jpn v PRC & Arg

From the perspective of the individual country team in a multi-scenario political game, the greater the number of issue-defined subgames in which its interests are involved, the greater the demands placed on the team members. One of the superpowers, for example, may have to be prepared to formulate and implement policies in a half dozen different issue areas. On the other hand, a lesser power may find only one issue of direct importance to it. Team size is thus quite variable. Not surprisingly, the American team is usually the largest. A need for internal organization manifests itself immediately and a division of labor along subgame lines is customary. With almost startling regularity, this leads to bureaucratization, however small the scale. Different perspectives emerge, each grounded in an organizational subunit. There may be coordination problems and turf battles. We have seen divisiveness in the US team in network games rivaling that attributed to the current US administration in the real "strategic arms subgame" (Talbott, 1985). Thus, multi-scenario games span two theoretical levels of analysis (Singer, 1961). Macro structures and processes generate micro structures and processes. International politics beget bureaucratic politics. For international relations scholars, this is reassuring. The bureaucratic politics model has had a profound impact on the way our discipline views foreign policy decision making (Allison, 1971).

Multilingual Gaming

English has been the lingua franca of most political gaming. The ICONS project has set a new precedent. Since the early nineteen eighties, network gaming exercises that it has conducted have been multilingual. The earliest experimental trials involved French, German, Russian and Spanish, in addition to English. By Spring, 1984, Hebrew and Japanese had been added. In recent ICONS exercises, all seven languages have been used, with native speakers on the Argentine, French, Israeli and Japanese teams sending their messages in their respective languages from Cordoba, Toulon, Jerusalem and Tokyo. While other country teams were staffed at American universities, they too sent their communications in the appropriate languages, especially German, Russian and Spanish.

Going multilingual has not been a casual matter, however. There is more to it than assigning people with appropriate language skills to the country teams. The ICONS project has been developing a carefully designed, systematic methodology, which integrates translation functions into the network. The idea is that, when a message is headed to a network address where no one can read and understand that particular language, it is branched to an appropriate translation node. There it is translated by humans (not machines) and sent on in either the original language, the target language or both (the choice is the recipient's). To provide an idea of the scope of the ICONS methodology, the above mentioned seven language political game generated over 4000 messages, a sizable portion of which had to be translated in this way. Without powerful new computers, that much information would bury one. With the POLNET system, it flows smoothly and is readily retrievable in part or in whole.

Thus far in the project, English has been the target language at all translation nodes. The methodology is general, however. A given translation node may work from any language of origin to any target language. It is particularly noteworthy that network translation nodes, like participating groups, are location independent. They may be anywhere.

One usually thinks of simulations as attempting to model reality. The ICONS network translation methodologies could well be an illustration of how simulation might serve as a model for reality. The city of Geneva constitutes a kind of paradigm for intercultural communications. Its conferencing capabilities are complete, versatile and neutral. All necessary interpretation and translation functions are performed there. What about "digital Genevas" -- electronic meeting places (Johansen, 1979)? The ICONS project is demonstrating their value as a means for assembling communities of scholars in international studies (Noel, 1979). The proliferation of microcomputer based multilingual networks beyond our esoteric field could contribute much to pluralization of communications, which all agree is essential for the establishment of a more just and balanced world information order (MacBride, 1980).

CRISIS GAMING VIA NETWORK: AN EXAMPLE

While the POLIS/ICONS methodologies evolved with fairly complex multi-issue scenarios, they also lend themselves to crisis gaming. As the term is used here, a "crisis game" involves two key sets of attributes. First, there is one primary issue; although there may be a number of ancillary issues which cannot be held altogether constant. A situation is defined in the scenario, and a catalytic event is hypothesized. An event that is created especially to evoke crisis must be perceived as threatening to one or more of the country teams' central values; it must require decisions almost immediately; and it may be a surprise (Brecher, 1979; Hermann, 1972). It occasions urgent reevaluation of policies; and it impels diplomatic, perhaps military, interaction. Second, crisis gaming exercises tend to be relatively short in duration, perhaps a few hours, perhaps a day. In a network context, this means synchronous interactions. Teams have to be at their computer terminals during the gaming period. The POLIS/ICONS software offers the investigator considerable flexibility in setting up communications structures to accommodate particular crisis game designs.

A concrete example of a POLIS/ICONS crisis game was prepared for ISAGA'86. It is called "The 'MUF' Crisis of 1986", "MUF" meaning (fissionable) "materials unaccounted for" in the arcane language of the American nuclear establishment. The central policy issue is the possible introduction of new nuclear capabilities into the matrix of conflict and cooperation in the Middle East. The event was drawn from the literature on nuclear proliferation (Spector, 1985). While this event stretches plausibility a little more than is customary in our scenarios, it more than amply fulfills the criteria for crisis games set forth above.

The main players in the MUF Crisis are: Iran, Iraq, Israel, Saudi Arabia, Syria, the USA and the USSR. These roles were taken by teams at several locations throughout the world. The exercise ran from 13:00 to 15:30 hours Greenwich Mean Time. The electronic mail capabilities of the POLIS/ICONS system made it possible for any team to send messages to any other team's "mailbox" at any time. The system's conferencing capabilities also made it possible to bring sets of teams together in realtime interaction. Table 4 shows some of the conferencing possibilities.

TABLE 4 Conferencing Possibilities in the MUF Crisis Game

team	plenary conference	bilateral conference	bilateral conference	bilateral conference
Iran (S. Calif)	X			
Iraq (France)	X			
Isrl (France)	X		X	
S.Ar. (N. Calif)	X			
Syria (Maryland)	X			X
USA (Maryland)	X	X	X	
USSR (Argentina)	X	X		X

The difference between the two modes of communication is that, in conferencing, teams need not observe addressing protocols; the computer both knows and shows who is commenting at any given moment. Also, one's comments show on the other conferees' terminals as soon as they are concluded. Let us turn now to the exercise.

*The MUF Crisis of 1986**

Introduction

*This is a diplomatic exercise, not a wargame. Although it deals with nuclear weapons, its focus is meant to be political, not military. The exercise is set in the contemporary Middle East and involves actual nations. The situation is purely hypothetical, however. Nonetheless, it poses problems which have great relevance for the nations of that region today. For our purposes, this setting offers interesting possibilities for diplomatic interactions. The principal country teams in the exercise are: Iraq, Israel, Saudi Arabia, the Soviet Union, Syria and the United States.*** *The matrix of conflicting and cooperative interests among these six nations constitutes a rich field on which some of the international political effects of nuclear proliferation may be played through.*

Scenario

This scenario makes no projections. The initial state of affairs is assumed to be just as it is today in the Middle East (June, 1986) and in the world generally. The same leaders are in power, and the same problems remain on the international agenda. Of particular note, there has been no progress in

---

Scenario and comments by Robert C. Noel.

**The play of all other countries will be represented by POLIS Control.*

resolving the conflict between Israel and the Palestinians and Arab states; the war between Iran and Iraq shows no sign of stopping; relations among Arab states continue to reflect both division and entent; and the interests of the superpowers in the region still add an East-West conflict dimension to virtually all regional affairs.

There is one new development, however. In an article which appeared in an authoritative Western newspaper yesterday it was reported that Western intelligence sources strongly suspect that several atomic bombs may have found their way into the clandestine arms market (Spector, 1985). They are believed to be uranium weapons with an explosive yield equivalent to the weapon dropped on Hiroshima. The article reported that the source of the weapons is not known, but that they are not believed to be military items stolen from one of the nuclear powers. This means that the so-called "back yard bomb," long feared by experts, may have become a reality.

According to intelligence sources, Iran, Syria and Libya have shown interest in the purchase of illegal nuclear weapons in the past. At a reported price of three hundred million US dollars ($300,000,000), it is believed that a number of non-state actors might also be interested in the weapons, for example, the PLO. This stunning announcement has precipitated a crisis in Middle Eastern capitals and in Washington and Moscow. It is precisely this kind of event that could ignite the Middle East and spread from there to a global conflagration. The news also points to a major failure on the part of highly touted intelligence agencies, which are only now scrambling to learn how many black market weapons exist and who has them. Clearly, this event constitutes a serious setback for the nuclear non-proliferation regime.

<u>Note to Participants</u>

The primary purpose of this exercise is policy analysis. Gaming actions and interactions are a testbed for creative policy ideas, not ends in themselves. The hypothetical situation with which you are confronted is rich in political and diplomatic implications. You should think first about your information requirements. Since the exericse takes "the facts" of late June, 1986, as given, you may work from common knowledge about your country and others. Additional information may be available from POLIS Control. After you have completed your information search, you should develop a set of policy objectives for the situation. Only then, should you consider specific moves to achieve those objectives. A useful device for getting the exercise going is for each team to prepare a statement expressing its reactions to the events hypothesized in the scenario and any public proposals it wishes to make.

<u>Situational Developments via Control Team Moves</u>

The following messages should be sent by POLIS Control after the exercise commences. The pacing of these inputs is a matter of judgement.

#1 To Iran (played by Control Team)

You should assume that your military have acquired one atomic weapon as described here.

yield *20KT*
fissionable materials *U235* *40kg*
gross weight *400kg*
dimensions *50cm diameter x 180cm high*
non-aerodynamic cylindrical gravity bomb; not designed for external mounting on aircraft; best dropped from cargo aircraft

#2 To Saudi Arabia

You should assume that your country has not been involved to this time in anything connected with clandestine nuclear weapons. Neither do you have any significant military nuclear program.

#3 To Iraq

(Certain non-state actors asked that you be given this secret message. POLIS Control)

Terms of offer:

An unspecified number of units comparable in effect to the Japanese original. Delivery to be effected in buyers country. Units fully assembled and ready for delivery. Testing (via simulation) will be effected in an agreed location in Europe in presence of buyer rep/experts. Sellers will arrange for transportation. Units have not been offered elsewhere. Net price in US Dollars: three hundred ten millions ($310,000,000) per unit. Deposit of one hundred fifty-five millions ($155,000,000). Balance due on delivery. You may respond via POLIS Control, Attention: Toy Department.

The Iraqi team should assume that Iraq does not have the money to make such a purchase. External financial support will be necessary.

#4 To USA

The US team should assume that the intelligence leak came from the CIA. The CIA learned of the possibility that a small number of atomic bombs may have entered the illegal arms trade from sources in Western Europe (Italy and France). The weapons are thought to employ highly enriched uranium (greater than 50% U235), which could have come from a number of sources. Design and fabrication was probably directed by people with some past connection with military weapons programs. Components for such weapons are known to have been diverted on past occasions. Their acquisition and assembly could have been done by a handful of skilled personnel at a cost of no more than a few millions (US Dollars, exclusive of the uranium). The weapons are probably large and bulky and deliverable only by something like a cargo aircraft or modified medium bomber. Their yields are likely to be variable and unpredictable.

#5 To USSR

The Soviet team should assume that the intelligence about the possible sale of atomic bombs came from Western sources, especially the American CIA. The KGB is working on the problem and will keep you informed.

#6 To Israel

The Israeli team should assume that the information about possible clandestine sales of atomic weapons caught Israeli intelligence by surprise. It came from the CIA. You should also assume that you have 25 atomic bombs at your disposal, some deliverable up to 1000KM via surface-to-surface missiles, some battlefield weapons, and some gravity bombs deliverable by F14's.

#7 To USA

American intelligence sources in Western Europe have turned up evidence, mostly circumstantial, that Iran may be a purchaser of clandestine atomic bombs. It is not clear how many or whether delivery has yet been taken.

#8 To USSR

Soviet intelligence sources in Western Europe have turned up circumstantial evidence that Iran may be a purchaser of clandestine atomic bombs. It is not clear how many or whether delivery has yet been taken.

#9 To Israel

Israeli intelligence sources in Western Europe have turned up circumstantial evidence that Iran may be a purchaser of clandestine atomic bombs. It is not clear how many or whether delivery has yet been taken.

#10 To Israel

Please prepare some comments for the press on reports that terrorists may have acquired atom bombs. How much does Israeli intelligence know? Where did the information come from? Are the Syrians involved? the Iraqis? the Libyans? the PLO? What are you going to do about it? (POLIS Control)

#11 To USA

Please prepare some comments for the press on reports that terrorists may have acquired atom bombs. How much is known about it? Why was it not learned earlier? Where did the information come from? Are American citizens in any danger? What are you going to do about it? (POLIS Control)

#12 To all (FIRST PRESS EDITION)

Washington, D.C. Both House and Senate committees have announced that they will immediately begin holding hearings on the matter of nuclear terrorism. The joint committee on intelligence will also begin an inquiry to determine how American intelligence agencies could have provided so little advanced information about the present situation.

Tripoli, Libya. Col Murmar Quaddafi, in a speech on Libyan television, heralded the news about nuclear devolution as the beginning of a new era of peace, in which the Americans and their proxies will be taught to respect the down-trodden and dispossessed peoples of the world. When asked if Libya had the bombs a spokesman for the Libyan Foreign Ministry said "..Libya will not be the last to introduce atomic weapons into the Middle East."

Tunis, Tunisia. Yasir Arafat was "..not available for comment.." about whether the PLO had bought black market atomic bombs.

Teheran, Iran. Sources close to the Ayatolla Khoumeni are quoted as saying that ".. the Imam just returned from a retreat in the mountains refreshed and invigorated, with renewed dedication and enthusiasm for the sacred cause upon which God's people have embarked."

END, PRESS EDITION

#13 To USA, Israel & USSR

Intelligence report.

*New information coming out of Western Europe indicates that the total number of bombs involved is three. If the Iranians have one, as seems to be the case, that leaves two unaccounted for.**

#14 To all (SECOND PRESS EDITION)

Paris, France. Usually reliable sources here report that new information available to Western intelligence agencies puts the number of black market bombs at three. The Iranians are believed to have one of them. The question is: Who has the other two?

15 To Iraq

Note to Control Team: This is a contingent move. It should only be made if Iraq comes up with the money to purchase one weapon (make the price negotiable from $310 million downward).

Send the offer to sell to Iraq (see message 1).

#16 To all (THIRD PRESS EDITION)

Tunis, Tunisia. In his first meeting with reporters since the MUF crisis began, Yasir Arafat laughed off the suggestion that the PLO might be interested in acquiring a black market atom bomb. "One doesn't liberate one's homeland by incinerating it.." he said.

Tripoli, Libya. Col. Murmar Quaddafi was quoted as saying to a group of Western visitors today that the Americans would lose more than a few jets if they attempted aggression against Libya again.

Note to the Control Team

The MUF scenario is short. But it is not simple in terms of its implications. Much depends on information. As the "oracle" in the exercise, POLIS Control is in a key position. The following comments are intended to be suggestive of some of the possibilities that are embedded in this scenario design.

**The idea here is to create uncertainty as to the whereabouts of the second and/or third weapons. This ambiguity should impact other teams. For example, are the Syrians getting a bomb? Would Syria want to give the impression that she might have a bomb? What about the PLO? In truth, known only to the Control Team, no one gets the third bomb.*

Clearly, the possibility of an Iranian bomb poses acute problems for Iraq, very serious problems for Saudi Arabia, serious problems for the United States, and could hardly be comforting for the Israelis. If all of these nations had identical interests in the area, solutions to these problems might be easier to find. But they do not. For example, the obvious solution for Iraq is to try to get a nominal deterrent vis a vis Iran. But, alas, what deters one Iraqi foe is certain to provoke another, far more dangerous one, Israel. The Israelis have already demonstrated how they deal with a perceived Iraqi nuclear threat. But, would an Israeli solution serve the interests, say, of the United States? The destruction of an Iraqi deterrent could lead to the ascendancy of a nuclear Iran in the Gulf. Compared to that, the prospect of a nominally nuclear Iraq may not be so bad. Surely the Saudis would press this view upon the Americans. Then there is the possibility that the Israelis might also dispose of the Iranian threat. This would require active American cooperation. But what about the Soviet Union? On the one hand, they could hardly be expected to view even a nominal nuclear nation of messianic fundamentalists on their Central Asian border with equanimity. On the other hand, the Soviets would find it difficult to sit idly by, especially in light of the position of their ally, Syria. What could Syria be expected to do in such a situation? Syria's support for Iran against Iraq is deeply rooted in the history of the two countries and of the Ba'th Movement. Would Syria consider a nominal nuclear option under such circumstances? The mere though leads us back to the Israelis. A Syrian bomb? Intolerable! So unthinkable, in fact, that one would expect the Israelis to act swiftly and decisively to nip the entire action-reaction process in the bud. The linkages implied in the scenario are indeed complex.

There is also a meta game, and it is a mixed motive game par excellence. Regardless of the deep differences between them, the two superpowers have very strong incentives to cooperate in controlling the kinds of situations that this scenario represents. Each may attempt to exploit the situation to its own advantage. But at some point, the overriding concern for avoiding direct confrontation in a volatile area such as the Middle East imposes the need for cooperation.

It may be that the Syrian and Saudi teams will not get enough stimulation from the scenario as it is written. The control team may need to help things along. That is why inputs are provided from the PLO and the Libyans. POLIS Control also represents Iran and the Press. These constitute useful devices to regulate the pace of events and direct attention to subtleties that may be overlooked by the participants. Iran can be made to behave in more or less menacing ways, not only toward Iraq, but also toward the Gulf States, Saudi Arabia and even Israel (via Shiite proxies in Southern Lebanon). Iranian moves do not have to appear entirely rational. It might make sense toward the end to offer the third bomb to Syria -- perhaps with a subsequent intelligence leak. This is delicate, however. The main purpose would be to evoke Soviet and American responses. Can the two superpowers keep their proxies under harness in such a situation? Witness the process of terminating the wars between Israel and Egypt in 1956 and 1973. With regard to the Saudis, the PLO might also be used to stimulate them, for example, by a request for one-time only supplemental funds in the neighbourhood of $300 million.

DISCUSSION

Communications and Interactions

One question comes up regularly in discussions of the POLIS/ICONS approach to political gaming. It concerns the written communications and the asynchronous interactions. Most people who have experience with simulation and gaming are accustomed to single-site exercises in which the participants meet face-to-face at least some of the time. The idea that representatives of country teams in diplomatic exercises never face each other seems unusual, if not unacceptable prima facia. To preclude the possibility of direct discussion and debate, of confrontation and compromise, is tantamount to throwing the baby out with the bath water. How can games of this sort involve sustained negotiations, let alone any excitement?

These differences in appraoch probably reflect underlying differences in purpose and perspective. From its inception at the RAND Corporation, political gaming has emphasized policy analysis. It has customarily involved highly structured forms of interaction, even when the exercise was held in adjacent rooms (Goldhammer and Speier, 1954). Teams met face-to-face, but interteam interactions were usually written and often indirect, through the submission of move papers to the control team. Recesses were even called when moves coalesced in particularly interesting ways, so that participants could think and write at greater length about the events. Such games were not without their own kind of intellectual excitement; but it was usually not of a kind that is clearly visible to the observer of the scene in the game room.

We too have been concerned with policy analysis, although our efforts also reflect strong interest in underlying questions of international relations theory. In our view, diplomatic interaction is not a rap session. Diplomacy reflects underlying configurations of enduring national interests, some compatable, some incompatable. It is our experience that scenarios which are structurally sound and situationally imaginative generate exercises which are substantively meaningful and rich in interaction processes, regardless of the mode of communication.

Although written, inter-team communications in POLIS/ICONS exercises are far less structured than they were in the RAND/MIT prototype. Country teams exchange messages directly; although in one variation messages may be routed to the umpires for plausibility rulings before they are delivered (the computer software manages the routing). In synchronous conferences when participants are on-line simultaneously, the dynamics of interactions can be both fast and powerful. For example, in one exercise a team that was involved in sensitive bilateral negotiations asked if something could not be done to degrade the performance of the computer. Spontaneity had burned them once before, and they did not want to give away the store in the excitement of the moment. In this context, it is instructive to note that the Direct Communication Link between Washington and Moscow (the "Hot Line") is not a telephone, as is popularly believed, but is intentionally asynchronous and written (Sinaiko, 1963).

Experience has also taught that there is no necessary loss of involvement either. The shared experience among team members is to be expected. What networking introduced is an element of identification with one's university, as well as with one's country team. But with global, multilingual exercises, what was only involving is now compelling. The authenticity of communiques and messages arriving in translation from Japanese and French and Spanish, the realization that real Japanese and French and Argentine players are gathered at their terminals in Tokyo and Toulon and Cordoba, this is the stuff of involvement in international relations gaming. If we may be permitted an anecdote, in one four week exercise, the American team brought their own bed to the university for use by their "duty officers" during late night vigils by the computer terminal.

There are, in fact, some distinct advantages to having political gaming exercises extend over fairly long periods of time with intermittant participation. Many of the people who participate in such exercises are difficult to schedule, especially foreign policy professionals, analysts, professors and the like. Getting them for one or two eight hour blocks is extremely rare -- even more so if travel to a single site is involved. It is far more convenient for such people to spend an hour or so daily from their own offices. They may enter and leave the interaction at their convenience. This feature may also work to advantage in educational applications. When participation in a political gaming exercise is an out-of-class assignment, a judicious combination of asynchronous message exchanges and synchronous conferencing can greatly ease scheduling problems among university students, just as it can among professional people. In fact, local networking can be as useful in this sense as distributed networking.

Reciprocity of Cultural Perspectives

The POLIS/ICONS research program has long been inspired by the possibilities for intercultural studies that are inherent in modern communications technologies. The purpose of the program has been to develop appropriate research and teaching methodologies exploiting some of those technologies (Dickinson, 1978, Wilkenfeld, 1983). The developments described in this paper are steps toward realizing the potential of these methodologies. We are nearing the time when network gaming and related forms of structured intercultural communications can take their place along side of more traditional social science methods as techniques, not only for analysis in comparative foreign policy, but for research on bargaining and negotiations, social attitudes, the social psychology of conflict and accommodation, and kindred areas of scholarly inquiry. The promise of these new methods is that the research process -- whether it be in the definition of problems for analysis, in conceptualization and design of quasi-experiments, in the composition of subject pools, in the study of intercultural interactions themselves, or just in thinking about problems -- routinely and cost-effectively -- may be enriched by multicultural perspectives. The pluralization of the social and behavioral sciences is a lofty goal toward which this work is but a modest step.

A LATIN AMERICAN PERSPECTIVE

As an intellectual cooperation process, ICONS is an original academic fact. Its interactive gaming exercises involve participants from North and South in the experience of collective creation that is implicit in simulations. Instead of discussing the reliability and validity of this tool, these comments will focus on the collateral effects that such multicultural creations have in the local context of Argentina.

More Equitable Use of the World Communication Infrastructure

Until now, the telecommunications network that links the developed countries with the less developed ones has been used primarily for the benefit of the former. This is another manifestation of the assymetry of relations among the developed and developing countries. From the perspective of the South, this unilateral pattern in the use of communication nets, instead of encouraging intercultural understanding, threatens cultural identities. Global gaming is a positive contribution that demonstrates the feasibility of using the world's network capabilities in a more democratic fashion, where messages move in equilibrium in both directions -- from South to North, as well as from North to South. It is also democratic, as it has been used pedagogically, in that it involves university students, who are ordinary citizens, in a worldwide dialog and exchange of ideas.

ICONS Gaming in the Argentine Historical Context

At a time when Argentinian society is making an effort again to exercise political freedoms, participation in intellectual activities dealing with political topics acquires special meaning. The first ICONS exercise occurred only fourteen months after the return to constitutional government in Argentina. For many participants, it was the first verification of the possibilities for intellectual development that are offered by a context of academic freedom. To be able to know and let themselves be known without censorship or auto-censorship was an original experience. This free communication, both locally and internationally, contributed to a strengthening of an incipient and still frightened consciousness of individual freedom.

There were other collateral consequences. The exercise facilitated a reencounter with the practice of rational political discussion in the university. The dynamics and requirements of the exercise created the necessity to redefine the style of autonomous decision making processes. Also, as the exercise progressed and contacts increased, there grew a sense of belonging to the world after a period of cultural confinement and chauvinism.

Concluding Comment

In the context of Argentine academic and informatic development, participation in the ICONS/POLIS research program is an historical event. It shows the great possibilities of the integration of communications technologies and simulation techniques. The positive effects of the experience allow us to think more broadly of the impact such practices could have if generalized.

Paradoxically, however, ICONS is also a manifestation of the "technological gap" in scientific work. In this it has a depressive effect. The unevenness overwhelms. There is a sensation of having "missed the train" forever -- a feeling that the importation of the scientific developments of foreign universities will drown the collective modernization efforts of the national university and people. On the other hand, this pheomenon opens ample opportunities for local applications which can be carried out almost immediately. The importance of the exercise in this sense consists in showing what can be done. In Argentina, it has pointed to the need on the part of the national university community for increased awareness of technological developments in communications that make posible appropriate applications to our academic needs.

REFERENCES

Allison, G. 1971, Essence of Decision: Explaining the Cuban Missile Crisis, Boston, Little, Brown.
Bloomfield, L. P. and Padelford, N. 1959. Three experiments in political gaming. American Political Science Review, volume 53.
Bloomfield, L. P. and Whaley, B. 1965. The Political-Military Exercise: A progress report. ORBIS, volume 8, 854-870.
Brecher, M. 1979. State behavior in international crisis. Journal of Conflict Resolution, Vol. 23, 446-480.
Campbell, D. T. and Stanley, J. 1966. Experimental and Quasi-Experimental Designs for Research, Chicago, Rand McNally.
Crookall, D. (ed). 1985. Simulation Applications in L2 Education and Research Oxford: Pergamon Press. Special issue of System, Vol. 13, No. 3.
Crookall, D. and Wilkenfeld, J. 1985. ICONS: communication technologies, and international studies. In Crookall (1985).
Dickinson, J. 1978. (POLIS) Gaming with telecommunications. In Bonham, G. and Meeth, L. R. (Eds.) 1978. Guide to Effective Teaching: A National

Report on Eighty-one Outstanding College Teachers and How They Teach, New Rochelle, N.Y., Change Magazine Press.

Goldhamer, H. and Spier, H. 1959. Some observations on political gaming. World Politics, Vol. 12, 71-83.

Greenstein, S. et al. 1972. Report on CCBS Text Analysis Project, Los Angeles, UCLA Center for Computer Based Studies.

Guetzkow, H., Alger, C., Brody, R., Noel, R. and Snyder, R. 1963. Simulation in International Relations: Developments for Research and Teaching, Englewood Cliffs, N.J., Prentice-Hall.

Helmer, O. 1960. Strategic Gaming, Santa Monica, The Rand Corporation, paper number P-1902 (21pp).

Hermann, C. 1972. International Crises, New York, The Free Press.

Hiltz, R. and Turoff, M. 1978. The Networked Nation: Human Communication via Computers, Reading, Mass., Addison-Wesley Publishing Co.

Johansen, R., Vallee, J. and Spangler, K. 1979. Electronic Meetings: Technical Alternatives and Social Choices, Reading, Mass., Addison-Wesley Publishing Co.

MacBride, S., et al. 1980. Many Voices, One World, Paris, UNESCO.

Noel, R. C. 1971. Inter-university political gaming and simulation through the POLIS network. Paper delivered at the annual meetings of the American Political Science Association, Chicago, Ill.

Noel, R. C. and Jackson, T. 1972. An information management system for scientific gaming in the social sciences. In AFIPS Conference proceedings, Vol. 40, Reston, Va., American Federation of Information Processing Societies Press.

Noel, R. C. 1979. The POLIS methodologies for distributed political gaming via computer networks. In Proceedings of the 10th ISAGA Conference, Leeuwarden, Netherlands, International Simulation and Gaming Association Press.

Noel, R. C. 1979. Global communication and the integration of international studies communities. Paper delivered at the annual meetings of the International Studies Association.

Shure, G. H. and Meeker, R. 1970. A computer based experimental laboratory. American Psychologist, Vol. 25, number 10, 962-969.

Sinaiko, W. 1963. Teleconferencing: Preliminary Experiments, Arlington, Va., Institute for Defense Analysis, Research Paper P-108, NTIS AD-601 932.

Singer, J. D. 1961. The level of analysis problem in international relations. In Klaus, K. and Verba, S. (Eds.), The International System: Theoretical Essays, Princeton, N.J., Princeton University Press.

Spector, L. S. 1985. The New Nuclear Nations: The Spread of Nuclear Weapons 1985, New York, Vintage Books (especially 45FF).

Talbott, S. 1985. Deadley Gambits, New York, Vintage Books.

Verba, S. 1964. Simulation, reality and theory in international relations," World Politics, Vol. 16, No. 3, 490-519.

Wilkenfeld, J. 1983. Computer assisted international studies, Teaching Political Science, Vol. 10, No. 4.

Wilkenfeld, J. and Brecht, R. D. (eds.) 1985. ICONS: International Negotiation Project, User Manual, College Park, University of Maryland, ICONS Project Documentation Series.

Winham, G. R. 1972. Complexity in international negotiations. In Druckman, D. (Ed.), Negotiation: Social Psychological Perspectives, Beverly Hills, Calif., Sage Publications.

Communicating about simulation design: It's not only (sic) pedagogy

Cathy Stein Greenblat
Rutgers University, USA

ABSTRACT: The process of designing gaming-simulations is complex, requiring that the designer engage in research on the system, develop a conceptual model of it, and translate that model into a gaming-simulation. Three communications problems attached to this enterprise are identified and discussed: (1) the nature of the process has not been well communicated to potential designers; (2) discussion of the underlying model is too often omitted from gaming manuals, leaving the operators and players to extrapolate it from the dynamics of play; and (3) persons outside the gaming community do not recognize the intellectual work of design, but focus on the use to which the products are put -- namely pedagogy. Since teaching tends to be undervalued in many institutions, we are told that our work is "only pedagogy". It is proposed that gamers must strive to communicate more effectively about the serious intellectual work of gaming-simulation design, but must continue to argue for the merit of the teaching enterprise.

KEYWORDS: Games; simulations; gaming-simulations; modelling; design; pedagogy.

ADDRESS: Dept.of Sociology, Rutgers University, New Brunswick, NJ 08903, USA.

INTRODUCTION

The process of design of gaming-simulations, as many of you in the audience know from personal experience, is a challenging one. The designer is forced to consider both the systemic character of the segment of reality to be included, and the manner in which these elements and relationships can best be communicated to the participants. That is, he or she must develop a way to translate that model into a gaming-simulation, or operating model. The game designer, then, engages in initial research on the topic of concern, and in subsequent exploration into styles of translation.

I would like to discuss today three communications problems I see as attached to this enterprise. First, the nature of this process has not been well communicated to potential game designers; indeed, little of the literature on gaming is devoted to systematic exploration and explanation of how to go about designing games. Second, those persons who have designed effective gaming-simulations often omit discussion of the underlying model from their gaming manuals. Hence the novice or advanced user of simulations is often left to extrapolate the model from the dynamics of play. And third, these problems in communication within the community of simulation afficionados have severe repercussions when we deal with those who are less knowledgeable, or worse, those who are skeptical or critical of our enterprise. Such persons often view the gaming-simulations we design and use as "only pedagogy". We are told that we have "just" designed teaching tools. The intellectual work of researching the real-world system and modelling it, which must precede creation of the simulation, is neither recognized nor acknowledged. The focus of their attention is on the use to which our simulations are put -- namely, pedagogy. And I am afraid that even in institutions such as my own, which claim they are striving towards excellence and committed to the transmission of knowledge, teaching is heavily under-valued.

Our work, then, has two dimensions. First, we must continue to argue for the merit of teaching as a valuable aspect of our work as faculty. Even if our simulations were most appropriately considered pedagogical creations, we must resist the concept of "ONLY pedagogy". But the second item on our agenda must be to communicate more effectively among ourselves and to others about the serious intellectual work necessitated by the decision to design a simulation.

SYSTEM RESEARCH, MODELLING, AND GAMING-SIMULATION DESIGN

Let me begin by setting out my understanding of the major steps in the design of a gaming-simulation. I will try to illustrate these by those examples I know best -- i.e. games I have designed. And due to the focus outlined above, I will place most emphasis on underlying research and the development of the conceptual model.

As those of you familiar with my own work and with Richard Duke's work already know, we have both argued for (and worked from) a conceptualization of the design process as consisting of four stages:

I. Setting objectives and parameters

II. Model development

III. Design of the gaming-simulation

IV. Construction and field-testing

Briefly described, the first step consists of careful consideration of the subject matter, purpose, intended operators and participants, and the context of use for the proposed gaming-simulation. The same subject matter might be developed into a gaming simulation for a 2 hour exercise by 25 secondary school students learning about the topic for the first time, or into an all-day exercise for 5 to 10 top-level executives to try out alternative strategies in a risk-free environment. Early decisions about which is the potential audience for your product, therefore, are essential to later decisions about what is to be included and in what form. At this point in the process the designer must also assess the constraints of time, money, and other resources available for both development and utilization. While all of us have made errors (sometimes gross errors) in our estimations of the time that will be needed for model and game development, failure to consider these resources will likely result in a bookshelf or file cabinet full of incomplete design projects, or in completed gaming-simulations that are rarely used because the problems of mass production and dissemination have not been solved or because users have insufficient time or money to employ kits that are available.

Assuming that these questions have been satisfactorily answered by the designer, he or she can move to the second step: developing a conceptual model of the system to be simulated. I consider this the most critical phase, and the point at which many potentially good games fall apart. Here the designer functions as a relatively conventional researcher. He or she must first spell out verbally (and graphically where possible) the system as he or she understands it from the literature and where necessary, from first-hand research into the system to be modelled. The designer here is a sociologist or an economist or a political scientist -- or all the above -- developing an understanding of the substantive content of the real world system. Roles, goals, activities, constraints, consequences, and external factors must be identified. Although psychological factors can be brought into the eventual gaming-simulation by players who are stressed in the same way their real-world counterparts are stressed, this necessitates that the most critical of these psychological factors be identified, and that the roles, resources, constraints, and contingencies that create them be identified as well, so the sources of the stresses can be simulated.

Let me repeat again that this step is essential if the model is to resemble the referent system and to operate as it does. UNLESS YOU HAVE AN UNDERSTANDING OF THE SYSTEM, YOU CANNOT SIMULATE IT, EVEN IN HIGHLY ABSTRACT FORM.

This stage of gaming-simulation design, then, takes one to the library, to technical reports, to case studies, and perhaps to direct field work. The game BLOOD MONEY, for example, was designed to simulate on an abstract level the general character of the experience of the hemophiliac and of those who provide him with care and blood. For the model development, John Gagnon and I read numerous documents about hemophilia and about health care delivery to these and other patient groups. These documents included social science studies, medical reports, technical documents about product development and availability, forecasts of change in social policy, and autobiographical accounts by hemophiliacs and their parents. We also had several all-day meetings with groups of experts on various aspects of hemophilia -- care, blood, insurance, research and development. We focussed our attention on the social and psychological conditions, paying limited attention to the more technical medical characteristics, as we were not physicians nor were we interested in teaching detailed medical facts.

This research enterprise led us to the development of a conceptual model consisting of two basic interacting role sets: the sick and those desiring to aid them. The first set consists of patients whose goals are to engage as fully as possible in the on-going "normal world" composed of school/job/family/friendship networks with their rewards of money, prestige, and social interaction. Patients are hindered in this by the periodic but unpredictable advent of attacks of bleeding, which require expensive and sometimes not readily available medical care and infusions of blood. The second set is comprised of those in one way of another concerned with the goals of provision of medical treatment. Some of these people gather and disseminate blood; others provide care and perform special treatments (operations); others are charged with helping pay for the services and product. The easy accomplishment of these tasks is hindered by unpredictable timing of demand, the limits of supply of blood and care relative to the demand, organizational problems that limit efficient delivery, and the absence of adequate information about patients. Several diagrams were developed to illustrate these relationships. For example, Figure 1 illustrates the character of the patient world, indicating the main sources of the psychological problems which were the focus of our attention.

This and the other diagrams were followed by one to two paragraph discussions of some of the details of relevant sub-topics in the figures, such as medical problems (sources, symptoms); medical treatment and costs; psychological problems; etc. After carefully reviewing this document with experts in health care, we identified the most important factors to be simulated, and generated a second list of factors we wanted to be emergent in most runs of the game --

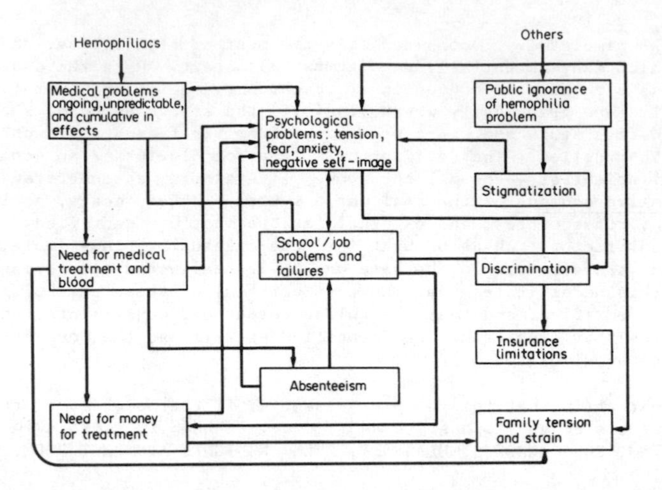

Fig. 1

i.e. factors that should develop through play. Finally, we made a diagram of the linkages between elements, viewing the problem as one of a continuing downward spiral, as shown in figure 2. At that point, we were ready to proceed to steps 3 and 4: design and then construction of the operating model - the gaming-simulation.

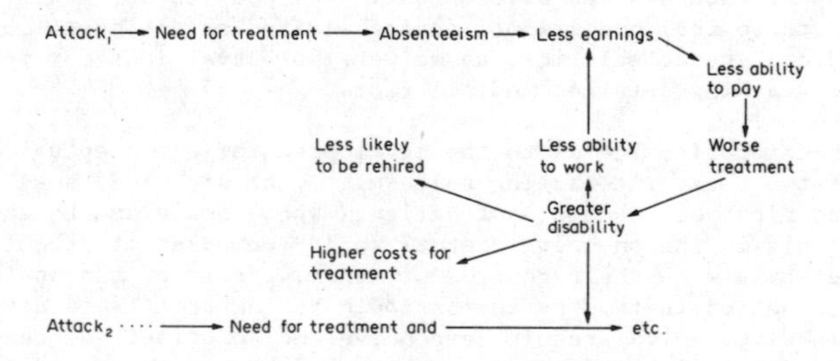

Fig. 2

My second example of the development of a conceptual model is from the design of CAPJEFOS, which some of you have seen operating here, and I hope many others have heard about during the conference. This gaming-simulation of the problems of development at the village level was designed in a workshop sponsored by UNESCO and the Pan African Institute for Development in Buea, Cameroon. Workshop participants were highly knowledgeable about dimensions of village life and about problems of introducing change in such contexts. In addition to tapping their expertise, we utilized research reports on village life throughout the less developed countries, field studies, case studies, training manuals, and direct research in two villages in the Buea vicinity.

CAPJEFOS is a quite elaborate simulation, with many submodels, so I can only hint here at the character of the conceptualizations that underlie it. Given the large size of the initial design team (25 persons), sub-groups worked separately on identification of the three major sets of information needed for subsequent translation to the gaming format:

Group 1 worked on identification of the major actors in the real world system. Their task was to define as fully as they could the main roles, and for each, their goals, activities, and resources.

Group 2 worked on identification of factors that promote and factors that hinder development in villages, and on the linkages between them.

Group 3 worked on identification of the natural and social characteristics (other than roles and development factors) in a typical village and the kinds of external events that might effect these.

Figure 3 presents a portion of the model of actors goals, activities, and resources, and figure 4 presents a portion of the model of factors fostering and factors impeding village development.

PORTION OF MODEL OF AFRICAN VILLAGE DEVELOPMENT GAME: ACTORS

ACTORS	GOALS	ACTIVITIES	RESOURCES
1. Village male farmers	a. increase cash crop production b. feed himself and family c. educate children d. increase family size e. enhance prestige f. participate as a community member vii. increase personal and family health	-farming -hunting -fishing -house construction -attending meetings -trading -participate in traditional activities -pursuit of pleasure -voluntary enterprises (e.g. road construction)	-land -farming implements -farming animals -time -fishing equipment -hunting equipment -friendship children
....			
7. Chief	a. Progress of the village b. Enhance his prestige c. Effective leadership d. Maintenance of law and order e. Maintenance of traditional customs	-preside over traditional meetings -settling disputes -representation of the village -collection of taxes -preside over ceremonies	-land -wives -children -animals -farming implements -time -wages given -friendship -authority

(cont'd)

Fig. 3

Later development of CAPJEFOS proceeded by a smaller group of 6 of us continuing the development of the conceptual model and then translating it into the gaming-simulation. Here several additional sub-models were created. Figure 5 illustrates our conceptualization of sickness and loss of productive times due to sickness in villages. Figure 6 is from a later stage of game development, and indicates how some of the conceptual model elements concerning farmers' time and capital allocations were to be translated into gaming terms.

There are many more examples that I could offer, and far more that could be contributed by members of the audience who have worked on different topics. Given the time limits today, however, let me leave this dimension of the talk in hopes that I have suggested something of the character of research that the

PORTION OF MODEL FOR AFRICAN VILLAGE DEVELOPMENT GAME:
FACTORS FOSTERING VILLAGE DEVELOPMENT:

Natural/Demographic

Balance of immigration with emigration
Existence of fertile soil
Good housing and resettlement
Clean streams with fish and sand
Adequate game/gaming
Use of minerals
Reasonable size of population
Balanced ratio of men/women
Birth rate renews population
Lopw mortality rate
Birth control and family planning
Existence of forests for lumbering

Social, Cultural, and Health

Cultural:
  Acceptance of changes in food habits
Social
  A good organisation of the population
  A good integration of ethnic groups
Health
  Nutritional education including use of local products
  Sanitary education

Political and Administrative

Solidarity and cohesion between different groups
Village Council is well run and represents all groups within the village
Dynamic leadership of village institutions
Awareness of the political aspects of under-development (How inequitable N-S relations are worked out on village level)

Economic

Use of appropriate technology
Use of natural resources
Adequate food production for domestic consumption
Maintain fertility of the soil
Accumulation of capital in the village
Invest surplus income in the village
Land tenure system, with land for all
Acceptance of recommended/improved production techniques
Local efficient marketing services
Suitable agricultural education of the people
Efficient transportation system

Fig. 4

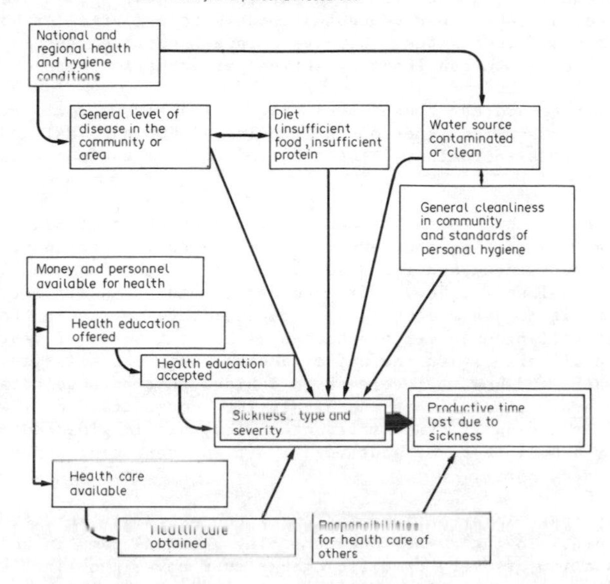
National and regional factors and resources
Community factors, cultural practices, and community and personal resources
National and regional health and hygiene conditions
General level of disease in the community or area
Diet (insufficient food, insufficient protein
Water source contaminated or clean
General cleanliness in community and standards of personal hygiene
Money and personnel available for health
Health education offered
Health education accepted
Sickness: type and severity
Productive time lost due to sickness
Health care available
Health care obtained
Responsibilities for health care of others

Fig. 5

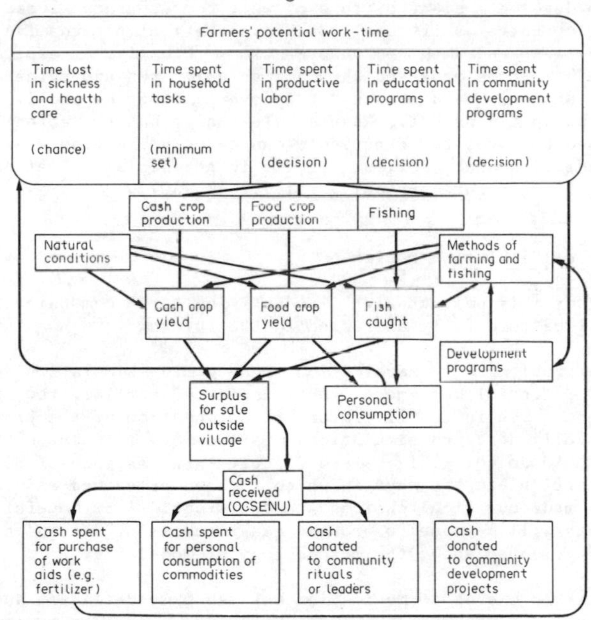
Farmers' potential work-time
Time lost in sickness and health care
(chance)
Time spent in household tasks
(minimum set)
Time spent in productive labor
(decision)
Time spent in educational programs
(decision)
Time spent in community development programs
(decision)
Cash crop production
Food crop production
Fishing
Natural conditions
Methods of farming and fishing
Cash crop yield
Food crop yield
Fish caught
Development programs
Surplus for sale outside village
Personal consumption
Cash received (OCSENU)
Cash spent for purchase of work aids (e.g. fertilizer)
Cash spent for personal consumption of commodities
Cash donated to community rituals or leaders
Cash donated to community development projects
(Social time, group meeting time, and community meeting time are dealt with in real time. Small supplements to farmers productive work time are given if they do not enroll older children in school.)

Fig. 6

serious designer must engage in prior to turning to the knotty problems of how to convert these understandings into an operating model. The stages of design of the gaming-simulation and construction and field-testing of it take the designer well beyond the standard research enterprises described above. Unlike our colleagues who can take their conceptual models to a journal or book publisher so others can learn by reading them, we have a larger purpose: turning them into a non-static form so people can learn by active participation with our models.

While I do not wish to address these later stages directly, I would like to quote from William Gamson, designer of the popular SIMSOC and WHAT'S NEWS, as he once described the intellectual excitement of moving from the conceptual model to the game design:

> I have found the process of continual modification of SIMOC a peculiarly absorbing and rewarding intellectual experience. It has a concreteness and immediacy which is lacking in much of the intellectual work I do... with the development of a simulation... it is as if I have a complicated Rube Goldberg device in front of me that will produce certain processes and outcomes. I want it to operate differently in one way or another but it is difficult to know where and how to intervene to achieve this purpose because the apparatus is delicate and highly interconnected. So I walk around it eyeing it from different angles, and imagine adding a nut here or a bolt there or shutting it off and replacing some more complicated parts.
>
> Each of these interventions must take the extremely specific form of a rule. To intervene, one must play a mental game in which the introduction of every specific change must be weighed in terms of how the whole contraption will operate. Such mental games force one to develop a clear picture of what the apparatus looks like and why it operates as it does. Each hypothetical alternative must be put in place and imagined in operation. Finally, an explicit choice has to be made and the game actually run under the altered rules, and one has a chance to discover whether he really understands the contraption or not. When a rule-change has the effect it is supposed to have, the experience can be very exciting - as exciting as predicting a non-obvious outcome in any social situation and having it turn out correctly. (1971: 306-307).

Communicating about Simulation Design

At the beginning of this presentation, I identified three communication problems which I hope will be more fully addressed in the future.

First, there are problems of communication <u>among</u> gaming simulation designers and between them and potential designers. As I indicated earlier, too little attention has been given in the literature to delineation of the process. Those who have successfully designed simulations must do more to pass on their experience to those who wish to do so. This is a difficult task, as many of us have found it difficult to articulate the ways in which we have proceeded and the criteria by which we have made our final choices and decisions. I am hopeful, however, that more attempts will be made to provide case studies of design and to offer experience-based guidance about the methodology of modelling.

The second problem is one of communication between game designers and the persons who employ the products we have designed. For years I have been dismayed about the low level of information conveyed in most game manuals about the nature of the underlying model. These manuals are often only practical guides to operation, dealing with role allocation, accounting mechanisms, etc. The operator is told little or nothing about the conceptualizations that (hopefully) preceded design and construction. We would be unlikely to adopt a textbook in our classes if we disagreed with the author's perspective, doubted his or her thoroughness, or thought the assumptions were unfounded or untested. Why then should we ask people to employ our gaming-simulations without telling them what <u>ideas</u>, theories, assumptions, etc. are embedded in them? I suggest that all manuals should have sections on the character of the conceptual model, and where possible, citations to the materials employed in the research stage of model development.

If these first two communication problems were more adequately addressed, some of the potency of the third problem would be reduced. Here I refer to the problem of communication between game designers and their non-gaming colleagues. While there are doubtless many versions of this problem, let me give a personal example of the kind of misunderstanding that currently arises. Last year I was considered for a "distinguished professor" award. Those nominating and supporting me described the design enterprise in terms similar to those offered above - i.e. as one requiring detailed analytic understanding of the system followed by construction of a working representation of that system as an interactive simulation. Several colleagues in sociology and other social sciences, however, and members of the administration, had a different view. They praised me as "a splendid teacher", "a distinguished and dynamic teacher, both in the classroom and at workshops and conferences", and as "an outstanding teacher who contributes to teaching beyond the classroom." These comments emerge from my general teaching evaluations in courses in which I employ no simulations, but, I suspect, are heavily based on reports of my simulation activity. And they are not based solely on reports of my <u>utilization</u> of my own and others' simulations; rather, they stem from a view that my simulation design is properly conceived of as a pedagogic enterprise, rather than as scholarly activity or research. This is seen in other comments, such as "she has developed simulation games that are effective teaching aids", found in the evaluation of my teaching effectiveness. The gaming-simulations I have designed are treated as teaching materials at best, but in another statement are essentially disregarded as my publication record is said to consist in large part of "gaming manuals". I wonder if those who have designed such computer programs as SPSS-X are evaluated on the basis of the SPSS-X manual rather than on the quality of the program itself, for which the manual is a guide to users...

A reasoned response to such comments (and I admit to having had many emotional as well as reasoned responses in the past few months) must focus on two issues. First, we must tackle the failure of persons outside the gaming world to acknowledge the scholarly, research, and intellectual character of our activities. In so doing, we can take a position such as that advocated by the systems theorist John Sutherland, who considers model-building to be the <u>essence</u> of science. Furthermore, we can employ his arguments for the merits of general system theory and its model-building enterprises as counters to the scientism and parochialism characteristic of so many of our disciplines. Sutherland argues as follows (1973: 188-190):

> ...if one were to search for a lesson from this survey of general systems theory philosophy it would be this: the ultimate ontological significance of the social and behavioral sciences will be inhibited by adherence to any epistemology or methodological platform which either dictates or unconsciously accepts a restricted interpretation of reality, and which thereby seeks to constrain the scientist to any single analytic or instrumental modality. The general systems theory approach most explicitly does not so constrain the scientist. In fact, as we previously suggested, the general systems theorist, if he had to choose among all thoughts which he could leave his colleagues, might only ask that they consider the suggestion that the only real truth about any phenomenon will be that found in the concatenative nexus which is formed between successively more

specific deductive interences - in short, at the point where percept and concept collapse and where logical and empirical, and inductive and deductive conclusions become indistinguishable.

Thus we find the general systems theorist; arguing for an end to academic parochialism and for adoption of interdisciplinary attack; arguing against simple statistical-mathematical models (e.g., the 'shock' models of the econometricians or the finite-state Stimulus-Response models of the behaviorist psychologists) and asking for more elegant and relevant formulations, even if these may somewhat delay the scientist's 'completion experience'; arguing against unwarrantedly deterministic paradigms and for paradigms which reflect the inherent complexity of most phenomena of any sociobehavioral significance.

The general systems theorist can argue in these ways because his commitment is not to any discipline or school, but to a philosophy; and his platform is one which lends him a perspective, not one which binds him to a constrained set of techniques or which seeks to algorithmatize his searches. As such, he is free to exercise his dedication to the subjects and to which he owes a responsibility; free from the ritualistic defense of his predecessor's positions and freed from the exegetical methodology which 'schoolism' implies; and he is free from the necessity to pay homage to any academic abstractions such as Freudian psychology, functionalist anthropology, or Parsonian sociology. Nor can he draw arrogance from such abstractions; rather, he can assume only confidence from his own achievements and humility from his own failures. He is autonomous man, free to attach his own discipline and free to advance the cause of science. For when one is captured by a disciplinary dogma, one ceases to be scientist and becomes evangelist, ceases to be investigator and becomes concept-mongerer. As Coleridge once noted:

> He who begins by loving Christianity
> better than Truth
> Will proceed by loving his own sect
> or church better than Christianity,
> And end by loving himself better than all.
> (Coleridge, From Aids to Reflection, XXV.)

The second issue we must address involves the nature and salience of pedagogy. I would hope that in our efforts to "set the record straight" about the scholarly character of our work, we will not join forces with those who deride pedagogy, but will continue to proclaim our commitments to the translation of knowledge gained by scholarly investigation into a form in which others can learn from it.

Beyond Pedagogy

Why is it, we might well ask, that esoteric formulations published in obscure journals, read by a handful of scholar colleagues, who are presumably "educated" by reading the theoretical and empirical work, are valued, but formulations which are equally serious, but cast in a form to have a wider audience and greater impact, are disvalued? What, indeed, do we mean by "pedagogy"? It seems that what my colleagues mean is that lower creatures (namely students or trainees) rather than a more elite audience, are the target. Or sometimes they seem to mean that the learning has practical consequences. They were informed that the work of the U.S. National Heart and Lung Institute was expanded, as a result of the hemophilia gaming project of which BLOOD MONEY was a part; the project was categorized as pedagogy. However, I wonder, would they characterize the design of gaming-simulations which the New York Times of February 12, 1986 describes as being regularly played by the National Security Council, the Joint Analysis Directorate in the Pentagon, the Center for Strategic and International Studies, the Federal Emergency Management Agency, and the State Department? In this article we are informed that "In this city, war games and political simulations are not games at all but deadly serious ventures seeking to prepare senior officials and military officers to deal with contingencies or to answer endless questions, all of which begin 'what if...'" (p.B8) I can only assume that the persons who have developed these simulations would be described as having developed pedagogical tools, and hence would be described as "good teachers", rather than "good scholars".

At these meetings three years ago in Bulgaria, Dennis Meadows conveyed to ISAGA members his new-born enthusiasm for gaming-simulations. He told us of his frustration at the limited practical consequences of his work with the Club of Rome. Although their reports gained widesprad scholarly and media attention, he explained, he felt that the material was not presented in a mode in which persons concerned with survival of our world could begin to think fruitfully about the implications of the findings for action to alleviate the impending disasters that would derive from continuation of individual and collective policies and actions. Gaming, he felt, provided the needed mode of translation of the conceptualizations to a form in which the scholarly work would have both educational and practical consequences. I am delighted that three years later, and after much assiduous attention to modes of game design, Dennis maintains this enthusiasm.

I am pleased and proud to be labelled a "splendid" and "dedicated" teacher. I make no apologies for the fact that people have learned from my work. But you and I must address our attention to ways of making the work we do recognized as <u>both</u> scholarship and pedagogy.

'Power' and 'realism' in simulation and gaming: Some pedagogic and analytic observations

W. W. Sharrock and D. R. Watson
University of Manchester, England

ABSTRACT: This paper examines some of the changes in educational orthodoxies, such as 'de-classrooming the classroom', which have facilitated the adoption and development of simulations and games as pedagogic devices, and also suggests that some of the claims made on behalf of simulation and gaming in education may not be sustainable. The paper also proposes a move away from the analytic evaluation of games and simulations in terms of their corresponding with or mirroring the 'real world', replacing this with a concern to examine participants' definitions of the realism of games and of the way participants build these definitions into their game conduct. This leads us to an examination of simulation/game participants' general cultural competences as deployed in the simulation/game.

KEYWORDS: Game; simulation; power; 'de-classrooming'; realism; sociology; pedagogy.

ADDRESS: Department of Sociology, University of Manchester, Manchester M13 9PL, England, 061-273-7121.

INTRODUCTION

As Professor Greenblat shows in her keynote address[1] in this collection, not all games are pedagogic and gamers' concerns are not all pedagogic. Games and simulations can, of course, be used as research devices and can themselves also be the objects of research. Indeed, the last-named concern is the one which interests us most. However, it seems to us that it is among educators that games have been most readily adopted, that the great majority of games in particular have pedagogic objectives and that most games are utilised in the formal educational sector (including Business Schools and the like).

It might make sense, then, to examine sociologically the place of games in that sector and to locate educators' concerns with games within relatively recent movements amongst them, or changes in their pedagogic orthodoxies. Our objective will be to subject games to analytic scrutiny and - though this is admittedly a secondary issue - to see whether educators' beliefs about the place of games can be entirely supported. After making some general observations on these pedagogic concerns, we shall move on to more analytic examination of some of the concerns and criteria gamers have with regard to the 'realism' of their games.

POWER

'De-classrooming the classroom'

During the liberalization of education in the USA, Britain and elsewhere which began in the early 1960s, traditional school classrooms came to be regarded, for a variety of reasons, as being 'out of touch with the real world', as in some sense not corresponding with reality beyond the classroom and as being discontinuous with the experience of that world which school students brought to the classroom. Some students' lack of motivation, involvement and achievement in the classroom was seen as a consequence of this discontinuity, and traditional classroom situations were seen as not adequately preparing students for life outside the school, in the 'real world of practice' where students had to make their living. Schools were increasingly seen as not serving the needs of society. We shall return later to these judgements of 'realism' and 'practicality'.

Secondly, the organization of the traditional classroom was seen as one in which the teacher possessed the authority and made all the choices. The traditional teacher-pupil authority relation fell into disrepute since it cast students into a passive mode, prevented them from communicating with each other, and prevented them from acting on their own initiative. The traditional classroom situation came to be seen as authoritarian and arbitrary, and as thereby closing off various opportunities for learning.

Simulations and games as pedagogic devices were seen as one of several ways of re-organizing the spatial and social organization of the classroom, and of bringing the classroom into close contact and correspondence with the practical realities of 'the world outside'. They shifted responsibility onto students to collectively find out what they needed to know and to find ways of gaining and establishing that knowledge. Here, guidance replaces didacticism on the part of the teacher, and where the traditional, highly asymmetrical and formal structure of the conventional classroom situation is supplanted by a large measure of self-determination on the part of the students, within which students can communicate relatively 'naturally', (i.e., according to the conventions of ordinary conversation as opposed to formal classroom talk). Games and simulations may be seen as mirroring natural, 'real-world' situations and as such may be seen as suitable for transmitting skills, the acquisition of which is not possible within the framework of the formally-organized classroom, e.g., real-world linguistic or other social skills. Games and simulations, then, are often seen as ways of 'de-classrooming' the classroom.

Along with all this has come a pedagogic concern with the motivational reorganization of educational settings. The putatively liberal educational programmes of the 1960s such as 'Operation Head Start' and 'Higher Horizons' in the USA, and 'Education Priority Area' schemes in Britain[2] were founded on - and indeed, served to disseminate - the premise that pupils' experience <u>of</u> and <u>in</u> educational settings should be continuous with and should build upon their 'background experience' outside the school. Games and simulations were seen as ways of using their broader experience as a positive resource rather than dismissing it as an irrelevant or even negative factor. The perceived continuity of students' intra- and extra-school experience was seen as potentially bolstering their motivation, as re-casting educational settings in terms of the students' putative interests in ways that maintain their ongoing involvement; games and simulations were seen as felicitious instruments for achieving these aims, where the traditional classroom situation was seen as involving imposed tasks, chores set and supervised by the teacher who was the ultimate arbiter and authority. Games and simulations were seen as devices for eliciting student involvement and for creating and sustaining student motivation;

motivation was seen as increasing in inverse relation to the degree of authoritarian control.

The introduction of simulations and games is not, then simply 'happenstantial' but is part of a process of educational change. In addition, it ushered in a concept of 'play' which was traditionally kept ecologically and behaviourally separate and distinct from the traditional classroom situation, where it was confined to out-of-school and schoolyard locales. 'Play' could now be seen as part of the 'business' of schools. This notion of 'play' and the re-definition of the 'work'-'play' opposition itself implied a motivational re-definition and re-organization. In passing, it might also be observed that insofar as simulations and games are played by adolescents, post-adolescents and adults in educational and training settings this introduction of 'play' as a legitimate rubric under which the educational institution's business could be transacted involved a re-definition of its appropriateness to different age levels, where previously 'play' had been conventionally predicated (so far as formal education was concerned) on the category 'child'.

One major pedagogic goal of many games, then, is to move away from the power or authority structure of the traditional classroom situation, along with all the motivation-deadening student passivity and the like which many educators associate with this structure. As an aside, we think it no accident that the increase in the popularity of gaming has coincided with the increase in the popularity of group therapy, which itself was seen as a liberalizing approach (particularly in the strict formal authority structure of mental institutions).

Some Reservations about Simulations and Games as 'De-classrooming' Devices

However, it seems to us that the extent to which games move away from this orthodox power structure in education can easily be overestimated. One precondition for the playing of many games is that prior to the game-playing, the teacher or game director issues a whole set of instructions which s/he insists must be strictly adhered to by the students/players. The issuing of these instructions seems to us to be often done with a didacticism and insistence which resembles the task-imposition usually associated with the power structure of the traditional classroom. These features become highly apparent to the detached observer if, for instance, s/he looks at videotapes of game-playing which also involve recordings of the pre (and post-) play situations as well as the actual playing of the game. Indeed, many of the practice-oriented observations made by gamers seem to disattend all but the actual playing of the game, omitting the flow of stipulations which create a framework that render this playing an *instructed* playing, and which furnish the game-director with resources for monitoring and arbitrating the process of playing the game.

To be sure, when the game is actually being played, some of the most visible, most obvious didactic and authoritarian features are relaxed, but the degree and nature of this relaxation is again determined and administered by the teacher/game director. For instance, the teacher/director typically controls game time, how game time maps onto real time, determines the periodicities of the game, not to mention deciding things such as when to break for lunch, what breaking for lunch means for (say) game options at this juncture, and so on. The teacher/game director also determines spatial co-ordinates, distances between (say) towns, nations, etc. In this sense, the game reality, the frame of reference which players must sustain when playing the game, is stipulated by the teacher/director and asymmetries of power are centrally involved in bringing off such a stipulation, and in rendering such stipulations non-negotiable.

In many respects we feel that simulation and gaming in education involves a *submerged* power authority structure, a power/authority structure which has less visibility but which is nonetheless highly present and diffusely operative;

after all, who is it that, ultimately, animates (to use Goffman's felicitious term[3]) the written textual instructions comprising the game rubric?

Indeed, the ways in which this power and authority structure is rendered less visible itself comprises a central topic for empirical research on gaming and simulation. Some of the enforcement or administration of game rubrics may, for instance, be effected during the game's playing by an intermediary, a game participant, a student who has been appointed the game-category of Chairperson, leader, president, chief, dictator or whatever, thereby rendering less obvious to other players the activities and interventions of the ultimate arbiter, the game director by ensuring that there is less direct intervention on his/her part.

In short, it strikes us that one could characterise the control/power/authority situation in games and simulations in terms of 'indirect rule', if we may borrow a phrase from Lord Lugard of the British imperial era, and, earlier, from Roman imperialism, the political technique of divise et impera for keeping conquered groups in a subordinate position. Instead of using a dichotomous 'we' and 'they' situation where 'we' have the power (and visibly so) and 'they' do not, these techniques deploy power and authority through one or more intermediaries. These intermediaries are, typically, selected from those who are ruled and possess some kind of priveleged status amongst the ruled (e.g., a traditional chief), upon which the power endowed by the ruler is superimposed.[4] This diffused power/authority structure can be particularly effective where subordinate groups are (or may be set) in conflict or competition with each other. Similarly, in many simulations and games some game participant may become a kind of conduit for authority that derives from elsewhere, as it were. Some observers might regard, then, such an attempt at 'liberalizing' or 'democratizing' the classroom situation as involving a significant element of manipulativeness in contrast to the frankly manifest dichotomous 'we' and 'they' situation in traditional classrooms. Moreover, the power/authority structure in games and simulations - that is, the organization of control over the game - may shift and transform somewhat, depending upon, for instance, the phase of the game (let alone the precise character of the game). For instance, the opening and closing down of games may involve more 'direct' and less 'indirect' rule by the game director when compared with, say, some ongoing intermediate juncture.

We believe that the introduction of considerations concerning power, authority and control procedures are warranted in view of our interest in the social or interactional organization of games and simulations; power is an organizational property integral to gaming and simulation and must be analysed as such. Moreover, we feel that power is a noticeably missing topic in the analysis of simulation and gaming. Simulations and games have often been designed to address issues in the accumulation, distribution and use of power (STARPOWER is but one example), but studies of gaming and simulation have seldom if ever examined the power of authority structure integral to the operation of simulations and games themselves. It is inappropriate to see power simply operating from 'on high', at, say, State level; power can be built into social-interactional settings too and this can be oriented to by participants in the setting. It seems to us crucial to inspect the control procedures organized into the course of the game. This theme, then, can be part of our treatment of simulations and games as deserving serious analytic attention, as objects to be examined in their own right.

REALISM

A pervasive concern amongst simulators and gamers, and amongst academic and other observers of games and simulations, concerns the 'realism' of games and simulations. This realism is typically assessed in correspondence terms, i.e., in terms of the correspondence (or lack of it) between the game and that facet of the 'real world' which the game purportedly models. However, as Cathy Stein Greenblat's keynote address implies, this correspondence assumes that we know in some secure way what that facet of the 'real-world' looks like. It is our distinct impression that establishing such correspondence is typically effected on the basis of no more than the designer's/observer's unexamined commonsense preconceptions about the 'real-world' setting being modelled; that setting is often characterized on the basis of what some analysts term 'face validity' alone. The detailed examination of what 'real-world' conduct and settings are like is relatively seldom undertaken by those with an interest in simulations and games, and in this regard Greenblat's keynote address constitutes a welcome conceptual explication of what is required if the establishing of correspondences is the desired objective (in particular, we regard her focus on 'translation' as a pivotal issue). These matters seem to be built in to what Bob Anderson, in his Rapporteur's paper in this volume, has treated as a set of paired oppositions (e.g., 'game'-'reality') in terms of which simulations and games are characterized and evaluated.

We want to focus, however, on the issue that judgements concerning the 'realism' of a given game, or any given occasion of playing a game, are not to be effected by the analyst or designer by reference to some standard derived from outside the game itself (e.g., some external standard involving some representation of the 'real world beyond the game'). Indeed, this allows us to sidestep the theory issue of whether even characterization of some 'real' setting based on research is adequate: after all, there can be many research characterizations of social settings, and any single characterization may well be treated as contestable; consequently, how can one warrantably use any given research characterization of a 'real-world' situation as <u>the</u> authoritative or definitive standard against which a game or simulation is to be judged? The degree and nature or realism in a game/simulation is, typically, to be established from <u>within</u> the game/simulation setting itself. Typically, it is game participants who effect and apply <u>in situ</u> judgements about "it's only a game", "it's not serious", "this must be what it's really like", etc. and build these judgements into their conduct, e.g., by establishing what Goffman terms 'role-distance'[5], a displayed separateness and aloofness of the player from the game role, or by showing embarrassment, etc.

Instead, then, of our judging <u>as analysts</u> the 'realism' of games against an extrinsically-derived standard of reality, we suggest that the proper focus concerning 'realism' is on <u>game participants</u>' judgements of the realism in (this specific occasion of playing) the game. What are participants' practices in jointly conducting the game as 'real-for-all-practical-purposs', in collaboratively shifting the 'accent of reality' around and in displaying their orientations to each other?

In other words, we are suggesting that rather than assessing whether a given game or simulation <u>mirrors</u> reality, we should, sociological analysts, treat games and simulations, or the parties to games and simulations, as <u>reality-producing</u>. How do parties to the game produce game realities? This, then, moves away from what we might loosely term a 'correspondence notion' of games/ simulations and some putative 'real situation', and we feel that this move is quite consistent with the move of contemporary philosophers of science away from correspondence notions of (scientific) concepts and 'reality' as reported in Anderson's paper. Instead, we treat game participants as 'pattern detectors' and as constructing game sense and realism-adequate-for-the-game.[6] Participants' practices in suspending and maintaining disbelief are themselves central data for this analytic approach.

Judgements concerning the 'realism' of games and simulations are for us, then, properly made by participants. It is the analyst's business to examine how these judgements are built into, and displayed in, participants' practices in collaboratively producing the game and jointly sustaining the 'game frame of reference'. In short, how do participants use their in situ judgements of the 'reality' or 'unreality' of a given game or simulation as part of their making sense of a game in the course of its production? This is our object of enquiry, and it implies an approach which allows us to avoid taking on the unnecessary burdens of 'correspondence' notions of the 'realism' of games and simulations.

In turn, this approach allows us to shift our analytic attention to other 'realities' than the one to which the game or simulation purportedly corresponds. For example, we can examine the ways in which game participants may 'import' their 'real-world' identities into the game situation, may orient to and indeed trade upon these identities to either sustain or subvert the game frame of reference; gender roles, age-grade roles, etc., and the actions typically associated with these roles, can all come to be built into the playing of the game, along with participants' cultural understandings of what these typical roles and typical actions characteristically mean in ordinary life and settings. Of course, one imported rule, outlined by James L. Heap, is of particular significance in the game is "treat the game/simulation situation as if it were real"; this rule constitutes for participants a procedure for displaying conduct as examples of game/simulation conduct.

Finally, this approach directs our analytic focus towards the general cultural competences which participants bring to the playing of games/simulations - competences which are unrelievedly used to produce and make sense of game/simulation conduct and of the game/simulation settings this conduct produces.

CLOSING REMARKS

We hope in this paper to have done enough to persuade the community of simulation and gaming practitioners that there might be some merit in the application of academic disciplines to the phenomena with which the community are concerned. This is not to say that our claims are incontestable; far from it - indeed many, perhaps most, of our fellow sociologists would not agree with our arguments. Be that as it may, the point is that the various academic disciplines, although they are dismissed in a rather wholesale manner by some in the simulation-gaming community as being too narrow, specialized and even sterile - in fact contain a variety of analytic resources which can cast simulation and gaming in not just one but a whole spectrum of new lights.

The potential insights and other contributions of these disciplines (which, usually, are more wide-ranging and internally diverse than narrow) should not be lightly disowned by practitioners in this field. This is particularly the case if practitioners in the sphere of simulations and games wish to avoid becoming a self-addressing audience, to avoid what Robert K. Merton has in another context called 'the fallacy of group soliloquy' and to avoid the marginality and even 'ghettoization' which some practitioners feel characterizes their specialism. Such group soliloquy can prevent the taken-for-granted assumptions that prevent gaming or any sphere of activity from being explicitly examined and subjected to critical scrutiny. We are not, of course, seeking to forestall a healthy and reasoned scepticism about the contributions of the academic disciplines - such scepticism is, indeed, sought for - but there seems little point in throwing the baby out with the bathwater.

FOOTNOTES

[1] C. Stein Greenblat: 'Communicating about simulation design: It's not only (sic) pedagogy', in this volume.

[2] It is, however, true that in Britain and some other Western nations,

these 'liberal reforms' have come to be regarded with suspicion by some elements in society, which has resulted in a reassertion, to some extent, of traditional values.

[3] E. Goffman: 'The Lecture', in his Forms of Talk (1981) Oxford: Basil Blackwell Publishers, pp. 160-96.

[4] For a discussion of this colonial strategy of indirect rule in maintaining power, and an application, by analogy, to the notion of Black ghettos in the USA as 'internal colonies', see S. Carmichael and C. V. Hamilton: Black Power: The Politics of Liberation in America (1965) Harmondsworth: Penguin Books.

[5] E. Goffman: Encounters: Two Studies in the Sociology of Interaction (1962) Harmondsworth: Penguin Books.

[6] Our general sociological perspective on game conduct, and indeed all conduct, is derived from Harold Garfinkel's Studies in Ethnomethodology (1985) Oxford and London: Polity Press.

REFERENCES

Garfinkel, H. 1985. Studies in Ethnomethodology. Oxford and London: Polity Press.

Goffman, E. 1962. Role Distance. In Goffman, E: Encounters; Two Studies in the Sociology of Interaction. Harmondsworth: Penguin Books.

Goffman, E. 1981. Forms of Talk. Oxford: Basil Blackwell Publishers.

The reality problem in games and simulations

R. J. Anderson
Manchester Polytechnic, England

ABSTRACT: This paper considers one conventional view of the relationship between models and reality, a view which is termed "the representational conception of modelling". This conception is found to be inadequate, first because it does not describe the essential and inherent discontinuities between games and reality, and second because it relies heavily on presuppositions about the use of models in science. The presuppositions have recently been under attack by various contributions in the philosophy of science. The implications of both of these arguments for the justification of the use of games are reviewed.

KEYWORDS: realism; games and simulations; philosophy of science.

ADDRESS: Department of Social Science, Manchester Polytechnic, Canvendish Street, Manchester, England. Phone: 061-228-6171.

INTRODUCTION

In his book on Margaret Mead, Derek Freeman (1984) entertains the possibility that what he calls Mead's "misconstrual" of Samoan culture, and in particular her claims about the unrestrained sexuality of adolescent girls, may have been the result of her failure to notice that she was being teased by her informants. Such teasing is, apparently, a prominent social practice among the young in Samoa (and elsewhere, no doubt), especially with regard to what are normally held to be tabooed subjects. Freeman tells us that many modern Samoans are convinced that this must have happened since what Mead describes and what they know, are so totally at odds. Be that as it may, the story serves nicely to remind us of the dangers of rushing to generalise about unfamiliar social groupings on the basis of very little experience of them. Ways of talking which appear, at first sight, to be obvious expressions of this or that attitude, may for the insider take on wholly different meanings. Indeed, as with Margaret Mead, there is always the possibility that certain ways of talking may be indulged in merely to provoke those few outsiders who happen to be around.

For my part, I will simply have to run this risk since, despite being an outsider, I want in this paper to discuss some of the ways we can talk about what goes on in games and simulations, and the sets of pre-suppositions which players and constructors of games appear to hold. My reason for doing so lies in an impression that some descriptions, definitions, defences even, of the construction of games and simulations are not just misdirected, though I feel they are, but also because they are unnecessary. They attempt to satisfy sets of constraints which are seen as external to and a priori for simulations and games. And yet the views which generate these constraints are by no means uncontested and uncontestable. Once game constructors, players and operators see that satisfying these constraints may not be obligatory, then wholly different styles of accounts, descriptions and explanations of what is going on in games and simulations, and hence of their value, may become available.

To bring this out, I will do three things. First, I will offer a brief sketch of what I will call "the usual way of talking". Next I will show how the usual way of talking is inadequate for the task it sets itself. Finally, I will introduce some ideas drawn (or poached, if you prefer) from recent developments in the philosophy of science which might provide the basis of an alternative way of talking about games and simulations. I will not develop this alternative, though. That task would be better (and more plausibly) carried out by those who do have an insider knowledge.

THE REPRESENTATIONAL CONCEPTION OF MODELLING

Let me start with an illustration and try to draw from it the general outline of the matters which I wish to discuss. In a paper which, by chance, came to hand while I was thinking about these questions, Wolfe and Roberts (1986) report the results of an investigation which they have carried out into what they term the "external validity" of games and simulations. External validity is a matter of some consequence, they feel, because users and proponents of gaming and simulations have defended the use of games and simulations in teaching, research, and so on by pointing to their "realism". In order to operationalise the notion, Wolfe and Roberts construe external validity as a symmetry, or congruence, between the set of skills required for success in any game and those skills required for success in that sector of life which the game simulates. The game which they chose to work on was Jensen and Charrington's THE BUSINESS MANAGEMENT LABORATORY. This was chosen because prior studies had found it to be an effective simulation. In order to demonstrate external validity, Wolfe and Roberts sought to correlate a number of variables indicative of success in the game and an equal number held to be indicative of success in business. Their hypothesis was as follows:

> A positive correlation exists between a student's economic performance in a business decision-making simulation and business career success as measured by salary level, salary improvement, number of promotions, and overall job satisfaction. (Wolfe and Roberts, 1986, p.51.)

It does not matter for my purposes that the results which Wolfe and Roberts report are inconclusive. As they make clear, the data could be used to support a case either way. Neither am I worried that their investigative strategy is somewhat hit and miss. They themselves summarise the major reservations one would want to express. What is important is their claim concerning what would have secured if external validity were to be demonstrated. If this had been the case, then there would have been a *symmetry* between the sets of skills required for high levels of performance in the game and those required for success in business. Because they could be mapped on to one another, to all intents and purposes they would be identical. It would not matter whether these skills were a partial set of other, more general sets, say skills fostered in and necessary for academic success. The mapping would ensure the justification of the use games as a teaching and research strategy.

Even though Wolfe and Roberts' findings are inconclusive, the very fact that they sought to measure external validity in this way raises the central issue which I wish to discuss, namely how is the mapping made possible? Or, to put it more correctly if somewhat less directly, how is it conceived that the symmetry between the simulation and that which it simulates will be achieved? Despite the apparent simplicity of this question, it is by no means easy to answer. In part this is because while everyone is at pains to stress the practical orientation of games, there is no consensus on whether this should be pedagogic, investigative, experimental, technical, or, indeed, all four. (See, for example, Elwood 1984, Gray and Waitt 1982, and Inbar and Stoll 1972.) However, this is a little room for manoeuvre here. Across all of the functional descriptions, we find constant reference to the importance of models. In some way or other, and just how is determined by the use to which the exercise is put, games and simulations are premissed in models of the processes which they articulate. So, the external validity of a game, then, will turn in the end upon the model which it incorporates and the systematicity with which it has been constructed.

Rather than begin by consideration of the somewhat daunting question "What is a model?", I propose to look at the process of modelling. Here is a relatively uncontentious view.

> The process of model development may be views as a process of enrichment or elaboration. One begins with very simple models, quite distinct from reality, and attempts to move in evolutionary fashion towards more elaborate models which more nearly reflect the complexities of the actual.....situation.
>
> (W. T. Morris, 1982, p.59).

This view seems to have three distinct components:

(a) the counterposing of "the model" and "how things (really) are";

(b) the counterposing of "simple" and "more complex" models and the inference that the latter are an improvement on the former;

(c) the intimation of a process by which simple models are elaborated into complex ones.

Lying behind these three, seems to be a fourth:

(d) a method by which the "actual...situation" is condensed, summarised, purified, idealised or mapped in the model.

For reasons which will become clear later one, I want to call this view "the Representational Conception of Modelling". According to this view, the picture or representation in the model is filled out step by step until it matches how things are, or as near as we can get to that given the state of our knowledge, techniques and other resources.

In games and simulations, the role allocated to models is central. They provide the premisses for the organisation of lists of possible elements, scenarios and interrelationships. They provide the grounding for rules of relevance and irrelevance, inclusion and exclusion. The models are, then, the stipulative framework on which the game is constructed and around which it is played out.

Although items (b) and (c) may be of prime interest to the makers and operators of games, I propose to set them on one side. For my present purposes, they are secondary to the more general issues raised in (a) and (d). If, as I will suggest, the attitude evinced in (a) and (d) could be re-thought, then it follows that those of (b) and (c) might similarly be re-considered. Let us take (a) and (d) in turn.

First, what is being said when we counterpose the model with "how things are"? The natural response is to say that the model is, in some sense or other, a picture, a representation, of how things are and, hence, that there is some one way that the world is and that the model is a picture of representation of it. A corrolary of this view is that we have a method for determining the "goodness of fit" between the picture and that which it pictures. This leads to a second and third question. What sort of picture is the model? And what is the method? The answers to these are tied together. The method for determining the goodness of fit is given by the form of the representation. A model may represent reality in a number of ways. The most prominent are probably by either more or less corresponding with it, or by being an analog of it. As is to be expected, correspondence takes numerous forms. Perhaps the most familiar are those where the model is a partial representation of the domain under study, e.g. a single dimension of cluster of properties, and those where the model is a putative microcosm of the totality of relevantly interrelated processes. In both cases, the models are purged of components or elements deemed to be irrelevant, ineffective, or unnecessary. In both cases too, the goodness of fit is secured by the systematicity with which the rules for determining relevance are applied and the recognisability of the correspondence between the model and those elements of how things are which have been selected. In analog models, correspondence is replaced by formal homology. All that matters now is the formal description of processes and outcomes. Thus, the universal Turing machine can be a model for all possible computing machines, including, it is alleged, the human mind, and biological homeostatic systems can serve as models of social processes. The degree of goodness of fit is now determined by observation of the requirements of formal reasoning.

If it is agreed that models are in some sense representations, the critical question, of course, revolves around the method for determining both the range of elements to be included and the loading which they are to be given vis a vis each other. The general approach is to adopt a modified empiricism. No-one suggests that all the investigator has to do is go out and look. Rather, the collection of information concerning the processes to be modelled has to be organised through concepts and theories. For the models used in games and simulations, it appears that the most effective general theory is system theory. Under this rubric, processes and elements may be displayed either as sets of linearly related components set out in flow diagrams, or as sets of functionally interrelated equations, or both. Arguments over the goodness of fit, the adequacy, veracity and validity of the model turn not upon the corrigibility of the theory but the systematicity and exhaustiveness with which it is applied.

Where lies the error in all of this? In my view, it is the centrality given to the reprsentational conception of modelling. To indicate just why this should be so, I will offer a series of observations concerning simulations of business and compare them to the actual practice of business life. My aim in doing this is not to show games to be inadequate but to try and clarify what it is that this sort of justification of their use is unnecessary. This can only be done if the question of symmetry can be reformulated in ways that do not bind it into putative tests of the veracity, validity and reality of games. The observations which I shall offer are not presented as discoveries. Indeed, to game players and constructors they will be news from nowhere. But, I will argue that the set of contrasts, the discontinuities between 'the game' and 'real life', mark endemic features of games and are not merely infelicities and minor flaws which might be overcome. As such they call the notion of symmetry severely into doubt. If a 1 for 1 mapping is what is required then the defenders of simulations are in trouble. But, and this is the second part of my attack on the usual way of talking, the supposition that such a mapping is required is a misconception largely derived from a narrow view of the nature of models in the natural sciences and a misplaced attempt to force a parellel between the use of models in games and their use in the natural sciences. Thus, and this is the crux, if models in the natural sciences do not seek symmetry with how things are in the natural world, and the defence of the use of games and simulations is the scientific character of the models they are premissed in, then the realism of the models no longer need be an issue, for scientific models are not to be defended on the basis of their realism. We can let go both of the argument for parallelism and of the argument for "realism".

Any description, whether formal or informal, of the place of decision-making in business life would have to begin from the observation that decision-making is built-in as a routine feature of bvusiness life. Although particular decisions may be taken with some care, after considerable thought and with great trepidation, as an activity decision-making is part and parcel of the executive's daily life. In this sense, decision making takes place *in media res* in the midst of the hurly burly of daily business life and is therefore connected to and located among all the other business activities which executives engage in; holding and attending meetings, dealing with paperwork, making phone calls, and the like. And, furthermore, decision-making, and business life in general, are connected to the rest of the individual's life. The business executive does not live a modular existence, shifting from one space-time capsule to another when he arrives at the office, goes out to lunch, phones his wife, arranges to play golf, or gets his secretary to type the minutes of the Parish Council Meeting. Business activities are continuous with other social activities and are entwined with them. This is precisely what is not the case with simulations. Here, game players do lead modular existences shifting in and out of the game frame, to use a term of Gottman's (1974), as required. The frame segregates business activities in the game from non-game activities, defines game relevant roles, and more often than not re-locates the players in space and time. Furthermore, for the purposes of the game, the real-time sequencing of activities may be re-organised, speeded up, or interrupted. The game has a start and an end. For the executive things are already in motion when he comes to them and continue after he turns away from them. Furthermore, only a finite body of information concerning the state of events at the start of the game is potentially available. For the businessman, this is certainly not the case. What the set of information to be collected is, what could be collected, what is potentially available, are, for him, matters to be discovered the course of making the decision. Thus, what for the busy executive may be major concerns, the relation of this decision to others, the relative need to make the decision now or to delay it, the history which the decision has, are purged from the game's scenario simply because the game players are game players and hence are not immersed, here and now, in the daily world of business. Closely related to this feature is the location of decisions in the flow of work which the executives achieve. Making decisions are not the only thing they do, nor are they isolatable in the flow of work activities. They are, so to speak, fitted in.

What we have been trying to bring out is the specific character of the social environment of business life and how this is to be distinguished from business life as simulated in a game. At bottom, the suggestion is that simulations are predicated upon a rational reconstruction of business life in which what are stipulated as non-relevant roles, activities and information are purged. The executive does not encounter the decision making situation through a rational reconstruction. For him it is a matter of discovery what is and what is not relevant, related, important and so on. Of course, in all simulations a little loose information is made available so as to make the taking of decisions problematic, but what that loose information is to be and how much of it is available are pre-determined. The 'rules of the game' specify what options are available, what proper plays might be, and who can be called into play. For the business decision maker all of these are matters for enquiry and interpretation. Whereas for the game player, there is a determinate set of relevances associated with the finite province of meaning which is the game, for the executive business life has an open texture of relevances where what things mean and what they can come to are discoverable matters.

MODELS IN SCIENCE

One way of taking observations such as this is to construe them as indicating the relatively undeveloped character of the simulation. In time, with enough attention and effort, the gap can be closed simply by elaborating the model upon which it is based. Such was the outlook espoused by Morris whom I cited earlier. This is, of course, no more than a reiteration of the representational conception of modelling and is associated with one particular philosophical story about scientific method. Hilary Putnam has labelled the general approach underlying both as "metaphysical realism".

> On this perspective, the world consists in some totality of mind independent objects. There is exactly one true and complete description of 'the way the world is'. Truth involves some sort of corresapondence relation between words or though-signs and external things and sets of things. I shall call this perspective the <u>externalist</u> perspective, because its favorite point of view is a God's Eye point of view.
>
> (Putnam, 1981, p.49).

Putnam contrasts metaphysical realism with what he calls "the internalist perspective. Of it, he says "...it is characteristic of this view to hold that <u>what objectives does the world consist of?</u> is a question which it only makes sense to ask <u>within</u> a theory of description." (1981, p.49).

In Putnam's view, it is precisely the advances which have taken place in the most developed and rigorous of the mathematical and natural sciences which press the need to abandon metaphysical realism. The discontinuities between the various descriptions of "how things are", both within the natural sciences and between them and other modes of exploring and describing the world, mean that it no longer makes sense to attempt to drive a wedge between "how things are" and the conceptual schemes we use to discover how things are. The two are interpenetrated.

To some this will appear to require the jettisoning of all claims to knowledge, the abandonment of reason, and the descent into solipsism and relativism. Either we can no longer say anything about what is real or we are confronted with the possibility that there are multiple realities, each of which is equivalent. If this is so, then any description of some phenomenon, be it natural, mathematical or social, is as good as any other. But this is not the implication of Putnam's view, at least as he sees it. I do not have the space here to provide an effective explication of Putnam's arguments and must retreat into recommending that those who feel they do imply some sort of Protagorean relativism should consult Putnam's own account. In any event, all I need to show is that metaphysical realism is contestable and that Putnam's alternative, his "internalism", while no doubt being equally contestable, can be just as stoutly defended. In other words, there is no need for us to take any metaphysics to be stipulative. Putnam opens up the possibility of letting go of the representational conception of modelling. We do not, thereby, have to become impaled on some other dogma.

To explore the possibilities which Putnam adumbrates in his contrast between internalism and metaphysical realism, I would like to introduce a couple of ideas set out in Nancy Cartwright's provocatively titled book <u>How the Laws of Physics Lie</u> (1983). In developing what she calls a "simulacrum" account of explanation and models in physics and in particular the physics of lasers, Cartwright employs a distinction between "phenomenological laws" and "fundamental laws". The phenomenological laws of physics describe the behaviour of matter under precisely defined conditions. These laws are varied, discontinuous and highly confirmed. The fundamental laws seek to explain and unify the phenomenological laws. They are pitched at a much deeper level, are more abstract, concerned with formal properties and relations, etc. In Cartwright's view

> really powerful explanatory laws of the sort found in theoretical physics do not state the truth.....The route from theory to reality is from theory to model and then from model to phenomenological law. The phenomenological laws are indeed true of the objects in reality - or might be; but the fundamental lws are true only of objectives in the model.
>
> (Cartwright, 1983, p.3/4).

Such models are, then, "works of fiction" (p.153). They compose possible properties, idealisations, and "properties of convenience", introduced simply to enable the application of specific domains of mathematical theory. Models are never realistic in the sense that they correspond directly with the situations they are models of. And no-one expects them to be. Different incompatible models are used for different purposes. In various essays, Cartwright examines a number of "problems" such as those of measurement, prediction, causality and explanation in Quantum Physics and comes to the conclusion that the difficulties these issues are said to present have their origins, first, in the discontinuities between the models utilised, and second, in the presupposition that a singular, unitary, consistent and integrated representation of natural phenomena is both possible and desirable. As Cartwright says, "This is a model of a physics we do not have."

Cartwright's is, of course, an argument with "realism" as a philosophy of physics and, perhaps, science in general. It is a demonstration that the attitude which Putnam called metaphysical realism does not capture the way in which some of the most significant parts of modern science can be described. So what? What is the relevance of all this for games and simulations? I think one possible implication might be something like the following. The defences which game operators and game constructors offer for the pedagogic and research relevance of their games turn upon the strength or adequacy of the model on which the games are based. Thus, games are defensible because they are "realistic" in some sense. This realism derives from the thoroughness and systematicity of the method by which the model is derived. The model genuinely represents how things are. The simplification and condensation embodied in the model are to be treated as temporary infelicities which, with closer attention to detail, will be ironed out. Any obvious disparities such as those noticed between business games and real world business life, are not fatal to the enterprise. In the long run, and with better techniques, more research, etc., they will be overcome.

And yet, the arguments of Putnam and Cartwright indicate that all this effort is misplaced. It is directed to justifying game usage in terms of a conception of modelling thought to be derived from the practice of physics and natural science and invoked as the template for methodological rigour, adequacy and exhaustiveness. In accepting that this conception of modelling was, indeed, the standard to which they ought aspire, game constructors and game operators have set themselves an impossible task. What Cartwright and Putnam indicate is not, as sometimes said,m that physics is a shambles but that there is no unified method exemplified there which other disciplines should impose upon themselves. Once free of the obligation to match up to and emulate what is felt to be the proper method for developing knowledge as that might be demonstrated in physics, the justification for the use of games and simulation can be couched, not of the realism of the models, but in other ways. One such might be in the nature of games themselves. The focus of attention will switch away from representation to what Ian Hacking (1983) calls "intervention" - the processes of constructing, planning, organising and playing games and simulations as means for reproducing specifically designed versions, simulacra, of facets of the natural and social worlds in which we live.

By way of a conclusion, let me return to the essential contestability of the views which I have been discussing. There is simply no agreement among philosophers and philosphically minded scientists either on the character of scientific endeavour (compare, for example, Husserl (1970) with a number of the contributions to Lepin's (1984) volume) or on what the relevance of various scientific methods might be for the social and humanistic sciences. Over the latter issue, the differences and similarities run in all directions. One has only got to try to reconcile Quine (1953), Putnam (1978), Schutz (1962), von Mises (1949) and Kaufmann (1958) to see just how complicated the questions are. It has certainly not been my intention to try to replace one handed-down framework with another. Rather, my aim has been first, to draw out the seriousness of what is at issue for the ways in which games and simulations might be described and defended, and second to indicate that there are no ready made and universally agreed principles to be relied upon. This makes it vitally important that those involved in constructing and playing games and simulations should work out for themselves what their responses might be.

REFERENCES

Cavlin, A. et al. 1982. Encyclopaedia of Enterpreneurship. Englewood Cliffs. Prentice Hall.
Cartwright, N. 1983. How the Laws of Physics Lie. Clarendon Press, Oxford.
Elgood, N. Handbook of Management Games. 1984. London. Gower.
Freeman, D. 1984. Margaret Merad and Samoa. Harmondsworth. Pelican.
goffman, E. 1974. Frame Analysis. Harmondsworth. Penguin.
Gray, L. and Waitt, I. (eds). 1982. Simulation in Management and Business Education. London. Kogan Page.
Hacking, I. 1983. Representing and Intervening. Cambridge. Cambridge University Press.
Husserl, E. 1970. The Crisis of European Sciences and Transcendtal Phenomenology. Evanston. Northwestern University Press.
Inbar, M. and Stoll, C. 1972. Simulation and Gaming in Social Science. New York. The Free Press.
Kaufmann, F. 1958. The Methodology of the Social Sciences. New York. Humanities Press.
Leplin, J. 1984. Scientific Realism. Berkeley. University of California Press.
Morris, W. 'On the art of modelling'. In Calvin et al.
Putnam, H. 1978. Meaning and the Moral Sciences. London. Routledge and Kegan Paul.
Putnam, H. 1981. Reason, Truth and History. Cambridge. Cambridge University Press.
Quine, W. 1953. Two dogams of empiricism. From a Logical Point of View. Cambridge Mass. Harvard University Press.
Schutz, A. 1962. The Problem of Social Reality. Collected Papers vol. 1. The Hague. Martinus Nijhoff.
von Mises, L. 1949. Human Action. Newhaven. Yale University Press.
Wolfe, J. and Roberts, C. 1986. 'The external validity of business management games'. Simulation and Games, 17:1.

ACKNOWLEDGEMENT

The writing of this paper was supported by ESRC grant number F00232213.

Language, computers and simulation: An introduction

D. R. Watson
University of Manchester, UK

David Crookall
University of Toulon, France

Our aim, in this section introduction, is largely of an orientational kind; so, rather than outline each paper individually, we shall confine ourselves to a more general commentary. This section covers a very broad area indeed, and it is not surprising that the papers here are not isomorphic in theme or analytic concerns. We cannot therefore, nor would we wish to, stipulate too closely, in advance, the characteristics of this set of papers. However, this is not to say that there are no unifying stands. It might be said, to use Wittgenstein's terms, that the papers are connected by 'family resemblances' or 'likenesses'. That is, whilst they do not share any criterial or centrally distinguishing feature, they might be treated as members of the same 'family', albeit an extended one. As Wittgenstein said of games, there is no central feature of, say, board and field games, but each type has some overlapping properties with some of the others, so that each is part of an overall network which we may identify as "games".

It is worthwhile citing Wittgenstein at length on this matter, particularly as his comments also have broad relevance for an understanding of what, as gamers, readers of this volume might hope to achieve by way of definition of their interests, i.e., the 'domain assumptions' of these concerns.

> Consider for example the proceedings that we call "games". I mean board games, card games, ball games, Olympic games, and so on. What is common to them all? Don't say: "There *must* be something in common, or they would not be called 'games'" - but *look and see* whether there is anything common to all. For if you look at them you will not see something that is common to all, but similarities, relationships, and a whole series of them at that. ... Look for example at board games with all their multifarious relationships. Now pass to card games; here you will find many correspondences with the first group, but many common features drop out, and others appear. When we pass to ball games, much that is common is retained, but much is lost. ... And we can go through the many, many other groups of games in the same way; and see how similarities crop up and disappear. And the result of this examination is: we see a complicated network of similarities overlapping and criss-crossing: sometimes overall similarities, sometimes similarities of detail. (Wittgenstein, 1986) [1].

The attraction of the notion of 'family resemblances' as a characterization of the activities of simulation/gaming seems clear. Indeed, the formal properties of games (each of which, in turn, takes on different manifestations in each specific game context) also raise fundamental issues of a far greater applicability than what we might commonsensically term the "game frame of reference". The notion of games as involving rules and strategies around and within which actions (which we may term "moves") are arranged is, of course, of generic interest to those involved in researching or teaching about the orderliness of social contexts. These actions may be seen not only as facilitated and constrained by the rules, but also as deriving their (shared) sense from players' co-orientation to the rules, their sense being established through members' interpretive work in treating actions as actions-according-to-a-game-rule.

Similarly, the sense of the actions is additionally established by conceiving of the game move as one in a sequence of moves; the move may be seen as occasioned by, and appropriate/inappropriate in terms of, the preceding moves, as well as by anticipation of possible future moves. Subsequently, the making of that move re-casts or re-organizes the field of possibilities for future moves. In other words, we have, in the game frame of reference, a manifestly temporal organization, a retrospective and prospective dimension to the game's management. All this having been said, when it comes down to actual instances of games, the substantive rules, the particular practices involved, the outcomes, etc., all manifest a very great diversity as well as, on occasion, similarities and connections at one level of generality or another. Games, quite simply, are not interchangeable or coterminous in any meaningful sense of those terms. One achieves little by introducing a rule derived from football into the game of tennis.

Likewise, a Wittgensteinian conception of games helps keep us away from the mentalism in terms of which many analyses of simulation/gaming seem to be cast. We are thinking, here, of the notion of motivation, learning, meaning and outcome which tend to be conceived purely and exclusively in psychological terms, as being events that happened, as it were, purely 'inside the individual participant's head', as private mental events. It seems to us quite ironic that, whilst most simulation specialists pride themselves, and rightly so, on the interdisciplinary character of their concerns, a few of these selfsame specialists de facto reduce the characterization of their activities to the terms of a specialized discipline, namely that of psychology - and to the most positivistic conceptions of psychology at that, which represents a further narrowing of concerns. Indeed, one might argue that the increased use of the computer in simulation/gaming has been largely based upon these already-established psychological reductions as well as upon positivistic approaches, and this has in turn greatly extended and institutionalized these reductions and approaches as the way to do, and to 'advance', simulation/gaming techniques. Psychologism and positivism have furnished many of the taken-for-granted, unexamined, and troublesome assumptions of the simulation/gaming sphere.

A Wittgensteinian frame of reference can help us avoid the mentalism and psychologism which, apparently, pervade some gaming concerns. His and his followers' emphasis on the public, shared dimension of mental predicates such as motives seems to possess major potential in the analysis, and indeed in the conducting, of games and simulations. Similarly, non-positivistic perspectives can also shed light upon the sphere of simulation/gaming, encourage the development of such approaches within the sphere, and make the perception of these approaches as just as legitimate as those derived from psychology and positivism. Indeed, we would see, as a major advantage of rendering the sphere of simulation/gaming a far 'broader church', the making explicit of what are presently tacitly-held assumptions.

Just as we cannot expect to find an easy, single, unifying criterial feature of games, so we should not expect anything of the sort in the papers in this section. There are good and healthy reasons why this should not be so; contributors' concerns may be predominantly theoretical (or analytic) or essentially pragmatic. Whilst a feature common to several (though not all) of the papers is an interest in computational work in simulations and games, the place of this work in the framework of each simulation is highly variable. In this respect, the papers which emphasize computer use are not to be understood as in any way interchangeable or continuous. Each must be inspected in its own terms, with careful attention to the detail of each individual account, rather than our seeking to impose an external framework for the comparison and evaluation of the entire set of contributions in overall terms.

It is, with great caution, then, that we should proceed with any general commentary on these contributions. One central theme, however, concerns the many and diverse ways in which language is both a contextualizing and contextualized medium, and as such shows its immense flexibility and versatility. The papers here presented show how communicative work is adapted, and takes on different forms depending upon the context in which it is located (where the form helps also to create the context as, for participants and onlookers, an identifiable state of affairs). The 'contexts' vary from highly specific interactional settings, to overall national or general cultural frameworks. Many of the papers address multiple contexts at varying levels of generality; moreover, they address these contexts in different ways, treating one (or some) of them as topics (e.g., modelling them) and counting on the others as essential background resources for the actual transacting of the pedagogic task. Consequently, they have a variety of orientations to language, sometimes 'simply' using it as a resource and at other times also treating it as a topic for examination in its own right. Each paper, of course, through its medium of presentation bears an essential relation to some natural language medium or other and often bears a variety of angles on that language. This, in turn, attests to the almost infinitely wide diversity of tasks which language can serve. It also attests to the way in which, if one wishes to analyse language use, one inevitably and simultaneously has to analyse the context of that use; and the converse too is the case, of course. To treat a context independent of its constitution in the language use of those producing that context would indeed be a strange task. [2]

Another strand which, in its varying manifestations, informs many of the papers is a broad set of pedagogic concerns. Here again, the conception of these concerns, the pedagogic object to be attained, the means favoured, etc., are highly variable. In various ways, these concerns are seen as being 'delivered' in a vivid and telling way by the introduction of simulation/gaming techniques; and the use of computers is, again, in its various manifestations, seen as a particularly effective way of 'delivering' the simulation or game into the teaching situation in a systematic, standardized and reproducible way, even when adapted to the individual's needs and choices. That having been said, the position of computer use is quite diverse in the various projects. Some simulations are entirely reliant on computers, whereas many teaching projects use computers in a way that may best be termed "computer-assisted". That is, computers are pressed into service to assist in the achievement of tasks which might eventually in some sense be achieved anyway without the use of computers. These tasks involve computers in a contingent, as opposed to a necessary, way. [3]

The reference in many of the papers to computational work raises another issue, one discussed by Berlinski (1976) in a brilliant and scathing attack on the use of cybernetics, information theory, systems analysis and computer modelling, particularly in the social sciences (and it is to be noted that Berlinksi addresses the models that underlie simulations, too). He points out that the employment of such cybernetic models in the social sciences may lend an apparent 'quantitative scientific', modern and technical cast to the disciplines, but in fact possesses a permissive, ad hoc and essentially woolly, indeterminate and logically incoherent quality. Cybernetics quickly takes on the soggy, poultice-like quality of the disciplinary framework into which it has become incorporated, claims Berlinski. This, we feel, is a contentious but serious point to be considered in relation to some uses of computers which are outlined in many of the papers in this section and elsewhere in this volume. Should it be assumed that computer models always add something to some pedagogical or analytic task in hand? Or, as Berlinski might lead us to believe, does the apparent technicality of these types of models detract in some way from the employment of the ordinary, less glamorous employments of the discipline's routine arts, crafts and conceptual devices of pedagogic and analytic work? Along with Berlinski, we can simply note here that there can be no _a priori_, 'disembodied' answer to this question; rather we have to examine the use of computer modelling on a case-by-case basis. Many of the diverse papers in this section offer us a useful opportunity to do just that, and to examine Berlinski's arguments critically as applied to the specific sphere of simulation/gaming rather than this or that discipline 'as a whole'.

Whilst there is a great deal of analytic attention paid to the actual playing of the game, there is quite frequently a lack of concern with - one might even term is a "casual attitude" to - a systematic evaluation of the game. Sometimes the outcomes of the game and their evaluation are only anecdotally rendered, with no explication as to _how_ the evaluation was conducted, _what_ assessment techniques were used, _whose_ evaluations these were, etc. We often have to take the authors' evaluation on faith, rather than being given basic details on the assessment procedure (let alone giving an independent basis for evaluation as is provided by some sociological perspectives which do not rely simply on evaluations based on our faith in the unaided, native intuition of the researcher).

Thus, whilst these papers may offer us no rigid set of similarities or criterial features, their rich diversity at least allows us the chance of bringing some general, 'open-ended' analytic considerations to bear, _mutatis mutandis_, on specific instances. Such an approach will not only inform our reading of this or that contribution but will also occasion our reflection upon the utility and limitations of the general considerations themselves. It is always important, though, to ensure that the general considerations can themselves encompass such family resemblances, and do not arbitarily 'include anything out' (as Samuel Goldwyn might have said). We feel that the reference to family resemblances, to various angles on language and computer use do indeed have this 'open-ended' quality, and will serve as useful orientational considerations to the papers in this section.

NOTES

[1] This quotation is reproduced and elaborated upon in an excellent article by Heritage (1978). For further discussion on Wittgenstein and games, Jones (1986) is to be recommended.

[2] For more discussion on language and communicational aspects of simulation, see Crookall and Saunders (1988) and Crookall and Oxford (1988).

[3] For a discussion on these and related issues, see Crookall _et al_. (1986).

REFERENCES

Berlinski, D. 1976. _On Systems Analysis: An Essay Concerning the Limitations of Some Mathematical Methods in the Social, Political and Biological Sciences_. Cambridge, MA & London: Massachusetts Institute of Technology.

Crookall, D., Martin, A., Saunders, D. and Coote, A. 1986. Human and computer involvement in simulation. Simulation and Games 17:3.

Crookall, D. and Oxford, R. (Eds). 1988. Language Learning Through Simulation/Gaming. New York: Newbury House - Harper & Row.

Crookall, D. and Saunders, D. (Eds). 1988. Communication and Simulation: From Two Fields to One Theme. Clevedon: Multilingual Matters.

Heritage, J. 1978. Aspects of the flexibilities of language use. Sociology: The Journal of the British Sociological Association 12:1, January 1978.

Jones, K. 1986. Games, simulations, Wittgenstein. Simulation/Games for Learning 16:2.

Wittgenstein, L. 1968. Philosophical Investigations (3rd Ed). Oxford: Basil Blackwell.

The ISAGA GAME: Inquisitive Speaking And Gameful Acquaintance: A mix of tongues and communicating across languages

David Crookall
University of Toulon, France

Alan Coote
Polytechnic of Wales, UK

Danielle Dumas, Alison Le Gat
University of Toulon, France

ABSTRACT: The ISAGA GAME was programmed at the start of the Conference, and all delegates invited to attend. As an ice-breaker and foreign language activity, its major objective was to encourage conference-goers to get to know each other early on and across language boundaries.

This account of that session looks at those objectives, gives an impression of how the game seemed to go, and contains an outline discussion on a few issues of language and behaviour as they seem to relate to the game and its objectives. For reasons of space, only the English part of the manual is provided here.

KEYWORDS: acquaintanceship, communication, game, intergroup relations, language behaviour, language variety, social identity, socializing, uncertainty.

ADDRESSES: DC, DD, and ALG: Université de Toulon, Av de l'Université, 83130 LA GARDE, France; AC: Dept of Management and Legal Studies, Polytechnic of Wales, Pontypridd, Mid Glamorgan CF37 1DL, Wales.

LANGUAGES AND RELATIONSHIPS

One problem which we had at the start of ISAGA'86 was how to break the ice in scorching hot weather. The natural way of tackling this seemed to be to conduct a plenary game on the first morning of the conference.

A title was found which produced the acronym ISAGA. As this was an international Conference with delegates from a large number of countries, speaking an almost equal number of languages, the major objectives of the game were to enable people to become acquainted early on, and to encourage people of different native tongues and foreign languages to establish communication with each other.

Moreover, as the theme of the Conference was "simulation and communication", it seemed that a game for communication among all conference-goers themselves was a good place to start. This was seen as even more important in an international conference, where there is always a tendency for people to gather in language-specific groups and to avoid those who don't speak their own mother tongue. Although the game had serious objectives, it also aimed to set a good-humoured and light-hearted atmosphere for the Conference by encouraging fun, and above all allowing as many people as possible to get acquainted at the outset and in a short space of time.

It is difficult to say how well the game went or indeed how well it managed to allow people to attain the objectives either during the game itself, or more importantly for the rest of the Conference. We would, however, like to say a few words on its flavour, before sketching some of the more serious issues underpinning it.

The title "Inquisitive Speaking And Gameful Acquaintance" attempted to capture the essence of the game and its objectives. It consisted of a manual (reproduced below) and a bunch of people who had all arrived with one thing in mind: to play, study and demonstrate games. But to do this we all needed to communicate, and overcome as quickly as possible those uncertain moments at the start of a conference when we first (wish to) get to know very different, but like-minded, others. Playing the ISAGA GAME would require some smiles and laughter gestures, but only when the jokes were in good taste. It had something to do with languages, but not your own, a sort of mini Tower of Babel to be conquered, with predictable chaos. Nothing could be more congenial to the ISAGA spirit.

The basics of the game were simple; first find a partner and establish his/her and your first language. If it was the same, find another language to communicate in, and discuss six topics. Thus we had two French people dredging their memories for school Spanish, to talk about Tchernobyl; an Italian and a Pole attempting German to give the definitive word on black underwear. And an Englishman and a Frenchwoman exchanging views in kitchen Italian on the vital issue of paper handkerchieves. Once the interviews were over it was on to another person and Hebrew, Spanish, who knows?

As an ice-breaker it appeared to be effective. Afterwords, encountering someone whose views on toothpicks you had exchanged in broken Portuguese, you felt little inhibition in getting properly acquainted, this time though in a more natural common language. But, it remains an open question as to how far people with no common native tongue (e.g., an English native speaker and a French native speaker, neither of whom spoke the other's language with great ease) were encouraged to maintain or even increase communication during the rest of the conference. But then, can one expect a few minutes in a fun game to radically change one's habits in social relations and in the choice of one's social partners?

Much research has been carried out into language behaviour in social and intergroup settings. Indeed it is not possible here even to outline the wealth of research, and the theories it has generated. But at least a glimpse can be given; we have chosen to take some extracts from various sources, and reformulate them with reference to games in general and the ISAGA GAME in particular.

Language plays a powerful role in the development and maintenance of relationships with others. Through language people try to reduce uncertainties and gain knowledge about themselves and others. Uncertainty can be seen as lying at the heart of speakers' problems in selecting, from a variety of alternative responses, those which will enable them to achieve their goals in social interaction. These problems are confounded when the situation is particularly ambiguous and goals unclearly defined, as is often the case at the start of a conference, where some people know each other well and know exactly "what it's all about", but where others are still in the process of assessing the general lay of the land, and seeking out new friends. Theoretically, a game with specific objectives and constraints should reduce this ambiguity and, at least for the duration of the game, provide clear goals.

In addition, individuals are uncertain about the ways in which their partners are likely to behave; and in initial encounters, they have very little information or knowledge about the various beliefs and attitudes that their partners hold, including language attitudes, which in multilingual settings, such as at ISAGA'86, gain particular significance and force. As a general rule, a game context encourages the reduction of uncertainty. A game provides a structure and a set of objectives known and accepted in common by participants; certain aspects of self, others and relationships are built into the game structure and objectives, which thereby provide a starting point for the attainment of further uncertainty-reducing knowledge. Moreover, the game structure narrows the range of possible alternative response behaviours, thereby reducing uncertainty even further. Uncertainty about the behaviour of partners is also reduced by the more bounded constraints of the same context; certain behaviours are commonly negotiated fairly early on as permissible within the game. With an ice-breaker, these aspects will be enhanced, because participants' common knowledge of the basic givens of the same situation include precisely those of an explicit aim of reducing uncertainty. In their initial encounters, participants have already accepted to adopt a basic common social rule which says "reduce uncertainty". This supposes that behaviours which, in other situations, would increase uncertainty, can be "laughed off", "excused as being outside the game", considered as "acceptable" transgressions under the circumstances, or passed over and off as otherwise not germane to the developing relationship.

Thus, the problem faced by participants, especially initially, is to reduce their uncertainties about each other and their situation in order to select messages which are deemed appropriate to the social situation. The ISAGA GAME provides at least some of the necessary certainties for a reasonably appropriate initial selection of utterances. The game context is an important element here; but the specifics of the ISAGA GAME also contribute. Each participant already has a set of given items to be discussed with his/her partner, and knows that his/her partner also has similar items. Uncertainty about the topic of conversation is thus considerably reduced, and its relevance to the situation is provided by the game itself. The game makes all required conversation appropriate, and that which is not is either not part of the game, or a misunderstanding of its rules, and therefore not critical to the ongoing social situation.

If uncertainty is reduced, the interaction tends to proceed smoothly; if not, interaction will be halting, and termination of the relationship is likely. Those encounters during the ISAGA GAME included an element which, with hindsight, was not conducive to uncertainty reduction. This was the requirement that partners did not use a common native language. An individual will employ a variety of linguistic strategies to reduce uncertainty, and the most powerful of these is the choice of language variety. In the ISAGA GAME, participants did not have a free choice.

Thus language can also create, and be used to create, uncertainty. If we recall that language plays a powerful role in the development and maintenance of relationships with others, then problems due simply to the choice of which language to use or to understanding, and expressing oneself in, a foreign language (such as were encountered during the ISAGA GAME) are likely to raise uncertainties by impeding knowledge about oneself, others, and our relationships.

Choice though is only one aspect. Language is above all a major feature in individuals' social identification of themselves. The particular way in which an individual speaks (intonation, accent, vocabulary style, for example) not only serves a direct communicative function, it serves also to signal to others the speaker's background and that person's self-image. These aspects became a major focus of the ISAGA GAME, and it is possible that they detracted from the ice-breaking objectives (reducing uncertainty, developing relationships).

ISAGA GAME participants were not able to use their native tongue with partners who also spoke that native tongue. This meant that either both partners used a foreign language, or one of the partners spoke the native language of the other, or they cheated.

Whether speaking one or five languages, all individuals belong to at least one speech community, all of whose members share at least a single speech variety and the norms for its appropriate use. In every society the differential power of particular social groups is reflected in language variation. Typically, the dominant group promotes its patterns of language use as a model required for social advancement; and the use of a lower-prestige language, dialect or accent by minority group members reduces their opportunities for success in the society as a whole. All too often, language-mediated barriers to communication are raised by negative cognitive representation of outgroups, particularly when language becomes a salient dimension of intergroup relations and of the ingroup's social identity.

Given this power of language over our individual and collective identities, it would not be surprising that the language constraints imposed during the ISAGA GAME, rather than reduce uncertainty, raised certain barriers to knowledge of self and others. This will have been confounded by the socially defined differential values attached to the languages represented at the Conference. Although all languages were encouraged during the ISAGA GAME, and many could be overheard, by far the most prominent one (during the game and during the Conference) was English. Although it seemed that the proportion of francophones speaking English was greater than that of anglophones speaking French (an unhappy reflection on the teaching of French in anglo-saxon countries, which in turn is a reflection of more diffuse language attitudes in society at large), it may have been that the dominance of English, due to the origin of the Conference population, had some of the more negative effects mentioned above, and this went counter to the ice-breaking objectives. When speaking a foreign language, we are forced to consider a variety of new social and cultural ideas and practices associated with that language, as well as to articulate the forms and logic of the language itself. It is doubtful that this socio-cultural content was very prominent during the ISAGA GAME, but it no doubt played a role in maintaining language group boundaries during the Conference itself.

In conclusion, the ISAGA GAME seemed to result in conflicting outcomes; the reduction of uncertainty by the use of an ice-breaker was offset by the rules which raised a consciousness of language varieties and their differential power and prestige. The ISAGA GAME, as an ice-breaker and general acquaintanceship game, seemed to work well, but as a means of encouraging communication across language boundaries during the rest of the Conference, the results seem to have been less clear.

BIBLIOGRAPHY

For further discussions on some of the above topics and issues, the following are to be highly recommended.

Atkinson, J. M. and Heritage, J. (eds). 1984. Structures of Social Action: Studies in Conversation Analysis. Cambridge: Cambridge University Press.

Berger, C. R. and Bradac, J. J. 1982. Language and Social Knowledge: Uncertainty in Interpersonal Relations. London: Edward Arnold.

Fraser, C. and Scherer, K. R. (eds). 1982. Advances in the Social Psychology of Language. Cambridge: Cambridge University Press.

Journal of Language and Social Psychology. Multilingual Matters, Bank House, 8a Hill Road, Clevedon, Avon BS21 7HH, UK.

Gardner, R. C. 1985 Social Psychology and Second Langauge Learning: The Role of Attitudes and Motivation. London: Edward Arnold.
Giles, H. and St. Clair, R. N. (eds). 1979. Language and Social Psychology. Oxford: Basil Blackwell.
Ryan, E. B. and Giles, H. (eds). 1982. Attitudes towards Language Variation: Social and Applied Contexts. London: Edward Arnold.
Turner, J. C. and Giles, H. (eds). 1981. Intergroup Behaviour. Oxford: Basil Blackwell.

GAME MANUAL

The English part of the manual is provided below. It is complete except for the three topics on which participants have to interview each other. One manual per participant is made, and then the three topics are written in by hand. It is probably a good idea to have one serious topic, one humorous, and one ordinary or everyday. The range of topics should vary as widely as possible, but some people can be given the same topics.

The ISAGA GAME: Inquisitive Speaking And Gameful Acquaintance: A mix of tongues and communicating across languages

David Crookall, Alan Coote, Danielle Dumas and Alison Le Gat
University of Toulon and Polytechnic of Wales

This game is in two stages - please do not read the instructions for Stage Two before completing Stage One. You will need a pen or pencil for this game. You will also need some smiles, with which to show your good humour; you may also deploy laughter gestures, but only when jokes are in good taste. Also, be warned that chaos is to be expected!, though not encouraged!!

The general pattern of this game is simple. However, there are a few small, but important, rules. Please make sure that you understand these and what you have to do before you start. So, before you begin, please read carefully all the rules for stage one of the game.

STAGE ONE

Basically, what you have to do in Stage One is interview people and be interviewed by others, on given topics. But there are restrictions on whom you may interview, and on who may interview you; or rather on the languages you may use in an interview. The basic principle is that you may not use a common native tongue with your partner. (For the purposes of this exercise, the various American, British and other Englishes count here as the same languages; and so do French, Canadian and other Frenches.)

Rules

The detailed rules concerning restrictions on the languages you may use in interviews are as follows:

1) *Case 1. If both you and your conversational partner speak the same native tongue, you may not interview in that native tongue. However, if you and your partner have another second or foreign language in common, you may use that language.*

   For example, your native tongue is Latin, and your partner's native tongue is also Latin; you may not interview in Latin. But if both of you speak Ancient Greek as a second/foreign language, you may talk in Ancient Greek.

2) *Case 2. You may speak your native tongue only if it is your partner's second/foreign language.*

For example, your native tongue is Latin, and your partner's native tongue is Ancient Greek; you may converse in Latin, Ancient Greek, or Morse.

3) *Case 3. Both you and your partner may use a second/foreign language that you have in common.*

4) *To summarize, Rules 1 to 3 state that you may use your own native tongue in an interview, ONLY if that language is NOT the native tongue of the other person.*

5) *You may, of course, speak to people in any language order to establish whether or not you can go ahead with the interview. Obviously, you may also say "hello", introduce yourself, etc., but this "polite conversation" should not be prolonged, if you are not qualified to interview or be interviewed.*

Procedure

1) *Find at least five different people to interview (one after another). Remember that people will also want to interview you.*

2) *Check carefully that you can interview them (see rules above).*

3) *If you are able to interview them, make a brief note of the following:*

   a. *their first names;*

   b. *the languages they speak and how well they speak them (e.g. native, fluent, second/foreign).*

   c. *what they think about the following three topics:*

   ____________________; ____________________; ____________________.

   d. *if you wish, talk for a short while about anything else, e.g., your/their countries, work, why you/they came to ISAGA'86, your/their likes and dislikes.*

You should now have at least five opinions on each of your three topics. Only when you have completed this cycle for five people may you proceed to Stage Two of this game.

STAGE TWO

If you are ready to go on to Stage Two, you should read the following carefully, and then get back into the game. You now need to spend a couple of minutes looking over the notes from Stage One you made on the answers people gave you. Turn these into questions.

For example, if, in reply to your question "What do you think of ISAGA'86?", someone said to you "I think ISAGA'86 is exquisitely fabulous", you can make a question such as: "Is ISAGA'86 exquisitely fabulous?".

Another example: if in reply to that question, someone said "I think ISAGA'86 is a pushover", you can make the following question: "Do you agree that ISAGA'86 is a pushover?".

You should have 15 questions (five people providing answers to three questions).

For this stage the following rules apply:

1) *All rules in Stage One remain valid (if necessary consult these in the instructions for stage one).*

2) *You may not interview, or be interviewed by, someone you interviewed, or who interviewed you, in Stage One of the game. That is, the people you talk to now must be different to those you talked to in Stage One.*

You should now interview five people on your three sets of five questions, as follows.

1) *Find five (different) people to interview. Remember that people will also want to interview you.*

2) *Check carefully that you can interview them (see rules above).*

3) *If you are able to interview them, make a brief note of the following:*

   a. *their first names;*

   b. *when they have replied to a question, you should also ask them "why?", i.e., why they agree or disagree, etc.*

   c. *if you wish, talk for a short while about anything else.*

If the game has not yet ended, please continue to circulate among the others, chatting with whom you like. Topics for conversation are entirely up to you, but the following two simple rules apply:

1) *you should try and continue to apply the language rules from Stage One of the game.*

2) *you may not talk with the people you interviewed in Stages One and Two above.*

Advances in the Development of Hand-Held, Computerized Game-Based Training Devices

Rebecca L. Oxford
Center for Applied Linguistics, USA

Joan Harman and V. Melissa Holland
Army Research Institute, USA

ABSTRACT: This paper describes advances in the development of two hand-held, computerized, game-based training devices. These devices are light-weight and battery-operated. They convert lull time into training time because of the attractiveness of their gaming features and their portability. Gaming is incorporated into the devices to create intrinsically motivating instruction. Effective gaming enhances learning by increasing the amount of time spent on the learning task, not only in laboratory settings but also in typical classrooms and in operational training settings. Current applications of the two devices are in basic skills and job training. The military services developed these devices, but many non-military applications are envisioned as well.

KEYWORDS: Computer-assisted instruction, gaming, job training, basic skills.

ADDRESSES: RLO, Center for Applied Linguistics, 1118 22nd St. N.W., Washington DC 20037, USA; JH and VMH, ARI, 5001 Eisenhower Avenue, Alexandria, Virginia 22333, USA, Phone: 202/274-5538, 274-5948.

INTRODUCTION

In recent years, the U.S. Army, Navy, Air Force, and Marines have developed and implemented increasingly complex and sophisticated weapons and communications systems. During this same period, entry-level military personnel have demonstrated declining reading, writing, and computing skills. The discontinuity between increasingly demanding jobs and decreasingly skilled personnel has created a substantial training problem for the military services. The problem is not confined to the military, nor is it restricted to the U.S. or even the "first world." The growing technological orientation of most parts of our global village has placed new demands on ordinary citizens at work and at home. Personal computers are found on the U.S. East Coast, the Ivory Coast, the Gold Coast, and the Côte d'Azur. Technology is flowing into Baja, Bali, Bataan, and Baltimore. Not everyone has the basic literacy and math skills nor the technical expertise to deal with these changes.

One means of addressing the disparity between technological demands and existing skill levels has been provided by the U.S. Army Research Institute for the

Behavioral and Social Sciences (ARI) in the form of two hand-held, portable, computerized training devices that teach technical, job-related subject matter and basic skills largely through the use of games. This paper describes advances in the development of hand-held, computerized training devices that stimulate interest via gaming. These devices are an example of how technology itself can help solve problems engendered by increasing technological demands on workers' skills.

ADVANTAGES OF HAND-HELD, COMPUTERIZED TRAINING AIDS

Most computer-based instructional systems consist of desktop microcomputers that are very costly and are almost literally chained to the classroom due to their unportability. Often such systems provide drill and practice in a rather unmotivating format, lacking the excitement and challenge of games -- although gaming is becoming more prevalent in computer-based instruction.

Light-weight, portable, batter-operated, computerized training aids can offer many of the advantges of computer-based training in a more convenient delivery system. When scaled down to the size of a looseleaf notebook, these training devices dramatically increase the opportunities for computer-based training. Advances in semiconductor technology are allowing the creation of hand-held devices that have the following advantages:

- Can accompany the individual in a variety of living and working areas.
- Can convert lull time into training time because of portability and gaming features.
- Can increase students' intrinsic motivation through the use of games.
- Can ease the teaching load by shifting certain areas of the curriculum, such as drill and practice, away from the teacher.
- Can provide more interaction and more on-task time than is possible in any instructor-centered classroom situation.
- Can provide more individualization of instruction than does the typical instructor, thus addressing individual differences in aptitude and learning rate.
- Can provide more detailed record-keeping than is possible for an instructor, who is usually busy with a large group of students.
- Can free larger, more sophisticated computers and simulators to train complex decision-making, problem-solving, and team performance instead of routine fact training or procedural training.
- Can support initial training.
- Can sustain or refresh previous training.
- Can offer "distance training" when the student is separated from the instructor.
- Can withstand the wear-and-tear of most field settings.
- Can provide training in a variety of technical specialities or basic skill areas through the use of substitutable, interchangeable, plug-in cartridges.
- Can offer audio effects and voice synthesis.

- Can provide graphics capability through the attached book and through the visual display screen.
- Can give an interface with a personal computer for the uploading of data for purposes of authoring or record-keeping.
- Can be widely used because of low cost.

We have indicated some of the advantages of hand-held, computerized training devices. However, such devices are only part of a total training system, which also includes large computerized training devices, simulators, individual microcomputers, local area networks of microcomputers, microcomputers linked to mainframes, print material, videotape and audiotape units, other mechanical devices, and, of course, "warmware"--instructors, curriculum developers, instructional systems designers, subject matter experts, evaluators, and students.

TWO HAND-HELD, COMPUTERIZED TRAINING AIDS

The two hand-held, computerized training devices to be discussed here are known as the Hand-held Tutor (to be called Tutor henceforth) and the Computerized Hand-Held Instructional Prototype (CHIP). Tutor was developed by Franklin

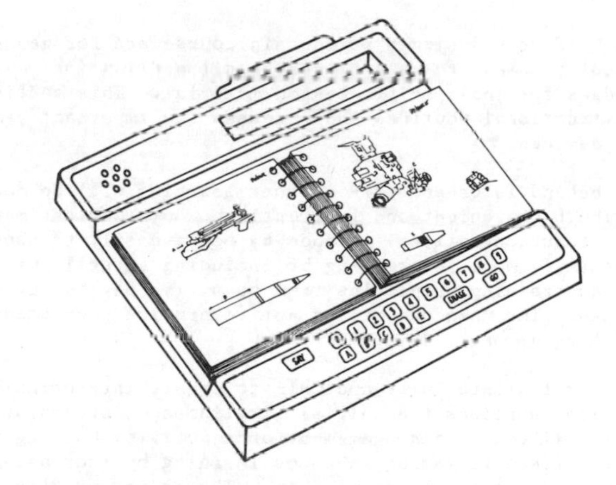

Fig. 1. Tutor

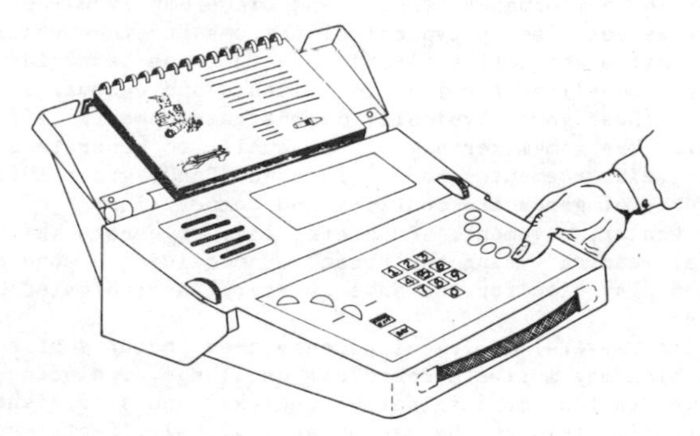

Fig. 2. CHIP with cover open

Research Center. CHIP is now under development by Technology, Inc. The development of both of these devices was planned and funded by ARI. Both devices are about the size of a three-ring, looseleaf notebook. Tutor weighs about four pounds or 1.9 kilograms, and CHIP weighs approximately five pounds or 2.3 kilograms. Both contain generic courseware for games and other instructional routines. Both have a place for a plug-in cartridge containing job-specific or basic-skill-specific training information. Both have diagnostic pretests; self-paced instruction compatible with initial knowledge levels, motivation levels, and learning rates; and frequent corrective feedback. Each device has a small speaker, a liquid crystal diode screen, keys for letters A-E and numbers 0-9, special function keys, and an attached book.

Tutor currently has three military applications: technical vocabulary training for the cannon crewman occupational specialty, basic mathematics training for combat engineers, and problem-solving for M-1 tank commanders. CHIP is now being developed for five job-related training applications: Army air defense, Army vehicle recovery, Air Force combat arms, Navy quartermaster, and Interservice explosive ordnance disposal. Many other applications have been recommended for both Tutor and CHIP in foreign languages, English as a second language, and job-specific areas for U.S. military installations throughout the world.

COURSEWARE

Both Tutor and CHIP were designed to contain courseware for generic instructional routines (including games) to be used across all applications, as well as special cartridges for job-specific lesson material. This section describes the generic instructional routines and stresses the important gaming features present in the devices.

The philosophy behind Tutor and CHIP is that learning will be more likely to occur if it is both convenient and interesting. Both devices make learning more convenient than conventional methods by being easily transportable. Both devices make learning more interesting by including several instructional routines based on training principles from cognitive psychology and video gaming techniques. The main purpose is not to provide recreational games but to provide teaching games.

Games are incorporated into Tutor and CHIP to create intrinsically motivating instruction, which increases the initial likelihood of student involvement and decreases the likelihood of disengagement once activity has begun (Condry and Chambers 1978). Effective gaming enhances learning by increasing the amount of time that is spent on the learning task. The amount of time spent on a given learning task is important. Increased task-related time is associated with better learning performance (Denham and Lieberman 1980) not only in laboratory settings but also in typical classrooms and in operational training environmentgs. Loftus and Loftus (1983) note that the best video arcade types of games are very compelling for most individuals and encourage hours of continuous play. These games typically demonstrate powerful effects of schedules of reinforcement. The gamemaker's goal is usually to generate a game in which a variable ratio reinforcement schedule is used (Pendergrass 1986). Other motivating features of games are conflict and competition; stimulus change (color, movement, sound effects); or events which derive their power from social meaning (being the "best" or receiving a monetary pay-off in terms of extended play time for the same quarter), as indicated by Pendergrass.

Malone (1984) cites several empirically researched features of successful learning games which may motivate learners: challenge, including a goal and an uncertain outcome; fantasy or imaginative context; and sufficient complexity to stimulate curiosity through the use of audiovisual effects, humor, and new informational content. Bobko, Bobko, and Davis (1984) cite evidence that destructiveness, dimensionality, and graphic quality are correlated with game popularity and hence motivational power. Keeping this gaming research background in mind, we now move to a description of the instructional routines contained in Tutor and CHIP.

Tutor

The major considerations in Tutor courseware development included multiple teaching techniques (gaming, drill and practice, etc.) to maximize a match with individual learning styles, initial knowledge levels, and rates of learning. Users can make selections from a menu of teaching and testing options that include gaming. The book includes many pictures and other graphics, and the computer provides both immediate and delayed visual and oral feedback to responses to multiple-choice questions.

Tutor coureware is divided into units that are sequenced from less to more difficult to promote an early experience of success by the user. Each unit consists of four parts: Pretest, Explanation, Picture Battle, and Word War. Users can choose any unit to work with and any component within the unit selected.

The Pretest is a short test that is intended to establish whether the user is knowledgeable about the subject matter being presented. Each unit Pretest contains multiple-choice questions with four or five answer options per question. Questions and response options appear in the book, and the individual responds by pushing the appropriate response key on Tutor. Upon completion of the Pretest questions, the person is given voiced feedback as to the number of correct answers. This number also appears in the display area. If all answers or all but one answer are correct, the user is permitted to move to any other component or any other unit or, if desired, to review the Pretest. If more than one response is incorrect, the user must return to the first Pretest item, review the test with accompanying corrective feedback, and then move to the explanation component in which the subject matter is taught. This component includes test questions as a check on the progress of the instruction. Rothkopf (1966) showed that testing can serve as a facilitator to learning, so the Pretest mode and the Explanation mode with its additional test questions are intended to provide direct instructional benefit.

The Picture Battle component is a game that requires matching pictures or graphic presentations with visual/oral stimuli. This component displays projectiles at each end of the display screen representing friendly and enemy targets. Correct responses result in movement of the friendly projectile toward the enemy target, and incorrect responses result in the same kind of movement of the enemy projectile. The objective is to destroy the enemy target before it reaches the friendly one. The impact with the enemy target is accompanied by a sound resembling an artillery shell exploding. The impact with the friendly target results only in both projectiles returning to starting positions to begin the game again.

Word War is a component of Tutor that is independent of the book. Both questions and multiple-choice answers are presented by the computer in the form of electronic flashcards on the display screen. The instructional method calls for drill and practice in an increasing ratio review format. That is, incorrect responses result in the question being presented again after one succeeding question, and once again after three additional items have been presented. Multiple-choice answers to questions answered incorrectly are randomly selected from other choices stored in Tutor's memory. Also, the position of the correct answer choice is randomly varied. The success of increasing ratio review has been demonstrated to shift learned information from short- to long-term memory (Wisher 1985). Siegel and DiBello (1980) found that the increasing ratio review was superior to conventional drill and practice, where missed items are not repeated until the end of the list.

Tutor courseware incorporates varying teaching techniques, presentation modes, and kinds of feedback in order to enhance acquisition and retention of selected subject matter. The courseware is heavily weighted with frequent, short tests to permit the user to monitor progress in acquiring needed information and to focus attention on the most relevant materials.

CHIP

CHIP courseware includes three games (Roll Call, Target Practice, and Mine Field) and one general instructional routine, sometimes called the Pregame. All the routines in CHIP are based on Tutor routines, except for Mine Field, which is a new job-step routine. The logic of all the CHIP routines (except Mine Field) reflects the logic of the Tutor routines, and the same basic learning principles apply to routines in both devices. For example, CHIP's Roll Call and Tutor's Word War both use increasing ratio review to facilitate transfer from short- to long-term memory. CHIP's Target Practice, like Tutor's Picture Battle, uses the technique of matching pictures with words and definitions as a multimodal way to enhance memory. The Pregame in CHIP is similar to the Pretest mode in Tutor in the use of delayed feedback.

CHIP's general instructional routine (the Pregame) offers both a warm-up and an explanation mode. The warm-up consists of multiple-choice of matching questions covering new material in the segment of instruction. If students miss questions in the warm-up or want further review of the material, they can enter the explanation mode. The general instructional routine does not have any gaming features, because they might distract students during the initial explanation process (Malone 1981).

Roll Call is a drill and practice gaming routine that requires students to match a list of ten words with their definitions, or vice versa. This routine is based on Tutor's Word War with an added gaming feature. Three choices of answers per word or definition continue to cycle on the screen, one at a time, until the student selects an answer. The goal of the game is to create on the screen a formation of ten soldiers, as at morning muster. For each correct answer selected, CHIP says, "Here, Sir" and adds a soldier to the formation on the screen; for each incorrect answer, CHIP responds "Missing" and displays a blank slot in the formation. If a question is missed, the cycle is repeated with the ten questions randomly rearranged until all questions are answered correctly. When the student scores 100 percent, CHIP says, "All present and accounted for, Sir."

A second game on CHIP is Target Practice. This routine is based on the Picture Battle routine of Tutor with enhanced graphics and gaming. Target Practice is used for vocabulary training or for general or multiple-choice matching questions. Typically, CHIP asks questions about a picture in the book. For example, CHIP might display the name or function of a part in a picture, and the student is required to locate the part and press the key corresponding to the appropriate label shown in the picture. The gaming technique has a friendly projectile and an enemy projectile at either end of the screen (much like Picture Battle). Each correct answer selected on CHIP moves the friendly projectile along a trajectory toward a target; an incorrect answer moves the enemy projectile. A new projectile is fired on each response, its distance depending on the number of correct and incorrect responses up to that time. The student's projectile can reach the enemy position after ten correct answers. On the other hand, the enemy projectile can reach the student's position after only four incorrect answers. If the student wins, the screen displays the words "You are a winner" along with the synthesized sound of an explosion. If the student loses, the display shows "Try again" and the game starts over.

The last routine is called Mine Field. This is a job-step routine that tests for correct answers to procedural steps. There are two principal modes: a job-step review that displays the steps of a job one at a time in the order they should be performed; and a game that provides drills about the sequence in which these steps are performed. The main objective of the game mode is for students to successfully progress through a mine field by correct responses to procedural steps. For each correct answer, a visual footstep successfully progresses through the mine field, and the student's score is increased at the same time as a chime sounds. For each incorrect answer, a footprint steps back one pace, the student's score is decreased, and a gong sounds followed by a spoken comment. The routine can also recognize steps that are impossible to perform and steps that are unsafe. In these cases, the game ends, and CHIP cycles to the job-step review mode.

To accommodate individual differences in learning and to provide a challenge to the student, each game has a basic and an advanced level. In the advanced level, responses are timed, and the student must respond YES or NO to every option; these two features increase the challenge to the student. In both levels, students control the order in which they work on the lesson segments.

EVALUATION OF TUTOR AND CHIP

In a 1984 field test, soldiers who used Tutor performed remarkably better than a group which used workbooks instead: 91 percent of the Tutor group completed all 20 units, versus only 58 percent of the workbook group. In terms of training effectiveness, the Tutor group again excelled, with nearly a 2:1 improvement over the performance of the workbook group. The control group, which had no special treatment, showed no significant difference from the workbook group. The Tutor field test also showed that soldiers enjoyed using Tutor and that they found it trouble-free and easy to use. There was a considerable difference between the fastest and the slowest time required to complete all units on Tutor, suggesting that self-pacing is a useful feature. In short, Tutor was found to be extremely effective in the field test. CHIP will be field tested soon.

SUMMARY

This paper has described two hand-held, portable, computerized training devices developed by the military services. One of them has been field tested and proven exceptionally successful in terms of student learning and attitudes. The other device will be tested this year. Gaming is one of the central instructional features of both devices and is used to increase student motivation and training effectiveness. These devices hold the promise of closing at least part of the gap between technological demands and existing skill levels.

REFERENCES

Asiu, B. 1986. Computerized Hand-Held Instructional Prototype. Unpublished paper.

Bobko, P., Bobko, D. J. and Davis, N. W. 1984. A multidimensional scaling of video games. Human Factors, 26: 477-482.

Bowen, K. C. 1978. Research Games: An Approach to the Study of Decision Processes. London: Taylor and Francis.

Condry, J. and Chambers, J. 1978. Intrinsic motivation and the process of learning. In Lepper and Greene (1979).

Denham, C. and Lieberman, A. 1980. Time to Learn. Washington, DC: National Institute of Education.

Gagne, R. M. and Briggs, L. J. 1979. Principles of Instructional Design (2nd ed.). New York: Holt, Rinehart and Winston.
Gagne, R. and White, R. 1978. Memory structures and learning outcomes. Review of Educational Research, 48: 187-222.
Goble, L. N., Colle, H. A. and Holland, V. M. 1986. Computerized Hand-Held Instructional Prototype (CHIP). Paper presented at the National Aerospace and Electronics Systems Society Conference of the IEEE, Dayton, Ohio.
Griffith, D. 1980. The Keyword Method of Vocabulary Acquisition: An Experimental Evaluation. ARI Technical Report 439. Alexandria, Virginia: U.S. Army Research Institute for the Behavioral and Social Sciences.
Hartley, J. R. and Lovell, K. 1984. The psychological principles underlying the design of computer-based systems. In Walker and Hess (1984).
Lepper, M. R. and Greene, D. (eds). 1978. The Hidden Costs of Reward: New Perspectives on the Psychology of Motivation. Hillsdale, New Jersey: Erlbaum.
Madigan, S. A. 1969. Intraserial repetition and coding processes in free recall. Journal of Verbal Learning and Verbal Behavior, 8: 828-835.
Loftus, G. R. and Loftus, R. E. 1983. Mind at Play: The Psychology of Video Games. New York: Basic Books.
Malone, T. W. 1981. Toward a theory of intrinsically motivating instruction. Cognitive Science, 4: 333-369.
Malone, T. W. 1984. Heuristics for designing enjoyable user interfaces: Lessons from computer games. In Thomas and Schneider (1984).
Oxford, R. L., Holland, V. M., Wisher, R. A. and Goble, L. N. 1986. guidelines for Designing Courseware for the Computerized Hand-Held Instructional Prototype. ARI Research Note. Alexandria, Virginia: U.S. Army Research Institute for the Behavioral and Social Sciences.
Pendergrass, V. E. 1986. Motivation and Representation through Gaming in Military Training. Unpublished paper.
Reynolds, J. H. and Glaser, R. 1964. Effects of repetition and spaced review upon retention of a complex learning task. Journal of Educational Psychology, 55: 297-308.
Robinson, C. A. 1986. A Hand-Held Training Aid in a Military Environment: Description and Proposed Evaluation. Unpublished paper.
Rothkopf, E. Z. 1966. Learning from written instructive materials: An exploration of the control of inspection behavior by test-like events. American Educational Research Journal, 3: 241-250.
Rothkopf, E. Z. and Coke, E. V. 1963. Repetition interval and rehearsal method in learning equivalences from written sentences. Journal of Verbal Learning and Verbal Behavior, 2: 406-416.
Rothkopf, E. Z. and Coke, E. V. 1966. Variations in phasing, repetition intervals, and the recall of sentence material. Journal of Verbal Learning and Verbal Behavior, 5: 86-91.
Siegel, M. and DiBello, L. 1980. Optimization of Computerized Drills: An Instructional Approach. Paper presented at the annual meeting of the American Educational Research Association.
Smith, E. E. and Goodman, L. 1984. Understanding written instructions: The role of an explanatory schema. Cognition and Instruction, 1: 359-396.
Thomas, J. C. and Schneider, M. L. (eds). 1984. Human Factors in Computer Systems. Norwood, New Jersey: Ablex.
Walker, D. F. and Hess, R. D. (eds). 1984. Instructional Software: Principles and Perspectives for Design and Use. Belmont, California: Wadsworth.
Wisher, R. A. 1985. The Development and Test of a Hand-Held Computerized Training Aid. Paper presented at the North Atlantic Treaty Organization (NATO) Symposium on Training Technology, Brussels, Belgium.

Simulation de dialogue pour l'apprentissage de la langue maternelle

Marie-Christine Paret, Michel Thérien,
Marcienne Lévesque
Université de Montréal, Canada

ABSTRACT: This study explores the possibilities for simulation in a pedagogical dialog concerning the teaching of language through the implementation of a tutorial which associates a theoretical model and the interactive, visual (colors, animation) and gaming capacities of the computer. The tutorial tries to reconstitute, by mean of an authoring language (PILOT), the elements of a pedagogical situation as presented in the Mastery Learning model of Bloom, which articulates the precise conditions for mastery learning, for which task analysis, formative evaluation, and individual correctives are essential.

KEYWORDS: Teaching, language, simulation, tutorial, mastery, learning, authoring-language, dialog, french.

La situation pédagogique se caractérise par les interactions qui s'installent à différents niveaux entre l'élève et le maître, l'élève et le groupe, le groupe et le maître; dans le cadre scolaire, le dialogue entre enseignant et apprenant est donc une des situations de base dans lesquelles se construisent les apprentissages. Depuis les origines de l'Enseignement assisté par ordinateur (EAO) on a tenté de reproduire ce type de situation, où l'enseignant était remplacé par la machine et présentait des connaissances à acquérir, au rythme de l'élève, par doses graduées et en fournissant immédiatement à ce dernier des renseignements sur ses succès ou ses échecs.

Si les critiques ont été et sont encore nombreuses envers l'EAO traditionne, tout le monde s'accorde pour dire qu'elles étaient méritées. Les tentatives actuelles de l'EAO de se rapprocher des systèmes intelligents (Sleeman and Brown 1982) en intégrant certains principes de l'apprentissage par la découverte dans une stratégie de tutorat, permettent de parler de "découverte guidée" ("computer coach instruction"). Mais pour l'instant, les acquis de l'intelligence artificielle ne sont pas accessibles au monde de l'éducation. Que faut-il faire en attendant? Peut-on s'appuyer uniquement sur les progiciels-outils que sont le traitement de texte, le chiffrier électronique ou la base de données et se priver de l'apport de l'ordinateur dans l'immense champ de la transmission de connaissances spécifiques? Il nous semble difficile de soutenir que ce soit là une solution socialement rentable. "Il faudrait plutôt permettre à tous ceux qui le souhaitent de créer des didacticiels. Pour cela, il faut lever la barrière informatique qui bloque l'accès des pédagogues au monde de la création du didacticiel." (Bibeau 1985). Un des moyens de le faire est l'utilisation de systèmes-auteur ou de langages-auteur.

A. QUELS ÉTAIENT NOS OBJECTIFS?

Nous sommes partis de l'hypothèse qu'il était possible de simuler, avec des moyens accessibles comme les langages ou systèmes-auteur, certains types de situations scolaires à partir de leurs éléments essentiels; autrement dit, nous voulions un modèle qui retrouve les conditions naturelles d'une communication pédagogique favorable aux apprentissages.

Plus immédiatement, nous voulions explorer un type d'outils dont pouvaient disposer les enseignants dans une direction peu empruntée - contrairement à la voie du traitement de texte - où les besoins sont immenses, et dans un contexte réaliste de moyens techniques et financiers limités. Nous avons donc entrepris l'élaboration d'un didacticiel sous la forme d'un tutoriel, à l'aide d'un langage-auteur (PILOT), puis d'un système-auteur (Micro-scope), de facon à connaître les possibilités conversationnelles offertes par ces deux systèmes pour le domaine de l'enseignement de la langue. Nous présentons ici ce qui a été réalisé à l'aide du langage PILOT et qui porte sur le discours narratif et le passage du style direct au style indirect.

La simulation de dialogue pédagogique repose sur la tentative de modéliser les démarches d'exploration de l'élève en situation d'apprentissage, soutenues et alimentées par les interactions avec l'ordinateur - pédagogue. Il ne s'agit plus, dès lors, d'une conception de l'enseignement comme accumulation pure et simple de connaissances mais comme appropriation progressive d'habiletés. La simulation implique l'élaboration de dialogues - entre l'ordinateur et l'étudiant - qui non seulement soient efficaces dans le sens qu'ils permettent de réels apprentissages, mais qu'ils le soient justement parce qu'ils sauraient retrouver les éléments essentiels de la réalité du dialogue pédagogique. Cela implique le respect de contraintes strictes en ce qui a trait à la qualité, au type et au style des interactions. Le système doit suivre une démarche qui permette la construction des apprentissages en partant de ce qui est acquis par l'apprenant, soumettre à sa réflexion des informations qui l'amèneront à une découverte personnelle, et dont on vérifiera l'intégration par un questionnement, mais aussi par des suggestions, des supplements d'information, des aides successives correspondant à ses diverses tentatives, et des retours en arrière. L'efficacité de ce dialogue repose avant tout sur l'analyse qui sera faite des réponses. Le traitement des mauvaises réponses devra pouvoir s'adapter étroitement au diagnostic des difficultés auxquelles se heurte l'élève car, comme le souligne Bork (1980): "Good dialogs often devote more of the program to responding to wrong answers than to the main line material".

Sur le plan pédagogique, nous avons choisi un modèle d'enseignement qui nous a paru apte à supporter de telles exigences et capable d'intégrer l'ordinateur sans subir de distorsions, soit le Mastery Learning de Bloom (1976) ou "Pédagogie de la maîtrise des apprentissages" (PMA), dont nous allons rappeler les grands principes.

Les différents points qui feront l'objet de la présentation sont les suivants: rappel des principes du Mastery Learning, choix du langage-auteur PILOT, PMA et simulation de dialogue pédagogique, structure et démarche du logiciel (didacticiel), discussion.

B. LA PÉDAGOGIE DE MAÎTRISE DES APPRENTISSAGES

Le modèle du Mastery Learning (ou PMA) est une pédagogie qui est en fait une philosophie fondée sur la croyance en la capacité d'apprendre de tous les enfants; elle critique la notion d'aptitude innée et rejette l'idée qu'il est normal que la majorité des élèves n'atteigne pas tous les objectifs que fixe l'école. Elle affirme que le niveau d'apprentissage n'est pas déterminé seulement par les caractéristiques de l'élève, mais aussi par celles de l'enseignement et que 95% des élèves sont capables d'apprendre à l'école ce que l'école a à enseigner.

Le modèle théorique a été développé par Bloom (1976) à partir des travaux de Carroll (1963) qui montrait que si la forme et la qualité de l'enseignement ainsi que la quantité de temps disponible pour apprendre sont appropriés aux besoins et aux caratéristiques de chaque élève, le rendement s'élève bien au-dessus de la distribution normale. Bloom isole les différentes variables qui déterminent l'apprentissage et précise leur définition. L'une d'entre elles, l'enseignement, devient pour lui une variable dominante, des recherches ayant montré que ce n'est pas tant les caractéristiques des enseignants ou celles de la classe ou de l'école, qui expliquent la réussite des élèves, mais celles de l'enseignement.

Cette variable, qu'il appelle qualité de l'enseignement, est définie comme fonction des quatre composantes suivantes: 1) les indications (ou indices) qu'on donne à l'élève pour qu'il mène à bien la tâche d'apprentissage, 2) la participation de l'élève, 3) le renforcement, c'est-à-dire l'approbation à sa réussite ou à ses efforts, et 4) le feedback informatif et correctif.

Dans une démarche de PMA, l'enseignant commence par définir des objectifs, c'est-à-dire des contenus à faire maîtriser; pour chacun, il fixe un niveau ou un critère de réussite. Ce système d'évaluation permet la promotion individuelle non comme résultat d'une compétition qui produit gagnants et perdants, mais comme la confrontation à des critères. Le cours entier est ensuite divisé logiquement en séquences ou unités d'enseignement et d'apprentissage avec aussi les objectifs à atteindre pour chacune. On planifie ensuite l'unité. Toutes les activités et toutes les méthodes peuvent être intégrées par le modèle. La planification doit inclure des procédés de feedback et des correctifs à la fin de chaque unité, c'est-à-dire des tests diagnostiques qui informent les élèves sur la qualité de leur performance, jouant un rôle strictement formatif.

Cependant, ce n'est pas tous les élèves qui répondent à l'enseignement initial de manière à maîtriser réellement l'apprentissage; la planification devra donc prévoir aussi d'autres procédés d'enseignement appelés correctifs qui sont définis pour chaque élève qui n'a pas réussi tous les objectifs en se basant sur les résultats au test formatif. L'élève, avec l'enseignant, seul ou avec l'aide de ses pairs, doit revoir les objectifs avant de passer au second test formatif. Après les correctifs, l'enseignant purpose un autre test diagnostique portant sur les mêmes objectifs tandis que des activités d'enrichissement sont prévues pour les élèves qui ont atteint plus rapidement l'objectif. Après plusieurs unités, l'habileté de chaque élève est évaluée par l'examen sommatif préparé au début du module.

C. LE CHOIX DU LANGAGE AUTEUR PILOT

Rappelons qu'un langage-auteur est un langage de programmation, mais conçu dans un soucis pédagogique, spécifiquement pour permettre la préparation de logiciels à contenus éducatifs, appelés encore "didacticiels". Il est destiné non à des informaticiens mais à des pédagogues. C'est un langage de programmation simplifié (chaque instruction correspond à plusieurs instructions d'un langage traditionnel) et auquel on a incorporé des fonctions utiles à l'enseignant, comme par exemple la possibilité de comparer directement une réponse donnée par l'élève et une ou plusieurs réponses jugées acceptables. Le vocabulaire et la syntaxe sont moins abscons que ceux des langages informatiques traditionnels; ils sont moins éloignés des langues naturelles.

Pourquoi avoir choisi un langage-auteur? Parce que nous croyons qu'il y a place pour l'ordinateur-enseignant puisque l'école reste un lieu d'appropriation des connaissances. Par ailleurs, dans bien des domaines, l'aspect systématique de la matière peut être favorisé par l'utilisation de l'informatique (sciences de la nature, langue, etc.). Parce qu'aussi le matérial pédagogique produit par l'enseignant lui ressemble, s'intègre naturellement à son enseignement, respectant son style et les approches pédagogiques qu'il privilégie. Enfin, aspect non négligeable, ce matérial peut évoluer, vivre en quelque sorte, selon les besoins des usagers.

Simuler un dialogue implique une grande souplesse dans les interactions, même si ces dernières sont orientées vers un but précis. PILOT semblait posséder des facilités particulières pour traiter les chaines de caractères. Et en effet, il s'adapte bien à l'enseignement de matières comme les langues ou les sciences humaines. Par ailleurs, il permet l'édition de texte en traitement de texte, le dépistage d'un bon nombre d'erreurs de programmation, et autorise même la poursuite de l'éxécution si le programme ne contient pas deux erreurs consécutives. Il possède une structure de base très simple qui correspond à une unité minimale d'enseignement: texte (explication et/ou question), entrée de la réaction de l'élève (réponse ou commentaire), comparaison avec le contenu attendu, stucture qui peut se complexifier et s'associer avec certaines commandes jusqu'à permettre une analyse très subtile des réponses. La commande "match", par exemple, permet d'associer plusieurs réponses, de les dissocier, de tenir compte ou non de l'ordre des réponses données, d'exclure de la réponse une chaîne choisie, d'accepter n'importe quel caractère dans un contexte donné.

Autres points forts, les facilités de cheminement dans le programme et de branchements qu'on peut toujours rendre conditionnels, les possibilités graphiques, le dossier de l'étudiant qui enregistre tous ses essais et ses réponses.

D. PMA ET SIMULATION DE DIALOGUE PÉDAGOGIQUE

PMA et ordinateur, un modèle et un outil destinés à se rencontrer dans la mesure où le coeur même de leur dynamique respective est le dialogue pour faire apprendre. L'efficacité de la communication pédagogique informatisée est fonction de la possibilité de se rapprocher d'une communication pédagogique réelle, c'est-à-dire dont on pourra simuler les composantes de base, soit une information pertinente pour l'accomplissement de la tâche, la possibilité pour l'élève de se concentrer sur cette tâche et d'avancer à son rythme propre, de savoir à chaque étape quelles sont les caractéristiques de sa performance et la possibilité de pouvoir la modifier pour finalement obtenir une évaluation de l'atteinte de l'objectif poursuivi.

On retrouve dans ce schéma les quatre composantes de la qualité de l'enseignement selon la PMA. En effet, la première étape consiste en la présentation d'indications (ou indices), c'est-à-dire d'informations qu'on donne à l'élève pour le guider dans les comportement à produire (les activités) pour atteindre les objectifs. Ces indications peuvent avoir différentes formes et par là des forces différentes. C'est un domaine où la technologie du micro-ordinateur avec ses possibilités grandissantes de jouer avec les formes, les couleurs, le mouvement, de rendre concret et de permettre d'observer en simulant, peut jouer un rôle déterminant. La PMA met de l'avant la nécessité de la structure en trois temps: informations à l'étudiant, réponse de ce dernier, feedback formatif, tandis que d'autre part, on ne cesse de répéter (Taylor (1980) que la plus grande force de l'ordinateur est sa capacité d'interactivité, qu'on se plaint justement de voir mal exploitée dans la plupart des didacticiels disponibles au niveau scolaire.

On sait qu'une des conditions de base de tout apprentissage est l'implication du sujet, sa participation ACTIVE; on a là la seconde composante de la qualité de l'enseignement. Les enseignants utilisent toute une variété de stratégies pour assurer cette participation; mais on sait que le manque de temps et les exigences de l'enseignement collectif empêchent l'enseignant d'observer chaque enfant pour être bien sûr d'une part qu'il est actif et d'autre part qu'il utilise au mieux les indications qu'on lui donne. Cependant, tous les élèves n'ont pas besoin de participer avec la même intensité pour apprendre. Pourtant, les conditions d'enseignement obligent les enseignants à faire comme si tous les élèves avaient besoin des mêmes exercises! L'intérêt que présente l'aide de l'ordinateur est d'une part de solliciter sans arrêt l'activité individuelle, d'autre part, d'individualiser les cheminements. Une de ses grandes forces est de permettre d'adapter la progression du didacticiel au rythme de l'élève et de tenir compte du type de difficulté auquel il se heurte pour lui proposer des contenus particuliers.

Le troisième élément qui contribue à la qualité de l'enseignement selon Bloom est le renforcement, c'est-à-dire les approbations et les encouragements prodigués à l'apprenant. Même s'il est clair que les renforcements donnés par la machine ne seront jamais de même nature que ceux que peut founir le maître, l'ordinateur a la possibilité de réagir avec assez de subtilité et de variété. De plus, il a l'avantage de réagir individuellement, ce qu'un enseignant qui a trente élèves dans sa classe n'a pas le temps de faire autant qu'il le voudrait.

Enfin, le dernier des quatre éléments pour un apprentissage de qualité selon le modèle de Bloom, et peut-être le plus important, est constitué des procédures de feedback et de correction. L'enseignant doit constamment s'adapter aux réactions de ses élèves; l'idéal serait qu'il s'ajuste aux réactions de CHAQUE élève, ce qui est évidemment impossible. La pédagogie de la maîtrise est une approche centrée sur le groupe mais avec correctifs individuels, l'ndividualisation étant un de ses concepts de base, tout comme il est une implication de la notion de dialogue. On se préoccupe particulièrement du feedback, notamment sous la forme de tests formatifs administrés à la fin de chaque tâche d'apprentissage et qui indiquent ce qui est maîtrisé et ce qui ne l'est pas. A partir de là, on peut décider des procédures de correction à suggérer à chaque élève dans lesquelles on va proposer des indications alternatives, différentes des premières, des démarches alternatives, du temps et des pratiques supplémentaires. Tout ceci joue une rôle capital pour améliorer l'apprentissage des élèves et permettre l'atteinte des objectifs pour 90 ou 95% d'entre eux.

On voit immédiatement comment de telles procédures sont tout à fait possibles dans un enseignement assisté d'ordinateur basé sur le dialogue pédagogique; l'élève peut évaluer son progrès, reprendre les notions mal maîtrisées et d'une façon pui peut lui être mieux adaptée personnellement, mettant ainsi toutes les chances de son côté.

A la suite de cette expërience d'ëlaboration de didacticiel, il apparaît clairement que l'ordinateur outil d'enseignement peut intégrer l'essentiel des principes d'une démarche aussi exigeante mais stimulante pour un pédagogue que la Pédagogie de la maîtrise, dans la mesure où ce modèle détermine les éléments de base de l'interaction authentiquement pédagogique, éléments à partir desquels peut s'élaborer une simulation de dialogue d'apprentissage par la découverte.

E. DISCUSSION

Mais dans quelle mesure, puisqu'on a affaire à des logiciels "à contenus", des didacticiels, est-il possible de maintenir une grande souplesse dans les interactions au point qu'on puisse parler de dialogue véritable? Il semble que la réponse s'articule essentiellement autour de deux points, d'abord la démarche pédagogique, qu'on a choisie de type inductif, c'est-à-dire où des données sont présentées à l'élève et où un dialogue guide l'observation et la réflexion pour en faire découvrir les caractéristiques. Ceci entraîne une concentration sur la tâche qui facilite les échanges puisqu'ils s'effectuent dans une domaine plus circonscrit. Le deuxième élément de réponse repose évidemment sur la capacité du langage d'analyser les réponses, qui est grande mais pas sans limite, puisqu'il n'est pas question d'analyse de la langue naturelle et qu'en conséquence toutes les entrées de l'élève doivent être prévues. Malgré tout, la vraie limite dépend surtout de l'imagination de l'auteur et, il faut bien l'avouer, du temps. On remarque également que ce qui semble favoriser les échanges, dans le matériel présenté, est la capacité de prendre en compte la variété des réactions de l'élève en même temps que la présence d'un soutien constant pour avancer dans la tâche. Ne peut-on pas alors s'interroger sur la complexité de la programmation et du temps que devrons y consacrer des enseignants qui ne sont pas des spécialistes dans le domaine? Il faut convenir que l'outil, même s'il s'agit d'un langage très simplifié, ne conviendra certes pas à tous, mais qu'il est étonnant de voir avec quelle ferveur des maîtres que nous formons ou perfectionnons à l'Université de Montréal se lancent dans ce travail.

REFERENCES

Bibeau, R. 1985. Les systèmes-auteurs: des outils pour les A.P.O. Bip-Bip (Ministère de l'éducation, Québec), 37-fév, 38-40.

Bloom, R. S. 1976. Human Characteristics and School Learning. Chicago: McGraw-Hill.

Bork, A. 1980. Preparing Student-Computer Dialogs: Advice to Teachers, in Taylor (1980).

Gagné, R. M. 1977. The Conditions of Learning. NY: Holt, Rinehard and Winston.

Sleeman, D. and Brown, J. S. 1982. Intelligent Tutoring Systems. Computers and People Series, NY: Academic Press.

Taylor, R. (eds). 1980. The Computer in the School, Tutor, Tool, Tutee. New York and London: Teachers College Press.

Ce travail a été réalisé grâce à une subvention de la Direction générale de l'enseignement et de la recherche universitaires (Ministère de l'éducation du Québec).

Computer-assisted language acquisition: Planting a SEED

D. Wells Coleman
Academy of Science of China, Peking

ABSTRACT: Language teaching computer-assisted isntruction (CAI) is almost entirely computer-assisted language learning (CALL), not computer-assisted language acquisition (CALA) (Krashen, 1982:10). If Krashen's distinction between learning and acquisition is valid, then what we should have is CALA, not CALL. Coleman (1985) suggested that this could be done with student-computer conversational simulations, since such programs "can simulate actual language use" by dealing "directly with linguistic, rather than metalinguistic skills." The presenter will outline a hypothetical programming language, SEED (Simulator of Environments for Educational Dialogues), which would facilitate the design of appropriate conversational simulations. A typical 'lesson' written in SEED would be able to simulate an artificial 'world' (via computer graphics), store and use 'knowledge' about this world, and 'discuss' this world in natural language (via keyboard input and text output).

KEYWORDS: computer-assisted language learning, CALL, second language acquisition, conversational simulation, SEED, computer-assisted instruction, CAI.

ADDRESSES: D. Wells Coleman, 1104 NE 3rd ST, Gainesville, FL 32601 (USA); Douglas Wells Coleman, Warsaw (Fulbright), Department of State, Washington, DC 20520 (USA).

The topic of this presentation is SEED (Simulator of Environments for Educational Dialogues), a hypothetical microcomputer-based programming language whose immediate purposes would be (a) to facilitate the creation of computer-simulated environments and (b) to facilitate natural language (e.g. English) student-computer dialogues about these simulated environments. The ultimate purpose of programs created with SEED would be to foster foreign language acquisition.

THE PROCESS OF FOREIGN LANGUAGE ACQUISITION

The model of foreign language acquisition I am assuming here bears a certain surface resemblance to Krashen's (1982). First, assumed is a distinction between language learning - "concious knowldge of a second language", i.e., "'knowing about' a language" - and language acquisition - "a process similar, if not identical, to the way children develop ability in their first language" (Krashen, 1982:10). I also accept, in general, Krashen's further arguments that learning "does not 'turn into' acquisition" (1982:83) and therefore plays "only a limited role in second language performance" (1982:16).[1]

The process of foreign language acquisition can be diagrammed as in Figure 1. First, target language input (input in the language to be acquired) is received by the student. Part of this input may be decoded, i.e. understood via use of the student's internal model of the target language. Part of the input may not be understood via use of the internal model; this part is what I will refer to as 'undecodable residue'. The portion that is decodable, taken together with extralinguistic clues, can 'fill in the gaps' - resulting in a 'semantic residue', i.e. meaning which cannot be associated with specific parts of the decoded input. The presence of these residues invokes the language acquisition device, which then attempts to modify the internal model of the target language. The goal of the modification is to define the relationship between the undecodable residue and the semantic residue, and in so doing, to attempt to make undecodable residues decodable.

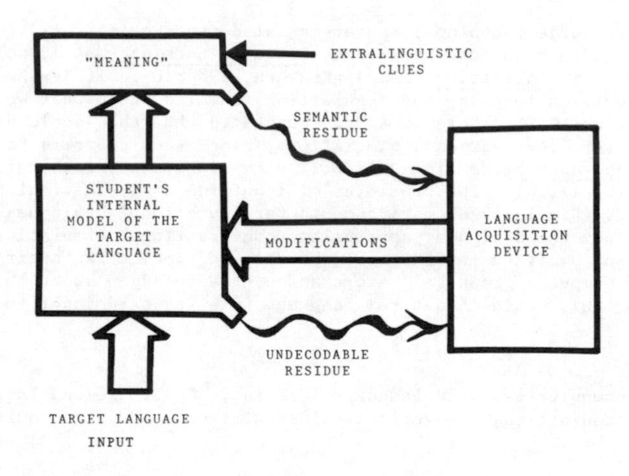

Fig. 1

WHAT DOES THIS MEAN FOR FOREIGN LANGUAGE CAI?

Foreign language CAI (computer-assisted instruction) which teaches students about the target language will be minimally useful, since it will tend to affect only concious learning and not acquisition. To promote acquisition, input must be provided which contains 'understandable' undecodable residues. The simulation environment provides the extralinguistic cludes needed by the student's language acquisition device for dealing with undecodable residues.

In Coleman (1985), I described TERRI, an example of a student-computer conversational simulation. A conversation with TERRI "can simulate actual language use" by dealing "directly with linguistic, rather than metalinguistic skills" (247).

Still, in light of the above view of second language acquisition, programs like TERRI have two serious limitations. First, TERRI-student discourse tends to resemble "teacher-student talk" in a negative way: the reality of the simulation is sometimes compromised by TERRI's comments about the student's linguistic errors (c.f. Dinsmore, 1985:229-230). Second, the student does most of the talking. A greater balance would provide more understandable undecodable residues. Hence, in my own mind, TERRI was less than a prototype; it was (or so I have thought) just a step in the right direction.

To realize more balance in student-computer conversational simulations, programs must have greater ability to engage in meaningful two-way conversation. Programs must be able to handle certain discourse functions, e.g. naming, attributing characteristics, managing conversational turn-taking, and so on. To accomplish this, SEED would need some sort of general parsing capability. I am assuming small-scale natural language processing capabilities are sufficient. Therefore, I will assume that the major features of an existing programming language like SNOBOL could form a basic part of SEED.[2]

FOCUSING ON THE MACROSCOPIC

Second, SEED would need to permit the programmer to define the environment at a relatively macroscopic level, something programming languages usually do not do very well, if at all.[3] Take for example the simple simulated environment in Figure 2: a ball rolling away, bouncing from side to side down a long corridor. This environment is so simple as to appear an abstraction. However, the programming task it represents can be a bit tedious.

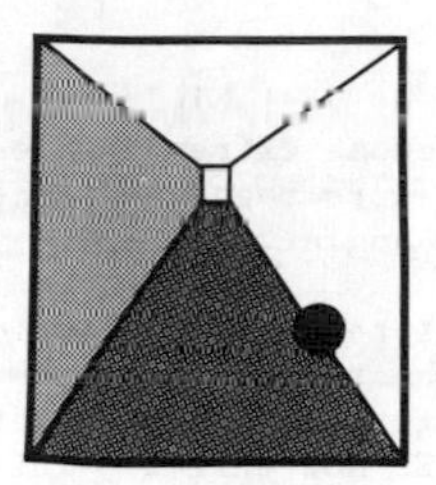

Fig. 2

The programmer is forced to focus on the representation rather than the environment, for example in the way the ball's movement must be dealt with: as changes in position on the screen display. What the programmer wants the student to perceive is a ball bouncing away from side to side down a hallway. What must be displayed is a disk shrinking in size, tacking from side to side up the screen, with a progressively shorter track between turns and with each turn forming a progressively sharper angle. Some programming languages commonly used for CAI even require that the movement of the ball against the background be explicitly regarded as a repetitive draw-erase-redraw operation.
To eliminate these problems, SEED should allow the programmer to write programs in such a way that an object in the simulated environment is an object in the program, an event in the simulated environment is an event in the program, and so on. This is what I mean when I say that the programmer should be able to define the simulated environment at a relatively macroscopic level. If this is taken in the strictest sense, it is probably a long way off, at best. But, there are certain things that can be done.

WHAT'S IN A NAME?

In the remainder of this presentation, I would like to focus specifically on the problem of naming in a simulated conversation. Naming, in this context, I will define as the association of character strings with variables or graphics in the program. The first step is the association of an object in the environment with an object in the program.

One already existing feature which helps a great deal is a graphics convention known as the movable object block (also called a 'sprite' on Commodore and Atari microcomputer systems). Movable object blocks (or MOB's), on an advanced system, are created with a sophisticated MOB editor which allows easy drawing of the MOB and which allows MOB's to be copied, rotated, inverted, and so on, to create other MOB's. Such an editor helps programmers feel they are creating images, not composing instructions. After using an editor to create an MOB, the programmer can refer to it by name (or number), thereby focusing wholly on the product - the desired image - rather than on the instructions needed to create that image. For representing discrete objects, MOB's have significant advantages over other types of graphics, such as customized characters. MOB's cover a larger area than a single character, so they allow unitary treatment of an image that might need to be composed of several customized characters. In addition, MOB's can be turned on and off; they can be expanded to double-size horizontally, vertically, or both. When the location of an MOB is changed, it moves nondestructively; that is, the background in its former location is automatically redrawn by the system.

If MOB's were treated as a type of variable, this would present some significant advantages. These advantages become extremely clear when MOB's are used to represent objects which require more than one image, as for example when a single object is viewed from different angles.

MOB arrays would greatly facilitate the animation of such objects. Suppose we want to represent a duck in a shooting gallery, swimming back and forth from left to right. The duck is to turn smoothly at the end of its track. First, a duck would be drawn with the SEED MOB editor. This graphic would be assigned to the first frame of the MOB array DUCK (Figure 3). The SEED MOB editor 'squash' function would then be used to alter the drawing so that it appeared as in frame 2 of Figure 3. The image would be successively 'squashed' until it appeared little more than a line (frame 5). Frame 5 would then be inverted to produce frame 6; frame 7 would be the inversion of frame 4, 8 of 3, and so on. After creating this array, the programmer could 'turn' the duck at the left end of the track by displaying each frame from 1 to 10 one at a time. SEED would contain a command (call it 'CYCLE') to accomplish this, e.g. "CYCLE DUCK 1 TO 10 RATE 6". To make the duck turn at the other end of the track, SEED would permit something like "CYCLE DUCK 10 TO 1 RATE 6".

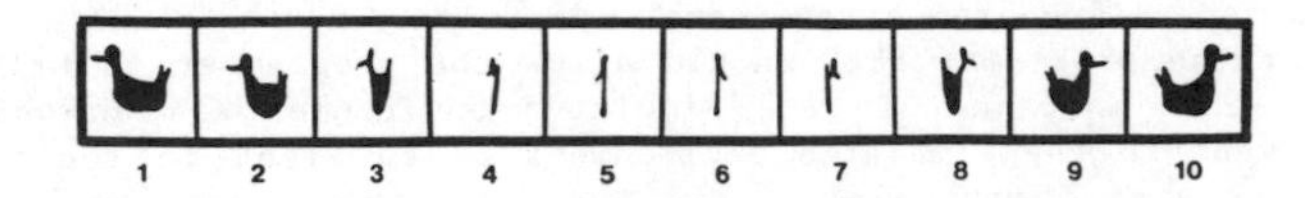

Fig. 3

Now suppose the duck were shot. When a shooting gallery duck is hit, it spins wildly. If the frames were displayed one at a time, forward, then backward through the array over and over, it would give the impression that the duck was spinning. The SEED command "CYCLE DUCK RATE 2" might do this.

This shooting gallery duck is perceived as one object in a simulated environment. Thus, the real advantage of MOB arrays is that they would allow the programmer to communicate with SEED about the animated duck as if it were one object. One object in the environment would become one object in the program.

Now, to completely deal with the problem of naming, SEED must allow the programmer to easily relate natural language elements (i.e. character strings) with objects in the program. Take a student question about whether a certain object in the environment possesses a certain characteristic, e.g. "Is the ball moving?" (See again Figure 2.) The character string "the ball" is associated with an object in the simulated environment. To the program, this means that it is associated with a graphic and/or an array enumerating the characteristics of the object. The string "moving" is associated with a certain value for a particular element of this array - the element defining the velocity of the object. The initial "Is" and final "?" are associated with a programmer-defined sequence of instructions which checks whether the named object possesses the characteristic identified.

Via parsing, "Is the ball moving?" is decoded to the following values (with strings converted to all caps for convenience).

OBJECT = "BALL"

CHARACTERISTIC = "MOVING"

FUNCTION = "IS"

Then the character string variable ACTION is formed this way (where '&' indicates catenation).

ACTION = FUNCTION & " " & OBJECT & "[" & CHARACTERISTIC & "]"

This yields the value "IS BALL[MOVING]" for ACTION. Suppose EXECUTE is a SEED command where

EXECUTE ACTION

Means 'take the character string ACTION and treat it as if it were a SEED instruction'. Now suppose (a) BALL is an array containing the codes for certain characteristics of the object, (b) MOVING references a subscript of BALL, and (c) IS is a programmer-defined subroutine which displays the character string "Yes." if BALL[MOVING] is nonzero and "No." if BALL[MOVING] equals zero.

At this point, SEED semantics would begin to resemble - at least a little more - the semantics of natural language. And if I can conclude here with a generalization, this will have to be a key characteristic of any programming language designed for conversational simulation in foreign language teaching.

NOTES

[1] Serious problems have been noted with Krashen's model. this point was mentioned by Douglas Morgenstern during the Conference. The model assumed here is based on a relational network approach to linguistic analysis (ultimately based on Sydney M. Lamb's Outline of Stratificational Grammar, 1966), however, not on Krashen's "Input Hypothesis". Unfortunately, those curious about details of this model will have to wait until the author's paper (currently in progress) "A relational network view of language acquisition" is available.

[2] As one participant pointed out, SNOBOL now runs on the IBM PC. However, neither SNOBOL nor any similar language is available for the vast majority of microcomputers currently used for computer-assisted instruction.

[3] One user of the FORTH programming language mentioned later in the Conference languages (like FORTH) which provide for "information-hiding"; such languages allow the programmer to build up program modules into more and more complex units until a relatively macroscopic level is reached.

REFERENCES

Bell, R. T. 1981. An Introduction to Applied Linguistics: Approaches and Methods in Language Teacing. London: Batsford.

Coleman, D. Wells. 1985. TERRI: a computer-assisted language learning (CALL) lesson simulating conversational interaction. In Crookall, D. (ed). 1985. Simulation Applications in L2 Education and Training. Oxford: Pergamon Press. Special issue of System, Vol. 13, No. 3.

Dinsmore, D. 1985. Waiting for Godot in the EFL classroom. ELT Journal, 39:4.

Higgins, J. and Johns, T. 1984. Computers in Language Learning. New York: Addison-Wesley.

Jakobson, R. 1960. Closing statement: linguistics or poetics. In Sebeok (1960).

Jones, K. 1982. Simulation in Language Teaching. Cambridge: Cambridge University Press.

Krashen, S. D. 1982. Principles and Practice in Second Language Acquisition. Oxford: Pergamon Press.

Papert, S. 1980. Mindstorms: Children, Computers, and Powerful Ideas. New York: Basic Books.

Sebeok, T. A. (ed). 1960. Style in Language. Cambridge, Mass.: MIT Press.

Internal audiences in computer simulations: Painting a moving target

Charles R. Ryberg and Edward B. Versluis
Southern Oregon State College, USA

ABSTRACT: During the short history of using computer simulations to teach communications, there has been a perceived paradox of relatively long standing. That paradox runs something like this: the more you use a computer to promote human communication, the less human communication will be promoted. We want to argue for the benefits to the student of a much higher level of computer intrusion into simulated communication situations, creating another more interesting paradox, the more extensive the intrusion of the computer into a simulated situation, the more invisible it will become. Perhaps the most important concept in grappling with this new view of computer assisted instruction is the audience within the computer simulation, the "internal audience," whose design really does constitute "painting a moving target."

KEYWORDS: audience analysis, communications simulations, computer assisted instruction, computer simulations.

ADDRESSES: Edward B. Versluis, Department of English, Southern Oregon State College, Ashland, Oregon 97520. (503) 482-6181 - School, (503) 482-3991 - Home.

During the short history of using computer simulations to teach communication, there has been a perceived paradox of relatively long standing. That paradox runs something like this: the more you use a computer to promote human communication, the less human communication will be promoted (Coote 1985). In its extreme form, this paradox predicts that extensive intrusion of the computer into a simulated communication situation will be detrimental to the student by drawing his or her attention away from other humans thereby creating a machine dominated isolation.

The authors disagree with this perceived paradox. Instead we want to argue for the benefits to the student of a much higher level of computer intrusion into simulated communication situations. Once the computer has been extensively involved in a communication situation, another more interesting paradox can be described. This new paradox can be stated as, the more extensive the intrusion of the computer into a simulated situation, the more invisible it will become, thereby concentrating the student's attention instead on the human aspects of communication. Perhaps the most important concept in grappling with this new view of computer assisted instruction is the audience within the computer simulation, the "internal audience," whose design really does constitute "painting a moving target."

Internal audiences in computer simulations can best be understood as arising out of the interaction of educational simulations, computer assisted instruction, and the teaching of writing for the professions (or, as it is sometimes known) Technical Writing. Though it is probably useful to remember that, "Simulations and language are virtually inseparable" (Jones), the class of simulations for communication we are discussing is more focused than that. This can easily be seen if we begin with a rather abstract description of a simulation, that, "A simulation mimics the interesting properties of a system" (Versluis 1984). The "interesting properties" in this class of simulation can be found discussed in the learned tradition from as far back as Aristotle's Rhetoric and coming down to the latest writing composition textbook for university students. In this tradition learning communication skills is equated with learning how to create connections between what the communicator knows and wants and what he or she knows about the audience.

How to mimic those interesting properties has been largely solved by all the many simulations whose operation involves communication skills. One area that remains to be explored is whether the audience for a given communication situation, always conceived of in terms of a cluster of traits, giving rise to a set of response probabilities, can itself be mimicked, and further, can that mimicry be made an integral and effective part of a computer simulation?

Few subjects are so intriguing as the role of the computer in our future. Indeed, the fact that computers seem so compelling as tools, toys, or even surrogate people, makes the discussion of their proper and probable roles particularly urgent. However, for some reason, actual computer assisted instruction is not quite that flashy. Computer software in the field of writing instruction has moved in the last few years beyond so-called "drill-and-practice" programs, each centering on the unrelenting repetition of some spelling or grammatical point. Today there are many programs designed to ask questions of a student before and during the writing of various kinds of essays. The content of those questions is derived from current theories about how writing takes place; the vehicle for asking those questions is the computer, guided by the computer program. Or, put more succinctly, today computers are routinely asking students pointed questions about written communication.

Whether computers are ever able to think exactly like human beings or not, what has gone unremarked so far is that we are already living in the presence of a powerful and frequent simulation of an instructor of written communication prompting a student before and during the actual performance of writing. The first and still most famous instance of a computer imitating a human being questioning another human being was produced and commented on by Joseph Weizenbaum (1976) at the Massachussets Institute of Technology. His ELIZA program "conversed" with its human audience. In one version of this program, called DOCTOR, Weizenbaum directed the computer to produce what he called a "parody" of a psychotherapist engaged in an initial interview with a patient. The fact that humans using this program tended to respond to the computer as if it were human deeply disturbed Weizenbaum, and governed much of his subsequent criticism of the possible dangers to society of computer use.

The ease with which the computer can be made to imitate human agents is explained by Alan Kaye's assessment of the computer's potential power.

> The protean nature of the computer is such that it can act like a machine or like a language to be shaped and exploited. It is a medium that can dynamically simulate the details of any other medium,

> including media that cannot exist physically. It is not a tool, although it can act like many tools. It is the first metamedium, and as such it has degrees of freedom for representation and expression never before encountered and as yet barely investigated. (1984)

If it would be desirable to mimic audiences, and if computers can be made to mimic human agents, we need only one more thing, some powerful means of conceiving of audiences as varied units, to enable us to design internal audiences for computer simulations. Technical Writing has grown in the last two decades to become a thriving speciality in both Writing and Business departments in American universities.

Every good Technical Writing textbook today puts a large part of its emphasis on audience analysis, defining who the people are that the student is addressing. For formal report writing typical distinctions are made among managerial responsibilities and levels of expertise in the subject matter. Distinctions are even made between primary and secondary audiences, those whose responsibility for action is immediate and those whose interests are less directly involved. For all writing, the "you attitude", considering what your audience's predictable interests and desires are, is strongly advocated.

What with its constant pragmatic emphasis ("Writing for results" is a popular title), Technical Writing has developed descriptions of probable writing situations, predicting both the writer's role and the audience's reactions. This approach has, naturally enough, had some influence on the more traditional forms of writing instruction. Cases for Composition (Field 1984) and Casebook Rhetoric (Tedlock 1985) are two text book titles, clearly reflecting a simulations approach to writing instruction, derived in part from the audience analysis work of Technical Writing. Significantly, both books are in their second editions in a highly competitive market.

Audience analysis, as practiced in both Technical Writing and the casebook approaches, involves predicting the likely differences in reponse of variously defined audiences. Typically such variables as how personally interested and how well-informed the audience is go into defining what the writer must aim at. In most simulated writing situations, each internal audience is described fully at the outset. But it is helpful to inexperienced writers if they can be reminded of particular audience demands just at that place in a written communication where those demands are most relevant. A friend looking over the student's shoulder can point out just where this or that concern for the audience should be dealt with. A teacher could be even more helpful than a merely sympathetic friend. But a teacher's time is expensive, and a simulation sometimes requires extensive repetition of a task to build mastery. Enter the computer with infinite patience and availability to comment on a student's writing choices.

A computer can easily be programmed to repeat information from a previously presented audience analysis or to comment upon the appropriateness of a student's writing choices. In fact the program can actually draw out the implications of a student's choices. By reviewing the audience analysis and extending the probabilities that follow from it, the computer's commentary creates a dynamic rather than a static view of the internal audience, in fact, a moving target.

In retrospect, then, we think those are the contributions made by educational simulations, computer assisted instruction, and Technical Writing to the theory underlying the computer simulation internal audiences we have designed. But theoretical possibilities should be bolstered by practical actualities, especially when an issue as ominous as the potentially dehumanizing effect of the computer looms over the discussion.

We have been designing, demonstrating, and discussing computer simulations for several years now. Since the American experience with computers in the Humanities has stressed growing databases for research and word-processing for writing, our examples will necessarily reflect our own rather narrow biases.

The first internal audience is still the dearest to us, Aunt Sadie. In 1979 one of us (Versluis) was teaching a regular first year writing course at our college. Bored by the usual "How I Spent Last Summer" and "My Most Exciting Moment" assignments, he wanted to give his students a writing task closer to what they would face in the real world, yet without the threats associated real world writing tasks. Writing a delicate thank-you letter to a relative seemed the ideal vehicle for the simulation. Thanking Aunt Sadie for a live 6-foot rattlesnake seemed the ideal element of fantasy.

So each student got a single piece of paper with: Aunt Sadie's kind letter announcing the gift, a short listing of background information, and this assignment, "Write a short and polite thank-you note to Aunt Sadie." As a paper and pencil exercise, the simulation is ridiculously simple. The letter and assignment sheet constitute the stimulus, and the student's letter back to Aunt Sadie is the response. But, as it turned out, there was more.

As each set of student letters was turned in, a pattern of responses emerged. Years of continued use as a paper and pencil simulation have confirmed and refined that first observed pattern. Not only was there a probable reponse predictable for Aunt Sadie, but even the ways students would address the problem in their response became predictable too.

When you can make good predictions about how students will respond to a simulation, you are on your way to designing a computer program. What was needed was some way to close the loop. Aunt Sadie would write to the student and the student would respond. But how could the student find out how successful his or her response was? Since Aunt Sadie was a finite quantity, defined and predictable, she could have "written" an appropriate letter back to the student in terms of the general path the student had chosen to thank her. That would have been a relatively simple and trouble free way to close the loop, but not very satisfying to a teacher.

Teachers, by their very natures, like to meddle in their students' lives. Writing teachers like to meddle in their students' writing. When we both sat down to computerize AUNT SADIE'S GIFT, then, closing the loop was obvious. Offer the student several choices at every stage during the composing of the letter, and comment on each choice as the student made it. This permits the computer to comment on each choice from the salutation, through the three sentences that comprise the first paragraph, through the subject of the last paragraph, to the closing. What the computer simulation is modeling is the segment of the writing process where three considerations bear simultaneously. First, in writing such a letter, we must constantly keep in mind what we've said and the audience's probable understanding and expectations up to that point. Second, we must consider what we are saying in the immediate present and how that will change the audience's understanding. Finally, what we have said and are saying must lead toward what we will say, the topics yet to be covered, the personal gestures of affection or politeness yet to be made. This continuous complex awareness is rendered more manageable for the student by dividing it up into small increments, each dominated by a sentence or a topic choice. The computer commentator is always there, supporting some choices and criticizing others. Eventually one of over 100 letters gets composed thanking Aunt Sadie for the snake.

Besides increasing the student's awareness of the progressive nature of the writing process, the simulation also sustains a sense of how the audience will read the developing letter. The fact that Aunt Sadie's probable responses can be discussed in the context of other writing considerations keeps her presence as an audience alive without allowing her distract from the task at hand. Actually, the simulation uses Aunt Sadie's letter and gift merely as a set of circumstances in which to write a thank-you letter. Such distractions as sorting out your childhood memories of her, or wondering what your Christmas present will be next year are reduced by the playfull circumstances of the simulation. While the frozen (now thawing) rattlesnake is the most obvious of those circumstances, the most persistant source of playfullness in the simulation is the computer commentary on each choice. A final benefit of the computer commentary is to concentrate the student's attention on process of writing, the act of communicating instead of simply on what is being communicated or to whom.

So far we may have made internal audiences in computer simulations sound like unexeptionable, even saluatary additions to simulations. But there are a couple of pitfalls, revealed to us by experience, and we should describe those hazards now. The first of these pitfalls predicably enough is an audience which excessively dominates the simulation.

A PRESENT FROM JASPER was intended to give the student another relative this time a shy cousin, as an audience for an even more challenging paper and pencil simulation of thank-you letter writing. One disasterous run through it with one group of students and the simulation was shelved. The problem was not with Jasper's personality directly. A rather shy cauliflower farmer (he regards the cauliflower as a "sympathetic vegetable") for years, Jasper has decided at last to send off a colorful present to his "favorite cousin," the student. The problem really begins with the present, a fully paid-up 12-month correspondence course in lapidary work and gunsmithing from the Ace School of Firearms Jewelry Design in Los Angeles. It turns out that a truckload of gemstones, which was deposited on the front lawn this morning was apparently quarried on an Indian reservation. A conservation group, which followed the truck from Montana, is now picketing the student's house. The Indian tribe involved has rented a local stadium. There a gathering of medicine men is working up a curse to shrivel the student's cat and clog the plumbing. Also the truckload of automatic riles being sent to the student was intercepted by the FBI, which has now mounted a 24-hour surveillance of the house.

Clearly the student was being asked to write a thank-you letter set against an outrageous background. Even so, a course could have been plotted between dishonestly suppressing all mention of the outrageous difficulties and brutally rejecting both the present and Jasper. But there was one more little complication. Shy, sheltered, and innocent of wordly vices, Jasper was about to be set upon by a horde of vicious relatives, the Gutheaps (distant cousins by one of his father's many ill-considered marriages). Since the student was the only person Jasper trusted and who understood the entire situation, the student had to write an entirely truthful and sympathetic letter about the present.

And so nearly every member of the class did. Being conscientous students, each one wrote a solemn and very detailed account of the present and its attendant difficulties. The disaster, of course, was that it was no longer a simulation of thank-you letter writing. Instead all the students produced long narratives of fictional difficulties. Jasper's circumstances invited the detailed, scrupulously correct accounts; the students' knowledge of the difficulties swelled those accounts. The entire intended function of the simulation was smothered in circumstantial consideration. That was the first pitfall.

This is not exactly an instance of Bonini's Paradox. It is not the case that too many details were supplied from the system being mimicked. Instead there was information about the internal audience and the problem to be solved which gave no greater insight into the task of communication. Much of that information was simply interesting in its own right (and hence a distraction) like casual gossip. That was the second pitfall. So the lesson to be learned about internal audiences here is that the essential design problem is to present just that information crucial to the simulation, and only then fill in with a minimum of other information for enhancement.

Besides deflecting the aim of the simulation, excessive information about the internal audience can also reduce the pleasure proper to the simulation. If you like, a third pitfall. The pleasure proper to a communication simulation comes from the anticipation of solving and then the recognition of having solved the problem of framing the communication well. In the case of AUNT SADIE'S GIFT the student's pleasure is ultimately derived from the knowledge that the issues have been addressed completely and appropriately.

Oddly enough, too little information about an internal audience is not much of a problem. In a complaint letter writing simulation of ours, BUCCANEER BILLY'S BAD BARGAIN (currently under development), the letter is to be addressed to a title, Complaint Manager or Customer Service Department. Here the typical function of the position (i.e., dealing with dissatisfied customers) sufficiently defines the internal audience. As in most business communication, the frequency and regularity of the problems to be solved permit a high level of formatting. In fact, in BUCCANEER BILLY we created a secondary internal audience, Billy himself, for added color. Billy, a small time entrepreneur and pirate, is dissatisfied with the pink wheels that Acme Cannon Company supplied with his Mk. IV Small Business Cannons. We smuggled in both Billy and an exposition of the problem to be solved by beginning with his virtually incoherent rough draft of a complaint letter. The student in this simulation serves as a cooler-headed friend, composing a more businesslike and effective letter for Billy to send.

For both of us, however, the most interesting design challenges lie in the area of human communications which are not simple business transactions, not easily formatted. It is at those moments when human feelings loom largest that communication skills are put to the greatest test. Then the conception of an internal audience and a set of delicate circumstances demands the best of the simulation designer and the eventual student user alike.

These letter-writing simulations are communications simulations in that they aim at motivating the student to consider the ways in which a message can be conveyed to an audience. The conception of their internal audiences provides a variety of targets for the student to aim at. Finally, the computer enables us to refresh, refine, and extend the view the student gets of any particular audience. The problems we have encountered and the means we have employed to solve those problems are largely a matter of keeping the student's attention concentrated on the essential task of conveying a message. Of course, the computer could be a distraction too.

Insofar as the purely machine operations of hitting the proper keys to make the desired changes occur on the screen draw the student's attention away from that essential task, the computer clearly gets in the way. We worked long and hard to reduce the complexity and difficulty of those machine operations. The problem we were trying to solve, however, was not one of the computer being a seductive, deluding presence. Instead the machine operations, stemming primarily from the need to employ a typewriter-style keyboard, are an irritating nuisance. In that way they are completely unlike the circumstances surrounding the arrival of Jasper's gift.

To the extent that the computer as nuisance can be reduced, what computer presence remains? The student sees an audience as a goal to be reached through writing a letter. The internal audience's presence is kept alive and relevant through the computer commentary that accompanies each of the student's choices. The computer commentary consists of the residue of what we know about communication and the likely effects of any particular choice. The computer as computer, not as keyboard or a collection of circuitry, primarily appears in the simulation as the letter to be read, as the choice to be made, as the critique of a decision. It is this transformation, or mimicking of communication, or of actually simulating the communication situation itself that creates the paradox of virtual computer invisibility. And the internal audience, displayed in different lights at different times, constitutes the chief evidence for the existance of this paradox and a still-intriguing moving target for us and for our students.

REFERENCES

Coote, A., Crookall, D. and Saunders, D. 1985. Some human and machine aspects of computerized simulations. In Van Ments, M. and Hearndon, K. (1985).

Field, J. and Weiss, R. (eds.) 1984. Cases for Composition. Boston: Little, Brown and Company.

Jones, K. 1982. Simulations in Language Teaching. Cambridge: Cambridge University Press.

Kaye, D. 1984. Computer software. Scientific American. 251: 3.

Tedlock, D. and Jarvie, P. (eds.) 1985. Casebook Rhetoric. A Problem-Solving Approach to Composition. New York: Holt, Rinehart and Winston.

van Ments, M. and Hearndon, K. (eds.). 1985. Effective Use of Games and Simulation. Loughborough: SAGSET.

Versluis, E. 1984. Computer simulations and the far reaches of computer-assisted instruction. Computers and the Humanities. 18.

Weizenbaum, J. 1976. Computer Power and Human Reason: From Judgment to Calculation. San Francisco: W. H. Freeman and Co.

SIMULATIONS REFERENCES

DOCTOR. Weizenbaum, J. 1976. In Weizenbaum (1976).

ELIZA. Weizenbaum, J. 1976. In Weizenbaum (1976).

AUNT SADIE'S GIFT. Ryberg, C. and Versluis, E. 1984. In Versluis (1984).

The educational potential of interactive literature

Bob Hart
City of Sheffield, England

ABSTRACT: Interactive literature is a new medium which is evolving through the development of adventure games and adventure generators. It has many applications in the development of skills in social interaction, communication, thinking, reading, research, and the creative arts.

In an adventure game students can take the roles of their own characters and enter into their own adventure stories which unfold according to the decisions they make.

The current paper examines two major adventure generator packages: THE TOMBS OF ARKENSTONE (for 8 to 14 year olds) and QUEST ADVENTURES, a powerful adventure generator (ages 10 to adult), which brings together the creative use of wordprocessing and data handling and enables the creation of a wide range of interactive scenarios. The author will introduce the software, discuss its educational potential and give an overview of some of the work which has arisen from its use in UK schools.

KEYWORDS: Interactive literature; interactive fiction; adventure game; adventure generator; education; primary school; secondary school; communication.

ADDRESS: Education Offices, Leopold Street, Sheffield, S1 1RJ, UK. Tel: 0742 735671.

INTRODUCTION

A computer adventure game is like an interactive book, into which the players/ readers enter in the roles of characters in the story. The player's decisions affect the subsequent unfolding of the plot.

Adventure generators enable users to create and then play their own computer adventure games, without having to descend into cryptic programming languages.

Over a period of four years, the author has designed and developed two adventure generating packages and through observation of children in UK Primary and Secondary schools and interviews with their teachers, has evaluated the educational value of students playing and writing adventures of this type. The two packages used were: THE TOMBS OF ARKENSTONE (Hart, 1983) and QUEST ADVENTURES (Hart, 1985).

THE TOMBS OF ARKENSTONE package includes a text-only computer adventure program with maps, a children's story book to set the scene and the authoring program "MAKE YOUR OWN ADVENTURE". It is designed for children aged 8 to 14 years.

Although the number of possible paths through the adventure is astronomical, the number of options players must consider at each decision point is small (maximum 5). There are 3 levels of play with more features and options being presented to the children as they gain experience and progress through the levels.

The adventure generator allows children to enter their own images - characters, locations, objects into the pre-defined logical structure of the sample game.

QUEST ADVENTURES is a more advanced adventure generator with much greater flexibility, designed to enable older or more experienced users to write adventures at 5 levels of complexity. The topological layout of locations and the relationships between characters can be defined by users who can also specify events that happen during the course of the adventure.

It is an integrated package which brings together the creative use of data handling (using the QUEST information handling package which is popular in UK schools), wordprocessing and optional teletext graphics. It includes a sample adventure based on the Arthurian legends - THE KNIGHTS OF CAMELOT. This is written at the highest level of complexity and demonstrates many of the features that children can include in their own adventures.

PLAYING ADVENTURES

Most teachers found it best to first introduce their students to playing one of the sample adventures. In this way the students could internalise the nature of the medium and the logical structure of the adventures and gain some understanding of the finished product before starting to write adventures of their own.

The tasks. The tasks are similar in both adventures - a search for a valuable object through a maze of locations involving encounters with other friendly or unfriendly characters.

In THE TOMBS OF ARKENSTONE, the Evil Dragonlord, Asgarn, has stolen the Elfin Ring. Without the Ring, the Land of Arken will remain in darkness forever. The adventures must go to the Tombs of Arkenstone to find the Dragonlord, survive the attacks of his monsters, work out a successful strategy to catch him, regain the Ring and save the Land of Arken from the Forces of Darkness.

The main task of THE KNIGHTS OF CAMELOT, is to find the Holy Graile. it can be useful to rescue the wizard, Merlin, to find the magical sword, Excalibur and perhaps rescue a maiden or two.

Selected random elements in the adventures ensure that each new group of players can have a different game but students are also able to stop playing, at any time, save their position and carry on later.

Group Role-play. In THE TOMBS OF ARKENSTONE, the children adventure in groups of three (or 6 if necessary), each child taking on a specific role and responsibilities.

The most important rule of the game is that all players should agree on every move. If however they can't agree, then they each hold the right of veto for their own areas of responsibility. The Hunter makes the final decisions on strategic matters and looks after the map. The Warrior decides when and how to fight monsters and the Magician makes decisions about Magic Spells. These roles are played "outside" the computer and are a device to stimulate cooperation and communication between the players.

In THE KNIGHTS OF CAMELOT, each player can take on the role and personality attributes of any of the 36 characters defined within the adventure. His/her moves are remembered and the consequences visited on his/her chosen character as represented within the program. Players can interact with other players within the program and can change characters when they wish.

THE EDUCATIONAL POTENTIAL OF PLAYING ADVENTURES

The notes that follow are based on teachers' observations of children using the two packages in schools.

In most cases, the children played the adventures over a period of a school term and went on to write adventures of their own during the following term. The main areas of value were in the development of skills in COMMUNICATION and SOCIAL INTERACTION, THINKING, READING, and CREATIVE ARTS.

Communication and Social Interaction. Players soon discovered the need to discuss each move. They had to clearly argue their viewpoints and explain their reasoning to fellow Adventurers, often under pressures imposed by the game - time or magical power might be running out, or a monster might be about to destroy them. Players are highly motivated in such circumstances to crystallise their ideas and present them effectively - not just for their own good but for the survival of the group.

These fantasy microworlde offer a safe environment within which students could argue out many moral issues; co-operation, consensus, democracy, freedom, responsibility, status and influence, good and evil. Teachers were able to extend these discussions with their students and apply them to real-life situations.

The adventures are designed to reward co-operative effort. In THE TOMBS OF ARKENSTONE, all three players must reach consensus for each move. The children gain Experience Points for successful hunting, fighting and use of magic. Each player's achievement benefits the whole group, who can rise through the experience grades to higher levels of play. By careful juggling of groups, teachers found they were able to encourage the integration of socially isolated children and ensure that girls had equal influence. "Less able" children are often on equal terms with their more academic peers in these new microworlde, where "streetwise" cunning is more valuable than academic analytical thinking. Teachers often found that a child of low status would make a useful suggestion which would win him new respect within the group.

Thinking Skills. Students' conversations often included phrases which revealed logical thinking; What if ...?; If we do ... then ...; I don't think so ... Why ...? Because ...?; Yes, but what about ...; But last time we did that ...; Look, we've got to work out a plan ....

For each decision the players had to consider their strategy and the topology of the plan. They had to weigh up all the factors of, for example; Magic, Time and Treasure, work out the possibilities, judge the probabilities, balance their current priorities and consider the consequences of each move. While in THE TOMBS OF ARKENSTONE children had only to choose between 3 options: whether to MOVE to another cavern, VIEW a cavern or BLOCK a cavern with the Circle of lightning, in THE KNIGHTS OF CAMELOT they could select from a large number of commands:

WHY	GO	READ	HIDE	OPEN	ACTOR
WHO	BACK	USE	SHOW	CLOSE	
WHAT	ESCAPE	GIVE	SIT	UNLOCK	COMMANDS
HERE	FOLLOW	ASK	STAND	LOCK	PRINTER
WHERE		EXAMINE	SLEEP	HELP	
EXITS			WAKE	FIX	PAUSE
CARRIED		FIND		ATTACK	
LOOK		CALL		HIT	
LIST	TAKE	SCARE		BREAK	FINISH
SCORE	DROP	BRING			
HOW IS	THROW	SEND			

Fig. 1. Quest Adventures commands

They also had to judge the reliability and value of information and make inferences from incomplete data. Scenes, actors or props often hold secrets that might be useful clues or misleading tricks. The maps give some clue to the landscape, but students may have to explore uncharted areas and discover secret paths. Teachers asked more experienced children to play without the map, so that they had to generate their own as they went along.

There are thousands of ways through an adventure and as many different winning strategies to be discovered.

Reading Skills. A 20-minute TOMBS OF ARKENSTONE adventure played by 9 year olds, involved the reading of some 4000 words and 800 numbers. They all had to be skimmed and sampled for relevance and their meaning extracted with speed and accuracy so that the adventure could proceed. This represents quite a task for children of this age, but in most cases this presented no problems to the children. They had a clear and attractive purpose that motivated their reading so that the volume of text decyphered and the level of comprehension was often much greater than teachers would have expected, based on their perception of the children's abilities as demonstrated on more traditional reading tasks.

THE CREATIVE ARTS

Creative writing. Teachers found that the children were well motivated to communicate the excitement of their adventures. They were happy to write about their adventures while the experience was fresh. Their involvement was intense and in many cases, the resulting work was surprisingly lively and vivid.

Arts and crafts. The project was a very rich source of inspiration for a whole range of art work - clay and plasticine monsters, puppets, paintings, prints, enormous paper serpents and tiny clay fantasy landscapes. Teachers found the lack of graphics on screen was an advantage in that it left the children's imaginations free to conjure up their own individual visual images.

Music, dance and drama. Depending on their age, children took their adventure role play more or less seriously. Some interesting situations arose as when a student had played as Sir Launcelot (a good guy) until he was in a favourable position and then was made to take on the role of the rival knight that he had just defeated in battle. This is rather like playing chess until you are about to win and then swapping sides with your opponent. There are great implications for the development of empathy and identification with others. Teachers were able to extend the potential of the role play into Drama and Dance. Groups of children also made instruments to accompany the dancers or wrote songs.

Cathartic fantasy. It is interesting that many children chose initially to cast parents and teachers as monsters: but it has also been noted that teachers often write adventures about education officers and advisers; advisers write about Her Majesty's Inspectors and H.M.I. like to write about government ministers!

WRITING ADVENTURES

Using the MAKE YOUR OWN ADVENTURE program, children can write their own adventures, set in any time or place, with characters of their own. These adventures can then be played on the computer within the logical structure of the original program.

Their ideas are captured in a series of on-screen cloze passages. They can insert missing words into paragraphs describing the scenes and characters in their adventure. They can also use or adapt a variety of plans provided which are topologically equivalent to the TOMBS OF ARKENSTONE plan.

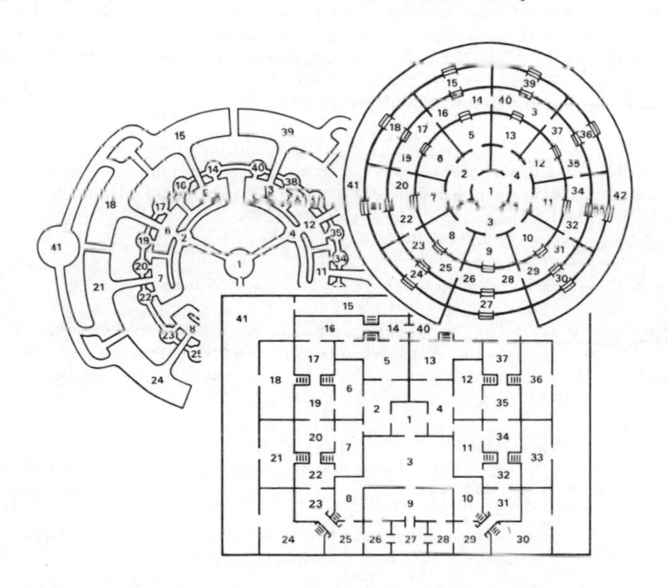

Fig. 2. Alternative TOMBS OF ARKENSTONE plans

Writing a QUEST ADVENTURE involves the use of an information handling package, wordprocessor and in some cases, teletext graphics.

Using the QUEST information handling package (The Advisory Unit, Hatfield) and prepared skeleton files, students complete three datafiles called: SCENES, ACTORS and PROPS.

The SCENES file must include such information as the name of each scene, its description and a list of other connected scenes. They write similar datafiles of ACTORS and PROPS. The program then compiles the three files into an adventure.

Adventures can start simple and be gradually enlarged and refined by adding more scenes, actors or props. By using additional fields of information within the database, more features can be added, for example; information or secrets can be held by a scene, actor or prop; starting positions of actors or props

```
---------------------------------------------
Record no 1
NAME       :Great Hall
DESCRIP    :A vast, high-vaulted hall,
panelled in exotic hardwoods and hung
rich in rare tapestries.  In the centre,
an oaken table, bearing a great feast.
GROUP      :2,3
---------------------------------------------
```

Fig. 3. Record for the Great Hall

can be specified, or they can be scattered randomly; actors can be given personality attribute scores - power, wealth, strength, wisdom and goodness or these can be ommitted for a simpler adventure.

The players can be left to their own aims, or the student can write a script of up to 20 EVENTS to make things happen to the adventurers. This involves writing a CUE to define the conditions which trigger an event (For example: The Adventurer is carrying Excalibur), a SCRIPT CARD of information which is displayed during the event and an EFFECT to define its outcome (for example: the Adventurer will be wounded and lose strength or the Adventurer will change into King Arthur's Dog).

Students can also change the names of, or omit, any of the 48 commands. They could, for example, write a peaceful underwater adventure, omitting the commands "HIT" or "ATTACK", and replacing "GO" with "SWIM".

THE EDUCATIONAL POTENTIAL OF WRITING ADVENTURES

Curriculum Applications. Teachers found applications throughout the Primary and Secondary School curricula, since children's adventures could model reality or fantasy and could arise from any personal interest or curriculum project in which they were involved.

With the TOMBS OF ARKENSTONE, children created adventures set in a pyramid in ancient Egypt, the catacombs of ancient Rome, in space, under the sea, in a haunted castle and on a pirate ship.

Furthermore they don't have to be restricted to the classic adventure genre. Here are some examples of students' and teachers' ideas for QUEST ADVENTURES:

My house, my family, my school.

A medieval village.

A walk around a museum.

Inside the human body.

My favourite book, film, TV programme.

Teachers were also interested in writing adventures to involve their students in, for example, a scene from the History syllabus.

An adventure might take place in a school or institution where players take on the roles of the current inhabitants and perhaps go back 100 years to interact with the characters' great great grandparents, or forward 100 years to meet their great great grandchildren.

Ideas could relate to the History, Geography, Mythology or the Religious heritage of any culture and could be as challenging and educationally rich for adult professionals, businessmen and women as for students and children.

Foreign language learning. The reading and comprehension skills involved in playing adventures together with the motivating force of the medium offer great potential in the teaching and learning of foreign languages. The level of student interaction is controllable (by restricting the commands available) as is the vocabulary used. Key phrases are repeated hundreds of times throughout a game and are absorbed almost unnoticed. The content of an adventure could also be designed to explore the cultural heritage of the countries whose languages are under study.

Organisational Skills. Writing an adventure is like writing a novel or a play. Much of the work is carried out away from the computer. Students had to organise their ideas, research the details and collate the information. This involved the following sub tasks: Deciding the subject of the adventure; Inventing or finding the location; Choosing the period; Discussing the story outline; Collecting information about the scenes, actors and props; Designing the plans; Dividing the plan into SCENES; Defining the connections between SCENES; Describing the ACTORS and PROPS; Writing the files on paper; Deciding how complex the adventure will be; Adding the EVENTS; Selecting the COMMANDS.

Students were then in a position to enter their information into the computer, compile it, play their adventures and subsequently refine them. Teachers found that students needed a surprisingly deep understanding of their subject and as they delved deeper into the task its complexity accumulated.

Cooperative Writing. Simple adventures can be written by individual children or students. For a more complex adventure, children found it useful to work in groups, each with a particular working role. Writing an adventure is similar to writing a novel or a play. Teachers found that students responded well to being given theatrical roles, for example: The Scriptwriters write the events; the Set Designers design the scenes; the Props Department create the props. There are potential roles for a producer, director and so on. This group organisation demands that students communicate effectively with each other as the work progresses.

Higher Order Reading and Research Skills. Teachers found that making adventures involved the collection of a large volume of information which had to be sorted and reorganised into a form relevant to the task in hand. This involved skills such as: Formulating open ended fruitful questions; Accessing library, catalogues and other media; Collecting a variety of suitable resources; Directed purposeful reading; Surveying techniques - skimming and sampling for relevant ideas; Reading for meaning, involving literal comprehension, reorganisation, inference, evaluation and literary appreciation. Study skills: Teachers took the opportunity to encourage students to explore new ways of collecting and organising information.

CONCLUSION

Playing adventures can stimulate a variety of valuable skills. It can motivate students to read accurately and think clearly. Children and adults alike can benefit from the social advantages of a task which demands effective collaboration and communication (or death at in the claws of a Dragon). Playing adventures on the scale described is not a matter of solving a serious of puzzles, but more of developing an understanding of a complex web of relationships, evolving strategies, sifting evidence, judging probabilities, predicting consequences and calculating risks.

Writing adventures brings together literary and artistic creativity, advanced reading, research and study skills, information handling and organisational skills.

Interactive Literature is a potent medium within which children and adults can safely encounter their internal heroes and villains or explore a microworld representing an aspect of reality. It thus provides a rich educational medium with potential applications over a wide age range, in many subject areas and in many cultures.

REFERENCES

Hart, R. 1983. An adventure in education. Micros in Action in the Primary School. Open University, UK.

Hart, R. 1984. I was chased by Anubis. Micros in Project Work INSET Pack. MEP National Primary Project, UK.

MEP National Primary Project. 1984. Talking Point (video). Language Development INSET Pack.

Hart, R. 1986. Interactive literature and the Ultimate Problem. Problem solving with a microcomputer INSET pack. MEP National Primary Project, UK.

COMPUTER SOFTWARE

THE TOMBS OF ARKENSTONE (BBC Microcomputer). Hart, R. 1983. Arnold Wheaton Software, Freepost, Leeds, LS11 5TD, UK.

THE TOMBS OF ARKENSTONE (Spectrum, RML 480z, Atari computers). Hart, R. 1985. The Arkenstone Project, 38 Hillcote Close, Sheffield, S10 3PT, UK.

QUEST ADVENTURES (BBC Microcomputer). Hart, R. 1985. The Advisory Unit - Microtechnology in Education, Hatfield AL10 8AU, UK.

QUEST information handling package (BBC, RML 480z, 380z, Nimbus, IBM computers). The Advisory Unit - Microtechnology in Education, Hatfield AL10 8AU, UK.

Artifice versus real-world data: Six simulations for Spanish learners

Douglas Morgenstern
Massachusetts Institute of Technology, USA

ABSTRACT: Concepts from general systems theory are posited to elucidate the relationship between the simulated and target environments. Four classroom simulations which make use of "creative abstraction" to achieve this correlation are described. These are NEW IDENTITY, a role-play with some simulation features, The ARBITRARY MARKETING SURVIVAL GAME, which demands strategic interaction under intentionally frustrating conditions, INFILTRATION PARANOIA, which calls for cooperation under difficult circumstances, and the TWO-COUNTRY SIMULATION, which tests participants' ability to negotiate. All of these are short-term simulations set in abstract or invented Hispanic countries. In contrast, ENCUENTROS and NO RECUERDO are computer simulations (still in development) which combine fiction with a real-world setting.

KEYWORDS: general systems theory, artifice, abstraction, informal discourse, Spanish, computer, interactive videodisk.

ADDRESS: 110 Union Street, Norwood, MA 02062, USA.

INTRODUCTION

The principal aim of this session was the presentation of several simulations designed for foreign language learners. A secondary goal was to touch upon a few aspects of an important and persistently troublesome issue, that of the realism of simulations. This theoretical consideration, slightly abbreviated during the session itself, will be treated first here.

Several simulation theorists, including Greeblat and Gagnon (1979), Jones (1982) and Crookall (1984), have concentrated on the relationship between external reality (or realities) and participants' perceptions and behavior during the simulation activity. More recently another approach, based on ethnomethodology and conversational analysis, focuses on the reality of the simulation without comparison to the outside world (Sharrock and Watson, 1985). By making the participants' interaction the object of study, the analyst can reveal new perspectives on phenomena such as role distance and topic organization. This approach is extremely fruitful for viewing and analyzing the simulation as an event, and has yielded significant insights, such as the fact that because in simulations the topic is stipulated in advance, negotiation and other topic managing strategies are not as necessary as in less structured environments.

This desire to "bracket" external reality has manifested itself in many areas, ranging from philosophy and literary criticism to experimental gaming. In the case of this last field, the relationship between subjects' behavior in laboratory games and in the outside world has been seriously challenged. Colman (1982) summarizes two positions on "ecological validity":

> Some have claimed that behavior in abstract labortory games is sufficiently interesting and important in itself to warrant continued research without making any claims about behaviour in naturalistic interactions... Most researchers, on the other hand, are in favour of generalizing from the laboratory to the cosmos in spite of the technical and theoretical problems which this entails.

The designer of a simulation intended for language learning or practice must perform the inverse task: adapt the cosmos to the simulation.

A simulation may function in a manner highly amenable to conversation analysis, yet it may result in a distorted representation (and internalization) of political, economic or sociolinguistic features of the target culture. A designer's (and evaluator's) perspective therefore requires attention to correspondence with the target culture.

Since simulations are by definition artificial - even though they give rise to conversational behavior patterns that are completely authentic, as Sharrock and Watson reveal - it might be useful to examine briefly the concept of artifice. In doing so, it is important to distance oneself, at least temporarily, from a rather restricted notion of authenticity characteristic of much of current language acquisition theory, which *a priori* casts artificiality in a negative light. One may move outside of the field. Simon (1981), utilizing concepts from general systems theory, posits several indicia which distinguish the artificial from the natural. Two of these are particularly relevant: (1) the artificial may imitate the appearance of the natural while lacking, in one or more respect, the reality of the latter; (2) artificial things can be characterized in terms of functions, goals and adaptations. We can combine these two characteristics for our purposes: a simulation whose goal or function is to provide an environment for language learning/practice will imitate (and lack the total reality of) various phenomena, including certain kinds of spontaneous conversations and commercial, civic, national and international transactions.

The correspondence between the target phenomenon and the simulated environment is based on abstraction. Again following Simon:

> The more we are willing to abstract from the detail of a set of phenomena, the easier it becomes to simulate the phenomena. Moreover we do not have to know, or guess at, all the internal structure of the system but only that part of it that is crucial to the abstraction.

The target phenomenon itself is usually a complex system. General systems theory (which abstracts from physical, biological and social systems) has advanced the concept of Near Decomposability, which is based on the filtering effects of hierarchy. Near Decomposability permits at least an approximation of comprehension:

> Subparts belonging to different parts only interact in an aggregative fashion - the detail of their interaction can be ignored. In studying the interaction of two large molecules, generally we do not need to consider in detail the interactions of nuclei of the atoms belonging to the other... In studying the interaction of two nations, we do not need to study in detail the interactions of each citizen of the first with each citizen of the second.

Yet because a simulation of the kind that interests us here is usually a reduced-scale representation of the target phenomenon, this hierarchical barrier may be diminished. Indeed, the individual, e.g. the individual citizen of a simulated nation, and his or her interaction with other participants, becomes paramount. The participant who is supposed to be a farmer must represent thousands of farmers in the target environment. This individual's particular behavior, which may be idiosyncratic, assumes a highly magnified degree of importance. From a language-learning perspective, the outcome of the simulation - what happens, who wins, etc. - may be less important than interactions among individuals. Paradoxically, in a successful simulation participants tend to be so highly motivated and caught up in the action that they feel just the opposite: the outcome is most important and individual interactions are means to an end.

Simulating (and not only understanding) a "nearly decomposable" complex system relies on abstraction. Abstraction provides a relational axis between artifice and reality. The simulation designer abstracts essential aspects of the target phenomenon. Jones (1982) notes that essence is more important than a wealth of detail.

These essential aspects may be derived from a single target event or situation, e.g. a business case study, a past or present conflict among (real) nations. Direct, one-to-one correspondence with real-world data does not necessarily lead to the most effective or even realistic simulations. For example, Sharrock and Watson (1986) and Cunningsworth and Horner (1986) note the predominance of formal discourse in extant simulations; the latter call for designs which foster more informal conversation. Now with direct correlation to real-world data, details are often lacking regarding these informal aspects of the target phenomenon. When participants are obliged to invent these details (the Prime Minister's breakfast conversation), they may enjoy themselves but they will end up play-acting and "reality of function" (Jones, 1982) will be compromised.

It is often better to abstract from several target environments and invent a new environment which, though imaginary, may be less distorted than a reduced version of a single real-world phenomenon. Common sense, however, dictates that such a process, which might be termed "creative abstraction," would not be desirable for highly contextualized, long-running simulations designed for language learning. This limitation will be treated in the section on "No recuerdo."

Following are descriptions of original simulations designed for learners of Spanish. The first four, which use creative abstraction, are classroom-based and have been run successfully at Stanford University, Harvard University Extension (adult classes) and MIT as a regular part of the curriculum. The last two combine creative abstraction with direct correlation to real-world data. Still in the design and development stages, these include both comptuer and classroom interaction.

During the ISAGA session, all of the actual simulation hand-outs, which are written in Spanish, were distributed and translated or paraphrased. These auxillary materials will be alluded to in the descriptions that follow; anyone interested in obtaining copies of the handouts in order to run the simulations can write to me at Foreign Languages and Literatures, MIT, Cambridge, Massachusetts 02062, USA.

NEW IDENTITY

This is actually a role-play exercise with some simulation features. It lasts from one to two hours, and can be spread over several classes. Each participant receives a hand-out sheet to fill in as an assignment or else at the beginning of the in-class session. Each person is to take on a new identity as an Hispanic residing in an unspecified Latin American city. The top part of the sheet gives four Hispanic surnames and four places of work (restaurant, clinic, store, bank); the participant chooses one of each, and then writes down details under the heading,"My characteristics." These include age, personality characteristics (the facilitator provides vocabulary as needed), aspirations, etc. The participants then sit together, grouped by surname; changes are made to avoid disproportionate size in groups. The task is to form a new family (not necessarily nuclear), based on the new identities chosen. The participants must negotiate and modify their original intentions, perhaps changing their ages to fit in with the overall plan, and selecting a different workplace (so that the workplace group overlaps as little as possible with the family group).

They then inform each other of their first names (invented), personal characteristics and workplaces. The middle part of the sheet is arranged for note-taking; each participant writes down data on his or her sister, aunt, cousin, etc. (the facilitator points out the prevalence of extended families in Hispanic society).

Afterwards the participants regroup according to the workplace chosen. A parallel process occurs: they negotiate, allocate power (who is owner, who employee), agree on a few basics (What kind of store?) and exchange information. Notes are written on the bottom third of the sheet.

Individual groups then present semi-improvised scenes of a few minutes duration. First each family group puts on a scene, with the others as spectators. They are directed to mention events and people in their workplaces. This creates an interesting audience-performer relationship, because members of the audience are referred to. These have the opportunity to incorporate their reactions when their family presents its scene, by giving their own perspectives or interpretations of events or motives. This reciprocity is maintained when workplace groups present their own scenes; now individuals refer to the behaviour of their family members, who now form part of the audience. Because spectators know that they soon will be interacting with performers, their attention to a current scene is not based solely on its entertainment value: they listen for references to themselves and others in their group, and strategize on how they will use what was said when it is their turn to perform. Debriefing occurs after groups of scenes. The facilitator gives observations on linguistic and cultural appropriateness of the performers' behavior, and a discussion often ensues.

This relatively simple role-play has two important features. The participants create the parameters of their own roles, so there is no role incongruence. Also, they negotiate when planning a scene and strategize when spectators. This involvement allows New Identity to partake of classic simulation characteristics.

Artifice is predominant in this exercise. It is unrealistic for people to have access to scenes of their interlocutors' families and workplaces, and to learn, as invisible spectators, what others are saying about them. But this privileged vantage point allows for creative interaction, and, within the framework of their new identities, encourages authentic expression.

THE ARBITRARY MARKET SURVIVAL GAME

This is a simulation which fosters cooperation and competition under conditions ranging from relative tranquility to uncertainty to turbulence. The exercise lasts approximately one hour. Participants receive written instructions (the facilitator is available for clarification) stating that they are citizens of a Latin American nation. Each will receive packets of cards representing products (meat, milk, shoes, etc.) and resources (petroleum, water etc.), but each packet is randomly incomplete, e.g. it may consist of several duplicates, be lacking in some products, etc. As in real life, some packets are better than others. The instruction sheet states that the goal is to obtain at least one of each of the twelve products. Also distributed (again, unevenly) are packets of money of various denominations, each containing a portait of the country's military ruler. The winner must be in possession of a complete set of cards. and have amassed more money than anyone else. Participants can trade with each other, buy and sell, bargan, and deal with the government bank (run by the facilitator), which opens periodically.

The hidden agenda is a special sequence of events during the course of the activity: (1) Normalcy. Participants trade. (2) Hard Times. Some needed products are not made available by the bank. (3) Frustration Onset: The government changes the rules, doubling the quota needed for certain products, then selling extras. Conversely, other products are crossed off the chalkboard master list, and the government buys them back from participants at low prices. (4) Frustration Peak: The government, now openly corrupt, restores or even doubles the quotas for products that were just eliminated, selling them at a profit. A revolution is announced. (5) New Government. The facilitator takes the role of a socialist leader and declares new policies. Products are distributed to the poor, the wealthy are taxed. The old currency is now invalid; citizens are prompted to exchange it for new currency depicting a clinched fist above the slogan "*Tierra y Pan*" ("Land and Bread"). (6) Facilitator as Instigator. The facilitator leaves the bank and circulates among the citizens, requesting support for a counter-revolution. (7) Coup/Elections. There are two alternate scenarios. In the first, the facilitator returns and stages a *golpe de estado*, immediately penalizing those who cooperated with the previous government. Those who hoarded the original currency benefit. However, international pressure leads to elections. In the second scenario, elections come during the socialist government. (8) Transfer of Power. Participants can form alliances, get a majority vote, and actually assume control of the government bank. During this final phase, the facilitator reassumes the role of instigator. The participants who gain control experience power and pressure; the others often experience betrayal. The final minutes before the end of the period find participants scurrying to form economic alliances and win the game.

During debriefing the participants recount their strategies and reactions to events. The facilitator contrasts the degree of control over one's life an individual has (or perceives as having) in native and target culture societies. A recent survey (reported in the June 30, 1986 issue of *U.S. News and World Report*) of 2300 young people indicates that almost half of those in the United States belief they have total control over their own lives, much more than the response ratio in other countries. Historical, economic, and political parallels to real-world events are drawn. The participants are shown a sample of currency from a renegade government in northern Mexico during their revolution.

This simulation is an example of creative abstraction. The country is not real; the survival rules are highly simplified. Abstraction from numerous Hispanic American events and situations allows for a convincing synthesis. The artificially accelerated time scale permits effective involvement. The structure obliges participants to use both formal and informal discourse. The result is a fusion of language practice and cultural empathy embedded in an authentically communicative situation.

INFILTRATION PARANOIA

This is a one-hour simulation intended for intermediate or advanced students. Participants are randomly assigned two groups, the Military and the Guerrillas. The Military receive a hand-out sheet which skews reality in favor of protecting the current Minister of Defense from terrorists, who will probably attempt to kidnap him before their incarcerated terrorist friends are executed. Details concerning the number of bodyguards available as well as the Minister's discreet visits to a female friend living in a poorer section of the city are provided. The military police must devise a plan to protect the general without interfering with his visits or causing publicity. The Guerrillas receive a sheet which provides the same basic information (the Minister's daily routine, his visits), but from the opposite perspective. The country must be liberated from an oppressive regime, and a plan must be found to kidnap the Minister and then release him only if certain conditions are met. Each group thinks it has a moral imperative to act. A common map of part of the city is given to both groups; it contains representations of the Ministry, the Minister's residence, his girlfriend's residence, and the locations of places whose inhabitants are favorably disposed to one or the other group, as detailed in the sheet.

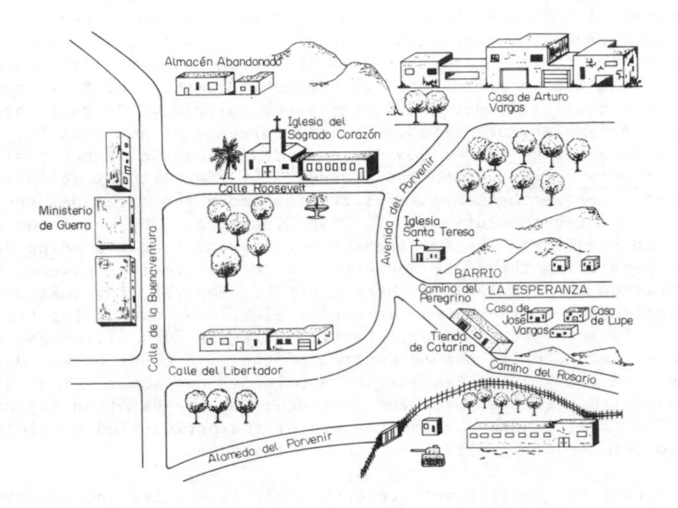

Fig. 1.

The Military are on one side of the room, the Guerrillas on the other. Each participant is given a slip of paper, labelled "militar" or "guerrilla," which must be kept secret. The facilitator creates one or more "spies" by giving out identity slips from the opposing group. There is a two-fold task within each group. First, the members work together, strategize and assign roles (which are never actually acted out). The Military tries to protect the Minister, assign decoys, determine positions, etc. The Guerrilla team studies different game plans, chooses routes and places for capture, determines who will be driving, who will be disguised, etc. The second task is that individual members carry out their own goals. The spies try to convince other members to accept defective plans without incurring suspicion; the non-spies have to ferret out the spies and be prepared to denounce them subsequently. The result is intense planning in a group format, with extra attention given to the language and motives of interlocutors.

There follows a side activity in which one representative from each group meet "secretly" (in a different part of the room) and exchange information about current plans. Because no one knows the true identity of anyone else, uncertainty can be either augmented or diminished. At the end of the hour, the facilitator stops all activity and stages two mock inspections. As a military investigator, he announces that a certain number of spies must be found. Within two minutes, intra-group accusations must result in discovery of the traitors; false accusations bring about the death of the accuser (the controller has a toy gun). The same procedure is followed with the guerrilla group. The team that "wins" is the one with the greater number of individuals alive at the end. During debriefing, students justify their individual strategies, referring to earlier conversations which led them to conclude that others were or were not spies.

Although the particular details are invented, the situation is abstracted from real phenomena and behavior patterns. At least one Latin American country experiences several hundred political and economic kidnappings per year. This activity gives extensive practice in several language functions, especially explanation and persuasion, and encourages informal interaction.

TWO-COUNTRY SIMULATION

This simulation provides an environment for decision-making under difficult conditions with high stakes. It lasts from one to two hours, and is designed for intermediate or advanced students. Participants are divided into two groups, each to inhabit an imaginary Hispanic American nation. The inhabitants of Sebastian are proud of their achievements despite lack of essential resources (water and fertile land), but feel threatened by their southern neighbor. The skewed hand-out sheet laments their dependence on Esperanza's exports, and vilifies the increasing numbers of Esperanzan immigrants. Esperanza citizens receive an equally one-sided perspective: their citizens are treated badly in Sebastian, and even imprisoned in the Isla of the Condemned without cause. Sebastian controls Esperanza's access to the sea, thereby imposing unfavorable trade policies. A map (see Figure 2, below) and an information sheet containing economic, political and military data for both nations are distributed in common to each group.

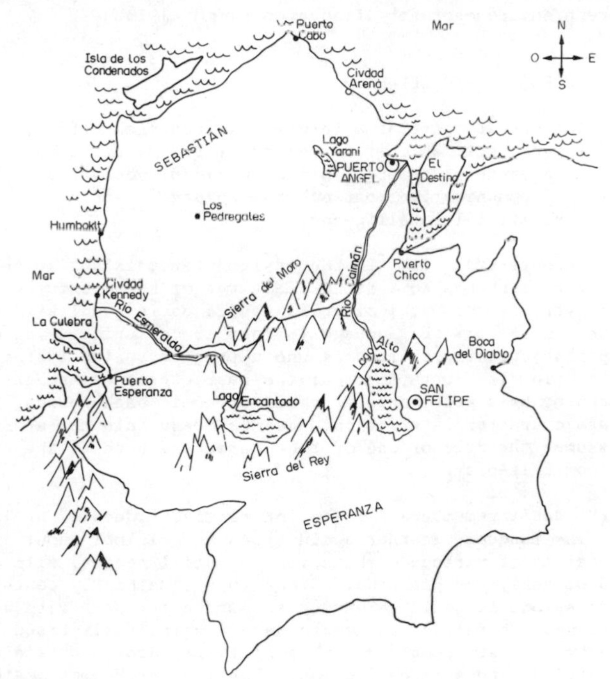

Fig. 2.

Participants discuss the needs and goals of their respective countries and designate representatives to make formal offers, counter-offers, complaints and threats to the neighboring nation. The facilitator then begins announcing a series of news events which test the participants' ingenuity in resolving disputes. These include the news of a deliberate fire set in an immigrant section of Sebastian's capital, rioting and brutal suppression in the island prison, and a press conference by the leader of Sangre Consagrada (Consecrated Blood), a Sebastian extremist group which takes credit for the fire and vows to drive out all Esperanzan immigrants. Comic relief from this heavy·scenario is provided again by the facilitator, this time in the role of a "mystic architect" who strives to unite both countries and form a theocratic technocracy.

The final news event is that a consortium (United States, France and Japan) wants to mine the extensive uranium deposits just discovered in the border area. Sebastian and Esperanza have to decide whether to co-operate, reject the offer jointly, or deal individually with the consortium. An optional scenario involves the addition of a mini-situation with sweethearts from the opposing countries (derived from Romeo and Juliet). Debriefing has proven difficult for this activity; participants tend to continue arguing with members of the opposing country.

A condensed version of this simulation was run at the summer language school of Middlebury College, Vermont, in 1985. The audience (mostly language professors) applauded when Esperanza declared war. The response was generally favorable, with the exception of one professor who objected to the use of invented countries instead of real ones, such as Nicaragua and Honduras. My response was that in a short simulation such as this, there would exist the risk of presenting a reductionist vision of Central American reality, thereby trivializing serious problems. The students who participanted in the activity also noted that they would have felt more constrained had the countries been real; one woman mentioned that she already knew her classmates' opinions about Nicaragua, but this simulation provided a fresh environment for discussion of moral issues.

ENCUENTROS AND NO RECUERDO

These two simulations are part of a large-scale research effort, the Athena Language Learning Project, which is directed by Dr. Janet H. Murray at MIT and sponsored by the Annenberg/Corporation for Public Broadcasting Project. Both seek to combine a communicative approach to language learning with technological capabilities in artificial intelligence.

Encuentros ("Encounters"), still in the design stage, is a text-oriented exercise for microcomputer which has some of the features of an adventure game. The adventure, however, is not mythical or fantastic as is usual with most commercial products. The aim is for the language learner to communicate via keyboard input with Spanish-speaking characters who appear to reside in the program. The setting is Colombia, South America; the characters are members of several families representing a wide range of socio-economic backgrounds. The learner first witnesses character interaction, becomes drawn into the story, and eventually assumes the role of one of the characters both at the computer and during classroom role-play.

No recuerdo ("I don't remembers") is an interactive videodisk project now in development. The language learner again types in original input in Spanish; the output consists of various combinations of still photos, film and video segments, audio, text, and graphics. Since this is a highly contextualized, long-term simulation, it would be unwise to burden the user with details of an imaginary setting. Therefore real-world data - principally based on explorations of Bogotá, Colombia - are combined with a fictional story containing elements of romance, intrigue and science fiction. The learner "communicates" with two protagonists and often sees scenes from their memories; as in Kurosawa's film Rashomon (1951), their versions sometimes conflict. The result is a system of "multiple realities," a concept explored by Greenblat and Gagnon (1979) in a different context.

The Athena Language Learning Project is developing other experimental prototypes for French, German, Russian, Japanese and English as a Second Language.

The reception of this presentation was highly favorable, although the amount of material and somewhat frantic pace may have burdened some of the audience. A few questions dealt with what particular language functions were being used in some of the classroom simulations, and with the (eternal) question of linguistic accuracy versus fluency. One person noted the rather obsessively political nature of the classroom simulations; another commented wryly that the sort of turbulence depicted occurred in Latin America, but not in his own country, Argentina. There were a few questions about the operation of an interactive videodisk system. At the close, everyone showed signs of Near Decomposability.

REFERENCES

Colman, G. 1982. Game Theory and Experimental Games: The Study of Strategic Interaction. Oxford: Pergamon Press.

Crookall, D. 1984. The use of non-ELT simulations. English Language Teaching Journal, 38.

Crookall, D. (ed). 1985. Simulation Application in L2 Education and Research. Oxford: Pergamon Press. Special Issue of System, Vol 13, No [illegible]

Cunningsworth, A. and Horner, D. 1985. The role of simulations in the development of communication strategies. In Crookall, D. (1985).

Greenblat, C. and Gagnon, J. 1979. Further explorations on the Multiple Reality Game. Simulation & Games, 10:1.

Jones, K. 1982. Simulations in Language Teaching. Cambridge: Cambridge University Press.

Sharrock, W. and Watson, D. 'Reality construction" in L2 simulations. In Crookall (1985).

Simon, H. 1981. The Sciences of the Artificial, sec. ed. Cambridge, Massachusetts: The MIT Press.

Simulation strategy and communicative approach in CALL

Pierangela Diadori
Scuola di Lingua e Cultura Italiana per Stranieri di Siena, Italy

ABSTRACT: Since the first use of computers in language teaching it has been shown that the best results are obtained if the teacher himself becomes the author of the courseware, while various problems arise if the work is entrusted to a technical expert. The teacher who wants to create a computer programme based on the principles of a communicative approach should not use structural or mechanical exercises, which are the easiest application of computer for CALL. He should refer the exercises to a context and favour a realistic use of the machine in the classroom. Furthermore, among the various CALL strategies (drill, test, inquiry, simulation and tutorial) he should prefer simulation, which is the less frequently used but offers the most interesting prospects for CALL.

An example of a simulation based on speech acts, which has been programmed using PLATO (TUTOR language) at CINECA (Bologna, Italy) as part of a courseware of Italian as a foreign language, is shown and discussed.

KEYWORDS: courseware design, authorising systems, CALL strategies, communicative approach, simulation, negotiation.

ADDRESS: Pierangela Diadori, via P.A. Mattioli no. 18, 53100 Siena, Italy. Tel. 0577/44004

Computer Assisted Language Learning (CALL) is a reality for a limited number of people only, but it is becoming an object of interest for an increasing number of teachers, also due to a rapid diffusion of this technological medium in the field of education.

As a matter of fact, knowledge of the possibilities and limitations of CALL will certainly prove useful both for those teachers who intend to implement their lessons with the use of specific courseware to appear on the market, and for those teachers who wish to become authors of their own programmes. Such a possibility should not be ignored, as it is becoming less and less remote. Since the first use of computers in language teaching it has been shown that the best results are obtained if the teacher himself becomes the author of the courseware, while various problems arise if the work is entrusted to a technical expert under the guidance of a teacher.

But how does a language teacher become an author of courseware? A good approach is outlined below:

a) First of all, he should familiarize himself with a text giving useful criteria for the evaluation of courseware;

b) He should also spend time revising existing programmes on various languages;

c) Then he should follow, at a specialized centre, a design course ranging from 1 to 3 weeks of intensive theoretical and practical study, depending upon the difficulty of the computer language to be learnt. It must not be forgotten that generally the language teacher has no technical background: he is not an expert, nor should he wish to become one. He only needs the more adequate means that will enable him to create on a computer those activities specifically connected to his own teaching subject. The difficulty is in finding a system that is sufficiently flexible and capable of adapting itself to the teacher's creativity and that, at the same time, will make programming as easy as possible.

   For these reasons, authoring systems have been created, providing environments in which the author can programme in the simplest and quickest way as compared to other systems, though with less freedom of choice. In fact he should limit himself to working with what the system offers. Among existing authoring systems there are those which are more flexible (and complex) and those less flexible (but easier to use). The former are called "authoring systems with computer language", the latter are called "authoring systems without computer language".

d) After having acquired the necessary skills for editing the programme in an authoring system, according to the logic of the computer (connection of displays, possibility or difficulty of applying certain commands, etc.), he will first conceive the various displays of his lesson and write them on paper ("script"). After that he will edit them on the computer.

CALL STRATEGIES

The "script" is probably the most delicate phase of the whole programming process, as it is the moment when the synthesis between the author's teaching approach and the characteristics of the computer takes place.

There are certain teaching strategies which can be used in CALL. Each of them can be more or less suitable to attain particular learning objectives. Such strategies are:

- drill: synthetic rules followed by texts, examples, questions and different explanatory feedbacks, according to the student's answer;

- test: various questions (open, true/false, multiple choice, matching, fill-in-the-blanks, etc.) followed by a numeric evaluation of the student's performance (score);

- inquiry: choice of options offered through a series of menus that will enable the student to have access to lists of useful information (words, texts, rules, data, etc.);

- simulation: a real situation reproduced on computer can be varied according to new instructions, choices or different parameters introduced by the student through the keyboard;

- tutorial: all the previous strategies are used to guide the student step by step till he is able to master the subject completely.

The majority of progrmmes produced for CALL use drill, test and enquiry strategies, whilst tutorial and simulation are applied less often. Tutorial is less frequently used, due to the impossibility of developing all the various language skills by means of a computer programme, though complex and exhaustive (4), and simulation because of difficulties in programming. As a matter of fact, simulation is probably the strategy that offers the most interesting prospects for CALL.

COMPUTER AND COMMUNICATIVE APPROACH

Both the first applications of computer for CALL and the description of the general characteristics of the machine given by technicians, might give the idea of a tool mainly suitable for mechanical exercises, rather than for activities related to real language use.

The teacher who wants to create a computer programme based on the principles of a communicative approach should consider the following suggestions:

- refer to a context: the manipulative exercises offered by drill strategies will be less mechanical if they refer to contexts, themes and whole sentences already presented in the class through written texts, audio or video-programmes. The same principle applies to more detailed explanations (tutorial) and to lists of data (inquiry) to which the student might have access in case of need.

- realistic use of computer: according to the various functions of computers in real life, the students should be asked to use it in class for similar tasks, like:

  a) language games to be done in pairs, in order to favour competition, motivation and active engagement of the students;

  b) problem solving activities, suitable also for groups, so that the common task will prompt oral interaction (which is very useful in multilingual classes);

  c) data to be found: the student will be asked to find out certain information among a series of data or to consult a memorized text.

- simulation activities: the possibility of giving an immediate feedback to the student's input, allows one to reproduce on the computer certain real life situations, whose development will vary according to the choices made by the student through the keyboard. Simulations are commonly used in mathematics, physics or other subjects that use numeric parameers and simple graphics, but the same principles could be valid also for language teaching. Given a situation (e.g., the beginning of a dialogue between two persons) the student will be asked to answer each cue according to the context. This would develop the student's ability of "negotiating", which means making choices that are adequate to the verbal stimula in the foreign language. This ability, which is typical of native speakers, contains all the various aspects of communicative competence, and is therefore one of the main learning objectives.

There are some operational difficulties, but it is not impossible to reproduce a simulation activity on computer, even using a very easy-to-use authoring system. The author might reproduce on the screen a kind of "animated cartoon", more or less realistic, according to the graphic editor at his disposal. The main problem is that it is impossible to foresee step by step all the different answers to a verbal stimulus presented to the student. In real life, the student's question: "What are you doing tonight?" might be answerd differently: a) "Nothing special"; b) "There is a good film on the TV ..."; c) "Why do you ask?", etc., and in each case the student should find an adequate feedback. On the computer, such structure would branch as follows:

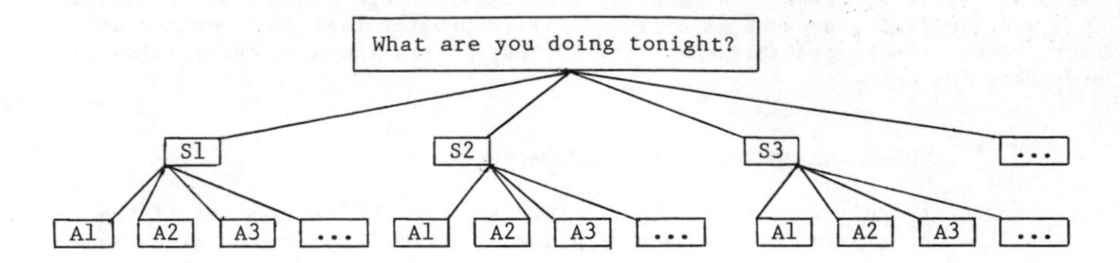

A possible simplification could be the introduction of a limited number of feedbacks. For example, each cue could lead to two alternative answers: one is wrong and causes a negative reaction of the interlocutor; the other is correct and allows the conversation to continue. This structure would be as follows:

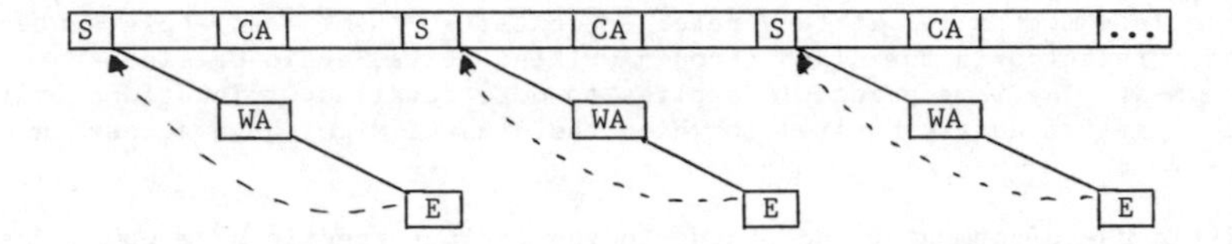

After having introduced the situation, the roles of the two speakers and the speech acts to be used, the cue of the first speaker will be presented with a choice of two possible answers (Fig. 1). If the student keys in the wrong answer, this will appear on the following display together with an explanation of the mistake and the reaction of the first speaker, that will end the conversation (Fig. 2). At this point, through the term BACK pressed on the keyboard, the student will automatically go back to the previous cue (Fig. 1). When the student chooses the correct answer, he will branch to a display where the conversation continues with a further cue and a new choice of answers (Fig. 3). The dialogue will go on for about ten or more cues, through the various phases of negotiation, till the final agreement (Fig. 4).

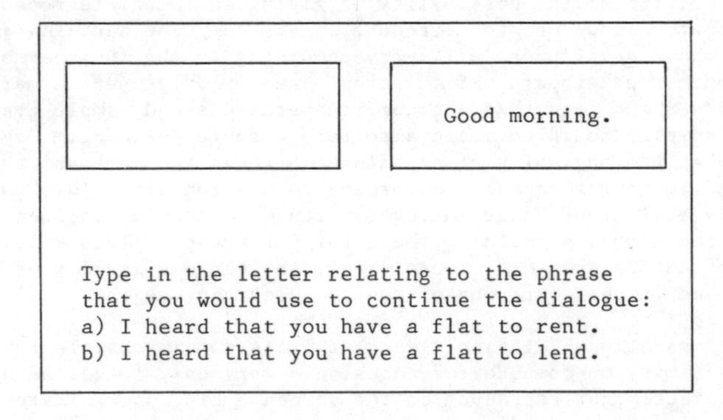

Fig. 1

I heard that you have a flat to lend.

You have been misinformed: it's for rent.

You have chosen the wrong phrase: lend means that you give something for free, whilst rent involves the payment of money, as is usually the case for an apartment.

press BACK

Fig. 2.

I heard that you have a flat to rent.

Yes. Are you interested?

Type in the letter relating to the phrase that you would use to continue the dialogue:
a) Yes, I should like to see it.
b) Yes, I want it.

Fig. 3

I'll do my best.

Here are the keys. I wish you a nice stay.

WELL! The first part of the dialogue is over.

Fig. 4.

Simulation/drama in the teaching of Asian history at a Canadian university

Ranee K. L. Panjabi
Memorial University of Newfoundland, Canada

ABSTRACT: This paper, presented at ISAGA 86, shared with colleagues involved in simulation projects, my experience in using this technique in my Asian History courses at Memorial University of Newfoundland, Canada.

I explained the methods used, gave examples of the types of simulation and showed how the dramatization of Asian history is very valuable in fostering inter-cultural awareness and understanding. Students stated that India and China "came alive" for them and that they were better able to understand the philosophy and way of life of the nations they were studying. Simulation stimulated student interest and also enabled them to participate actively in groups which researched together and planned the presentation. The advantages and benefits of this technique are stressed in my paper as well as the significance of the instructor's role as advisor and guide to each student group.

Within the normal requirements of historical research and presentation, i.e., thoroughness, objectivity, clarity and precision, simulation/drama projects can be as useful a teaching technique as more conventional methods like term papers and written exams. The combination of simulation and conventional methods of teaching offers a new and very rewarding approach for university history professors and students.

KEYWORDS: Simulation, drama, Asia, history.

ADDRESS: Department of History, Memorial University of Newfoundland, St. John's, Newfoundland, Canada, A1C 5S7. Tel: 709-737-8419.

A subject like Asian History, when taught in a North American university, requires a creative, even an innovative approach. The subject is exotic and out of the ordinary.

Hardly anyone today would argue the need for better understanding between Asia and North America. It is in our Universities that we can break down some of the misconceptions and self-imposed barriers and hope to create a mutual awareness which can only be beneficial for both cultures. The real challenge for the teachers of Asian History lies in arousing the interest of the student by emphasizing the political and cultural differences while simultaneously stressing the similarities between the peoples of Asia and North America to give the subject a relevance which justifies the hours of study and research which the student will devote to the subject.

One of the real challenges facing the techer of history in our modern world lies in the need to make a subject which deals largely with dead people come alive in the minds of young students. While we who teach history naturally have an intrinsic fascination for it, communicating that enthusiasm to others exclusively through traditional methods of education can be quite a challenge. It is certainly possible but today's students have so many mental diversions available to them that in order to compete with these we need additional methods to attract their attention, arouse their curiosity and sustain their interest. Largely in response to this, a number of history professors and teachers around the world have experimented quite successfully with the idea of involving students actively as participants in the historical process. The results of such endeavours go variously by names such as simulation, role playing and dramatization. The aim is to make history "come alive" for the student, to give him a "feel" for the era he is studying, to convey to him an idea of the complex choices available to real historical figures, to give him a sense of identification with personalities from the past and to make history a compelling, lively and exciting subject to study.

The literature on the qualitative and quantitative impact of this method is now quite extensive. The process of assessing its worth is still continuing. I would like here to recount in detail the manner in which I have used the method and give some idea of its impact. I must emphasize that the method involves a great deal of extra time and commitment both from the professor and students but the results seem well worth the effort.

While simulations are now popular in any number of academic fields and are used in a variety of learning situations at all levels, when adapting this method to my university history courses, I set for myself and for my students some very specific goals to keep the method viable and relevant to the aims of the course.

The dramatization (a term I find more applicable to my version of this method) of history has to be fitted into a tightly scheduled semester filled with activities and assignments. It forms only one assignment out of a number. Students also write research exams, term papers, and so on. The time constraints have always been the biggest problem. Class enrollments can be a crucial factor. Dramatization is only feasible when classes are not unduly large. It is just not practical to have too many groups performing in the limited time available. In recent years, increasing enrollments and large classes at Memorial University have made implementation of this method an impossibility.

I would strongly recommend that any instructor considering using this method persuade but not force students to participate. Some students are too shy, perhaps too diffident to speak publicly and they should be given the option of writing a term paper instead. I have invariably found that the overwhelming majority are interested in doing the dramatization. One student voted to write a term paper instead of doing the dramatization and later wished she had participated in the dramatization after she saw her fellow students performing. She returned to another of my courses and participated in the dramatization. Indeed, by giving the student a choice in the matter, the professor guarantees a degree of commitment and dedication he could not hope to achieve by making participation mandatory.

It is also imperative that the professor and students both have a shared mutual understanding of the objectives of this assignment. Each participant will need advice from time to time regarding his/her role and the professor often becomes a resource person assisting in this process.

Essentially, a dramatization of history involves one or more (usually a maximum of 6) students in an authentic recreation of the past in the class-room before an audience of peers.

The standard and level of information to be imparted is the same as that required for a conventional assignment like a term paper. Grading is based not on acting ability but on clarity of presentation, historical accuracy and evidence of careful research. Each student hands in a bibliography. Given the nature of the assignment, students are urged to read primary and secondary sources. If the historical character represented by the student has written books expounding his/her views, these automatically become required reading along with analytical or critical secondary sources.

The student's task is to portray a person from the past as accurately as possible. He has to know the life story of the person, his views, his contribution, his foibles, weaknesses and fit this information into the context of the era covered by the topic. The process of research and presentation varies, depending on whether it is a solo performance or a group effort.

For a solo performance, the student does the research and then focuses on a particular aspect he wishes to dramatize. For example, a rather shy young woman decided to study non-violence and gave an excellent speech as Gandhi. This particular student went on in life to become a very effective television journalist. Another student presented the class with a sermon as she portrayed Martin Luther King Jr. Both students were careful not to rehash actual speeches but to present a rendition based on their research of the historical figure. Both stduents had an obvious interest in the subject of non-violence and both expressed their interest in the rather intense class discussion which always follows each dramatization.

While some solo presentations have been very effective, dramatization functions best as a group effort with about four students. The shared activity generates friendship, and the joint participation makes the assignment both a learning experience and a lot of fun. The sharing of responsibility also serves to lessen the individual's burden in "pulling off" a good presentation.

After students select a subject, they usually meet with me to discuss it in detail. This first briefing is crucial. The topic has to be narrowed to fit into a limited time frame. Each student has to be assigned a role. Students have to decide whether to portray an actual event involving all the characters or whether to expound individually on a topic which was of common interest to all the historical persons. If the presentation is to focus on a philosophical issue of historical relevance rather than an actual event, then it has been most effective when portrayed as a discussion of the issue between the characters. This takes a certain licence with history but as long as the entire class knows ahead of time that the actual discussion did not take place but that the discussion in class will expound on the actual views of the historical persons, the students' interest is aroused and there is no chance of them being misinformed. A really good dramatization involved four students representing four Chinese philosphers who lived at different periods of China's history discussing a topic which was of mutual interest. The very lively discussion which ensued revealed not merely an understanding of each philosopher's views but a keen critical analysis of the weakness in his thought. Each student defended "his philosopher" and attacked the "other philosophers". I could not but be impressed by the apparent insight these students had gained in what would normally be regarded as a complex subject.

I leave it to the students to decide whether to write a script or to speak extempore. Those who write scripts are free to use them provided they do not simply read the entire presentation. Most students prefer to settle the broad areas to be covered and then speak in class without a script. Many rehearse at least twice or thrice in the evening on or off-campus and such meetings also encourage the collegial spirit of the whole enterprise. The rehearsals also highlight any weak areas which require more research. The combination of students in each group ensures that the "A" students set the pace and motivate the others to extra effort. This peer persuasion has propelled some students out of the rut of "C" grades into becoming more ambitious and striving for better marks.

Students have told me that this assignment requires considerably more research than a term paper because each student has to acquire a thorough knowledge of the life of the character and his or her personality in order to project it successfully to the class. No detail is too minor to be overlooked. to interact in a dialogue, students have to search for different historical views to arrive at the truth.

I receive a progress report from each group mid-way through the research. These reports also explain the directions being followed by the group. If students have any questions they generally ask. Normally, once they have done all the reading and had one rehearsal they have a much clearer grasp of the subject. Their questions tend to be more of detail relating to peripheral matters.

Because the event is enacted, students also have to acquire ancillary information on mannerisms, traditions, attitudes and ideas current at the time so that there are no inadvertent modern insertions into the dramatization. The professor can be a very helpful resource person in answering such questions.

Students are left free to decide whether or not to include props and costumes as part of their dramatization. Most appear to enjoy the idea of doing the assignment authentically. Students have often been inventive and ingenious in their use of props. Class-room furniture, skillfully draped with material raised from home provides just the right atmosphere. Classroom chairs have turned into very convincing thrones, even carriages. My lectures witnessed gruesome scenes of massacre as it was skillfully turned into a guillotine. The blackboard becomes a "newspaper" announcing significant events of the era. Posters decorate the room and sometimes fact sheets are handed out by the group to inform the audience before the dramatization begins. In one instance, a group of my students, obviously dedicated to authenticity, borrowed a truck and transported some beautiful antique furniture out of one very indulgent parent's home to our classroom. The furniture was from the era being portrayed and it was apparent that the students had verified this before bringing it to class.

Music authentic to the era and the region being portrayed is often used. Ethnic community groups can be extremely helpful in providing this.

The process of hunting for such props is itself an education and students acquire an insight into a variety of areas of knowledge by studying costumes, customs, art, furnishings, and so on.

This is particularly useful when dramatization is applied to the study of Asian history. The identification of the student with the historical character he has chosen to represent is obvious. His involvement with the way of life and the era in which that charcter lived forms a vital part of his growing awareness of the nation being studied. Dramatization involves the student personally in Asian History. For the duration of the assignment he participates in the process of being an asian or being a foreigner who was actively involved in Asian History. The dynamic nature of this type of assignment, the public projection of a point of view quite new and unknown to the student before he joined the course, provides a personal insight into the subject.

It is very useful to have a class discussion after the presentation. It helps to clarify issues which were raised, to clear up any confusion in the mind of the audience and also to highlight the overall significance of the episode or personality portrayed. By answering questions in the same way as a professor would following a lecture, the student becomes both learner and instructor and I have found that student responses to class questions are generally to the point and sufficiently detailed for the level of their achievement.

After I began using this technique I felt it was important to gain some feedback on it. What follows is a distillation of student comments on the assignment. This information makes no claims to being scientifically based nor does it purport to represent any cohesive analysis of student opinion on dramatizations of history. Most students admitted to being initially wary and apprehensive about the project and then to relaxing after they had begun to rehearse with their group members. A number indicated that shortage of time was a crucial problem. It required an amount of class time in group activity which was sometimes difficult to organize. If the project is announced very early in the semester, students can set aside a certain amount of time each week for their group meetings.

Most students said that they learned not only history but a great deal about human motivation from studying these varied historical personalities. Generally, the amount of research required was thought to be as much if not more than that necessary for a term paper. The discussion after the presentation also required preparation as students could not be sure what the questions would be.

They were more conscious of the need to fare well in this assignment. The group could not be let down by any one member so there was more incentive to work. The need to be concise, precise, to include historical fact, to retain continuity of thought and action and do this in the form of a dialogue was the real challenge to some students.

Occasionally, a very real competitive but friendly rivalry emerged between groups. I noticed this particularly in one Indian History course. A number of very good students had joined the course. Most had taken my courses before and many had done this type of project in their earlier courses with me. There was a definite competitive spirit which inspired each group to attempt to excel in its presentation.

Students generally praised the group nature of the activity which they claimed made research less a lonely affair, more interesting and more fun because others were involved. There was more zest in seeing how other students presented their topics and therefore more incentive to do well in one's own project.

My experience has been that dramatizations of history involve and awaken an interest in history among students who are history majors as well as those who major in other fields. This type of assignment has a very special appeal for all types of students. Dramatization of history is an important activity in which the student becomes the essential participant. Dramatization also provides a most significant means for encouraging the student's personal awareness of history and of historical personalities. It is also a means of communication which attracts attention, evokes interest and generates a response. In the process, it became an excellent learning device. Far from being a frivolous activity, it is a serious absorbing educational process. It is a means by which a student can extend his experience and acquire a firmer more personal grip on his subject. As an activity, it is both meaningful and creative. As such it evokes more involvement by each student in the learning process. It creates a situation in which students learn from each other.

The assignment is a challenge. It challenges the student to research extensively and it challenges the student to reproduce this research in an interesting manner to his peers. It involves both individual study and co-operative learning, personal research and teamwork. It brings the student a feeling of achievement and pride in his work. It evokes a positive attitude and outlook towards their work among students. It enthuses them to react to and become interested in history and historical figures. It encourages an enthusiasm in history and historical figures. It encourages an enthusiasm in history and world affairs. It encourages them to explore the historical process in a more involved, more committed, more personal manner. Essentially each student must become his own discoverer of the processes of history. By becoming a participant in the dramatization, he embarks on this personal voyage of discovery. The fact that students have retained vivid memories of the roles they played years after graduating, testifies to the impact the method has on them.

Finally, I have found all of the methods mentioned in this paper particularly useful in the teaching of Asian History. The introduction of a new area of knowledge requires a certain innovativeness on the part of the professor who has to excite the interest and arouse the curiosity of the stduent while recognizing the need to give him a thorough factual and analytical grounding in the subject. It seems only logical to use all the modern aids available to educators to make an ancient culture come alive and to make its past part of the mental riches which the student acquires in his university education.

While in his later life all the historical facts may not be remembered, the sense of adventure with which he approached the subject will not be forgotten. Even if the emperors and rebels, the philosophers and writers remain only as dimly remembered names, the stereotype images of Asia will probably never return. The open mindedness will hopefully surve as will the appreiciation for that which is different and unique. A sense of tolerance, of universality, of broad-mindedness is the lasting gift a study of Asia brings to the students of North America.

ICS: Interactive Communication Simulations

Clancy J. Wolf
The University of Michigan, USA

ABSTRACT: ICS is a role-playing exercise that immerses participants in the dynamics of political affiars. The objective is to give students a profound understanding of the complexity of political reality and to introduce them to the potentials of computer conferencing. The format of ICS is a general one and can be used to study politics at both national and international levels, the focus of the discussion is on the Arab-Israeli Conflict Simulation. The whole communication process is mediated by a control group located at the university. The three month long simulation includes a preparation period, simulation period and debriefing period. New developments are also discussed.

KEYWORDS: Telecommunications, role-play, Middle East, microcomputer, simulation, social studies, politics.

ADDRESS: Clancy J. Wolf, Room 1225 School of Education, The University of Michigan, Ann Arbor, Michigan 48109, Phone: (313)-763-5950.

The Arab-Israeli exercise has been the mainstay of the ICS program at The University of Michigan for the past three years. ICS, the Interactive Communication Simulations, is a growing collection of role-playing simulations moderated through Confer II, a computer conferencing system based at the university. The Arab-Israeli exercise is an educational program designed to combine research, discussion and hands-on computer experience in a political and diplomatic exercise. A major goal of the exercise is to give students an understanding of the complexity of international politics while introducing them to the potential of computer networking.

Large simulations, as used in ICS, can only be undertaken by linking participants from different schools via computer technology. Confer provides such links and enables groups of people geographically separated from one another to be in continuous communication.

Furthermore, individual teams never need to be available to each other at the same time, thus avoiding the problem of coordinating multiple school schedules. All that a participating school must have is a microcomputer adapted to perform as a computer terminal. When participants in the simulation connect the terminal by telephone to the Confer system, they enter the simulated world of international politics.

The simulation is a result of a collaboration between The University of Michigan's School of Education and the Department of Political Science. The simulation has been a part of a political science course on the Arab-Israeli conflict at the university for over a decade, although not in the computer mediated form. In 1981, the simulation's designer, Edgar C. Taylor, took the simulation to other campuses. Through the efforts of Robert Parnes, who designed the Confer system and Professor of Education Frederick Goodman, the simulation was presented to secondary schools in 1983 via computer.

In the five terms since the Fall of 1983, several thousand students from ages 10 to 64 have been involved from 94 schools in 11 states (and 4 countries) in 28 separate exercises. Very few schools have decided not to continue with the program, while more than half of the schools have already participated more than once. ICS also appears to be becoming institutionalized at many schools as the program continues even after the initial facilitator leaves the school.

STRUCTURE

In the simulation, participants are organized into teams, each team being assigned the roles of five political leaders. Each team represents one of the principal political factions in the Arab-Israeli conflict. The twelve teams currently represented are: Egypt, Israel, Jordan, Lebanese Christians, Lebanese Muslims, PLO Fatah, PLO National Salvation Front, Syria, Saudi Arabia, USA, USSR, and the West Bank. Individuals assume the roles of presidents, kings, ministers, opposition figures, private envoys and others who form the nucleus of leadership in each group.

The roles are all real ones so participants must immerse themselves in the personalities and politics of those they play. Careful preparation and a variety of research activities are required for the students to become effective in their roles as international and national leaders.

Each country group is located in a separate school. A neutral, university-centered Control Group of faculty and students from various schools at the university administers the overall operation of the simulation. Each team also has a team facilitator, usually a teacher, who is active in guiding the students as well as working with the other facilitators to maximize the students' experiences.

The fifteen week simulation is divided into three main parts: preparation, actual simulation, and debriefing. Preparation for the actual simulation generally extends for a period of seven weeks. Extensive materials are provided by the university to cover the basic technical and substantive information. Participants focus their efforts in six areas:

- Participants read and discuss selected role and country profiles. Further research into the specific roles and country background is encouraged.

- Participants respond to daily informational items submitted through the computer network by the control group.

- A pre-simulation introduces students to the structure of the simulation as well as the complex substance of the subject matter.

- Participants exchange comments on the credibility of what was exchanged during the pre-simulation.

- Each country/team submits a paper outlining its members' individual and collective goals.

- Participants learn how to manipulate the microcomputer, establish a connection with the university's computer, and function within the Confer environment.

The actual simulation lasts for a period of five to six weeks. This segment begins with a scenario, entered by the control group, designed to prompt action from each group. The scenario places the participants in a hypothetical state of the world in the not too distant future.

During the actual simulation, students pursue the personal and collective goals that they developed in the pre-simulation. By signing-on to the university's mainframe on a daily basis the students are able to pursue these goals in a number of fashions:

- Participants are allowed to send private communications, or messages, to other individuals. A communications matrix is imposed upon these communications designating a sub-set of participants who may communicate with one another directly. This communications matrix is designed to reflect constraints on communication that exists in the "real world."

- By subamitting press releases, students are able to make public comments to the entire simulated world. Typically each team has its own press agent assigned the task of submitting these press releases. Press agents are admonished to remain within the constraints of the press in the country they are representing. A daily International Wire Services Release is generated from these submissions by the control group.

- Teams may submit action forms to initiate economic, political or military actions. Each proposal includes analysis of the issue they are addressing, possible alternative actions, why they have chosen this specific action, who is the major proponent and opponent within the team, what the possible side effects are and what is the likelihood of success. Based upon the preparation demonstrated by the team, the control group will issue appropriate press releases indicating what level of success the action achieved.

- Participants are able to organize and conduct multilateral meetings.

- In addition, participants discuss options and policy with members of their own country team.

Debriefing lasts approximately two weeks. Participants initiate the debriefing by outlining what they tried to do, how well they think they did it, what problems they faced, and what assistance they received and opposition they encountered. This phase highlights the specific knowledge students have gained about the Arab-Israeli conflict as well as about national and international politics in general. The final portion of the exercise is to tell who they really are, where they live and what their interests are.

GOALS

The support and enthusiasm for the simulation at the schools is undeniable. We believe that there are a number of benefits for the participants. Preliminary evaluations support our belief in the following benefits:

- Participants demonstrate high levels of critical thinking.

- Participants achieve understanding of the dynamics of political affairs.

- Participants learn to look at reality from different geographical and political perspectives.
- Participants acquire detailed knowledge of a particular political problem area.
- Participants gain valuable experience with computer communications technology.
- Participants practice the writing of clear, forceful and effective prose.
- Participants make important decisions and experience the consequences of the choices they have made.

Several concurrent simulations are conducted during any specific term of the academic calendar. Parallel to these games, each one a separate unity, is another forum in which the teachers associated with all of the games may exchange ideas. This forum, known as the facilitators' conference is the gateway that all teams must pass through to get into their respective games. Originally intended as a way for the game directors to disseminate information, the teachers themselves use this forum for a variety of purposes. For example, when the last set of games was initiated, several facilitators posted announcements stating what computer hardware and software they were using, offering assistance to any others with similar equipment. A similar exchange is underway with specific activities used by the facilitators to prepare their students for the simulation. The forum has also been used to coordinate bus schedules for face-to-face workshops held in Ann Arbor at the close of the exercise.

One of the university's goals with the ICS exercise is to extend the resources of the university to the classroom teachers as well as the individual students. The most obvious way this is achieved is through the development of the suppoprt materials. Each term, a variation of the simulation is conducted on the university's campus within a political science course. The university students must prepare their own role and country profiles, as well as establish which are the important roles, rather than start with the kind of materials provided to the middle and secondary school students. The results of these students' research, combined with the previous simulation materials, are used to generate the role and country profiles for the following term at the pre-collegiate level.

A second approach to disseminating the university's resources is in the use of graduate and undergraduate students in the control group. Students from various schools within the university who have demonstrated a level of expertise in the subject area are invited to act as monitors and guides for the schools. At least one controller oversees each individual game. This individual is responsible for the maintenance of the Wire Services Releases as well as providing a credibility check of the participants' actions by reviewing their communictions and negotiating their action forms with them. Other students are assigned the role of National Security Advisors. NSA's follow the actions of a particular country in each of the different games. The NSA for Egypt, for example, would follow each of the schools acting as Egypt in however many games are being run. The NSA's role is to act as a resource, when sought, as well as step in and perform subsistence activity for schools having technical or other difficulties. The controllers and NSA's meet on a weekly basis with the ICS support staff as well as actively participate in the facilitator's forum.

DEVELOPMENTS

Finally, the resources of the university are being used to develop other simulations that can be conducted in the ICS format. A new simulation, the US constitution Exercise, will undergo its experimental run in February through April of 1987, timed to coincide with the 200th anniversary of the US constitution. Three other simulations are also under development. One is designed to have participants examine alternative views on drug and alcohol use, tentatively titled "Miami Advice"; another is a United States Public Policy Simulation; and a third is a Global Environment Simulation designed to fit into the science curriculum. These additions are all being developed with continuous guidance from experts in the associated fields from throughout the university community.

ICS has proved to be a dynamic experience both at the university and pre-college levels. The simulations are constantly changing based upon our experiences and ideas generated from all quadrants. Students, teachers and administrators in the schools have made unique contributions to the present simulations as have undergraduate and graduate students from various parts of the university community. We look forward to many more years and many more exercises to add to the ICS repertoire.

Make a mask! The face you save may be your own!

Karin Blair
Ecole Supérieure de Commerce de Malagnou, Switzerland

ABSTRACT: The same speed which has taken a few of us outside of earth's pull has taken many more beyond any homogeneous cultural gravitational field. In such circumstances how can I keep my sense of self and my sense of direction? Such questions are addressed through Cubal Analysis, which is based on a vision of all people sharing three inescapable dimensions of time, space and value. As a thinking being I am always making judgements about the young/new or the old, the male/masculine or the female/feminine, the good/desirable or the bad/ undesirable. As a living being I am at the intersection of these three axes, which are in constant motion. The eight apexes of the cube, where one term of each set of polarities joins, provide character types which can be used to animate a variety of classroom games, all of which invite us to find that place in ourselves from which we can define our own "ups and downs". Although each type of activity is self-contained, it can also be related to the conceptual model, serving as a point of reference in dealing with a variety of contexts including foreign languages and foreign cultures.

KEYWORDS: cultural diversity, masks, cubal analysis, matrix of games, zero-gravity.

ADDRESS: 8 av. Adrien Jeandin, 1226 Thonex-Geneva, CH, (22)49-11-01.

The model underlying the following game activities, some of which were played at ISAGA '86, is based on a value neutral model I call Cubal Analysis (1983). It is based on a vision of all human beings sharing three inescapable dimensions of time, space and value, each with its own sets of oppositions: young/new versus old, masculine versus feminine, good versus bad. Through this hypothetical structure we are in contact with a conceptual matrix that functions independently of any particular context. It can therefore provide a game structure applicable to a wide variety of situations. Looking through this mathematically symmetrical model can enable insights into the possibilities of everyday life. We can see the fact that we make these judgements is distinguishable from how we actually apply them in any given circumstance. Therefore we have a hypothetical play world or a matrix for games which can mediate between the individual and the wider community.

By focussing on the corners of the cube where one pole of each of these three oppositions join, Cubal Analysis provides eight structured but blank character types. On each corner of the cube the conjunction of one aspect of each of the three axes produces: the bad young man and the good young man, the bad old man and the good old man, the bad young woman and the good young woman, the bad old woman and the good old woman.

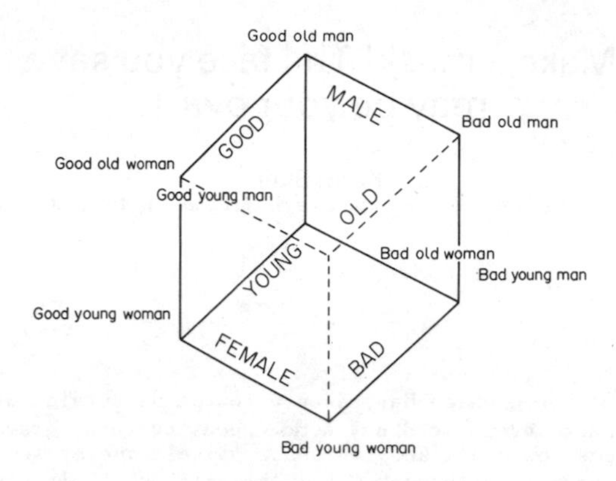

Fig. 1

As a three-dimensional object the cube when rolled as a die could only turn up one surface: one term of each pair of oppositions. The sides of the octahedron, however, are analogous to the corners of the cube, and the corners of the octahedron to the sides of the cube. By means of an octahedronal die (see Figure 2), the player can throw one character type to be animated with his or her own interpretations which can vary with changes in time, space and context. By adding masks to verbal interpretations one adds a visual focus which frees the participant from reliance on notes while speaking; also after we have made a mask we are obliged to clearly see that no living person could look like it any more than s/he could live out our generalizations or stereotypes. Using the Cubal die also engages the element of randomness. In this context the player is encouraged to go beyond seeing chance as fateful and impersonal. While the cubal world arrays eight character types which are equally available, the specific one I threw may be particularly important to me. Only I can say.

Basic procedures include throwing the Cubal die so that each person has a character type, thereby generating some questions to answer about each character to be animated. Basic materials for making a mask include paper plates and magic markers (or coloured pencils or crayons). In addition a variety of odds and ends are sometimes handy, such as string, yarn, coloured paper or cloth which can be glued on to the masks for beards, hair, hats, scarves or other decorations. Even after little study, participants can answer questions on behalf of the character such as the following: Name, Age, Country of origin, Job, Hobbies, Family, Friends, Neighbors, Enemies. More advanced tasks would include: Describe his or her neighborhood. Or: What does s/he look like? How does s/he dress? What is s/he like? What is s/he buying at the market? Where is s/he going (at the railroad station)? What does s/he always/often/sometimes/rarely/never do? What would s/he like to do? What is s/he going to do? What would s/he do if ...? What will s/he do when ...? What did s/he do yesterday/last year/on the National Day? A good final task is to ask each student to find a characteristic phrase for his or her character: "Go for it!" or "Keep praying!" for examples.

Once the questions are answered and the mask made, there are several options for processing the material. The players can take turns introducing their characters

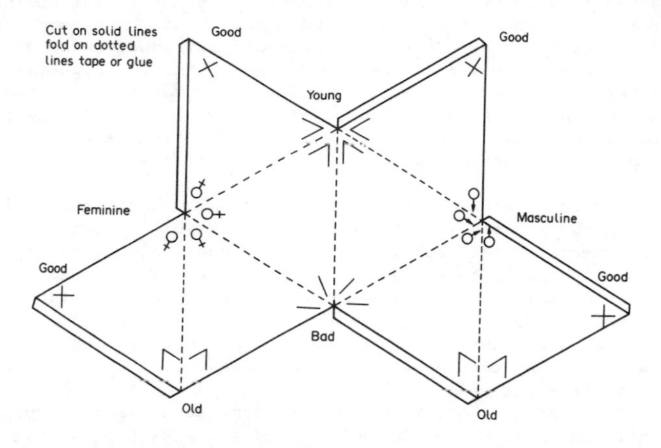

Fig. 2

to the rest of the group; they can take turns asking others, one at a time, questions about their character type; as their English improves they can try to hide the identity of their character types that the others must try to guess; or they can flesh out their character types in response to the questions of their peers.

Asking students to come to the front of the class gives the feeling of being "on stage", which encourages them to remain in the world of the game. In the ISAGA playshop we did a basic interview exercise appropriate for beginners, plus one of the following which can be used for those who do not need to be "on stage" all the time. First of all individuals can work in pairs or small groups to compare their character types. With a larger group, the leader might post on the wall the eight types so that each participant can find the appropriate place where others with the same character can gather. This resulting mini-group should share individual ideas about their shared character and agree on a profile. (Questions to keep in mind: have all three axes been taken into account? How do they vary with context?) Create a typical scenario which dramatizes the characters' qualities. Create another scenario which changes a positive character to a negative one and vice versa. Go to the group with the opposite character type and imagine a social setting which can accept and valorize the qualities of everyone's character. How can various types cooperate? How can they relate peacefully? Here self-identification is not by exclusion (I know who I am because I am not you) but by complementarity (since we are different we need each other). One can end by passing each mask around and have each person say something s/he has learned from it.

Given the impact of speed of transportation and communication, no one today can ignore cultural differences and the need for ways to accommodate them. Since cultural differences are dependent on human intangibles, I have been looking for more concrete analogies. The experiences of astronauts adapting to the absence of gravity and to the consequent absence of a predictable physical orientation provides an interesting perspective on the problems of coping with the lack of a predictable social orientation such as characterizes cultural diversity.

In a paper by Lichtenberg _et al_. (1984) the phenomenon of space motion sickness - or "space adaptation syndrome" - was studied by four specially trained crew members on a ten-day flight of the Space Shuttle/Spacelab 1 launched in 1983. Space sickness was observed in relation to 'provocative' situations in which 'rules' defining the 'normal' relationship between body movements and the resulting feedback to the nervous system have been upset. The most distressing disruptions resulted from the absence of a clear visual 'down' and the presence of others whose physical positions were askew in terms of your inner sense of down. After the initial few days, however, everyone became adapted to weightlessness and one who had earlier been sick said working upside down "was a lot of fun". Congruity between inner and outer orientation was still possible but no longer necessary. What had been a provocative experience - the disjunction between the two - became pleasurable play, once one had an inner sense of orientation.

The relevance of these studies of the experience of weightlessness in the present context lies in their analogy to the effect of multiple cultural orientations. "I felt like I needed a real visual 'down', ... and I didn't really have one of my own," sounds analogous to someone outside a home culture feeling the lack of familiar environmental reinforcement of others behaving predictably to reinforce a personal sense of identity. In a recent interview a young American woman in Geneva for a few months described how everyone she met from different cultures seemed to call her own sense of identity into question. Just as seeing another astronaut in a different relationship to shared space often makes one question one's own orientation, without stability of background, seeing another person act differently may make you wonder about yourself. With or without the problems of "other languages" there is always the problem of "other people". The wider the gap between people's assumptions and expectations, the less room for play there is in the relationship.

Csikszentmihalyi (1975) has described the zone of "fun" or play as that which lies between boredom and anxiety, both of which seem hard to avoid in cross-cultural situations. The woman who has to invest great amounts of time learning how to shop in a new cultural environment every three years is both bored with the task and anxious about what is going to happen in this strange place, for example, how will she find the right kind of vendor for what she needs?

The foreign language classroom is in particular the interface of at least two cultures where students frequently confront anxiety and boredom. On the one hand they cannot express themselves at the level of their thought and are bored, on the other they cannot be sure of the meaning of what they say and can be anxious. In addition the usual packaging of teaching materials can be seen, like the foreign language itself, as prestructured, abstract and closed to the student. However telling the student to "Express yourself" or "Create your own characters" leaves the terrain too open. Extreme openness like extreme closure is uninviting; the one provokes anxiety, the other boredom. A game structure, non-threatening, non-judgemental, hypothetical, relevant in any context, can provide shape, orientation and space to project our own meanings, reducing anxiety and increasing self-awareness.

Clarity of background structure within which one can actively participate is an important element of play. In making a mask of a fictional character type as well as the scenarios for interactions with other characters, the participant controls both the foreground and background first by animating a character type and then by modifying its "meaning" by placing it in different scenarios. Hence games of this type can offer a way to grasp and play through some of the difficulties inherent in foreign languages and foreign worlds, much as living with zero gravity became playful to astronauts.

From here I wish to return to the Cubal Analysis model which has served me as a reference point and analytical framework in multicultural circumstances where diversity of social orientation parallels diversity of physical orientation in outer space. Just as we have access to our bodies for full three-dimensional physical movements, so we have the potential for a fully three-dimensional psyche. The psyche as that by which we take in, digest and orient ourselves in relation to our life situation is itself ageless, sexless and without culture. It is like that source of creativity Lessing (1975) describes as "immensely ancient[...

while being born] neither male nor female, and]...having] ... no race or nationality .... It isn't far off that creature or person you are when you wake up from deep sleep and think: Who am I?" Having heard the question, we must realize that it is up to us to answer it. We are the only ones who can follow the promptings of our inner gyroscope.

When these promptings reach our consciousness, however, we find we are always making distinctions: the vaguest of dream figures begins to be graspable when we can say is "it" is more or less young or old, masculine or feminine and attractive or not, in other words when we can locate it somewhere on the cubal axes. Similarly a new-born child is always identifiable by gender, time of birth and nationality perhaps even before acquiring a name.

At first glance my conceptual cube functions in terms of mutually exclusive opposites: one cannot be both masculine and feminine, young and old, good and bad at the same time. And at first there are advantages to this, such as its practical utility, which right now provides the eight character types about which everyone usually has something to say. Identifying oneself solely through mutually exclusive oppositions, however, leaves one bound to one place and one set of unquestioned definitions. Cultural diversity makes it hard to live unreflectively, to assume, for example, that evil is the absence of good, that my foreign neighbor is supposed to change personal habits from simple exposure to my "superior" culture. We cannot expect others to live out our stereotypes of them any more than we could expect them to look like our masks. Such an attempt would in any case produce people who are only adapted but not necessarily adaptable, a quality more relevant today.

On a second level of reflection one sees that the cubal model, viewed from the inner intersecting axes, is in motion and thus more concerned with reciprocity and the relative weighing of oppositions. Since two poles are needed to structure an axis, one is led to see that both terms of an opposition are indispensable, just as opposite character types needed each other. Furthermore any expression of value is relative to the time and space through which it is manifest. With a gyroscope I know where I am because of where I have gone and where I am going; so with this model I can know who I am because of who I have been and who I want to be. Thus games played in such a hypothetical world can let us try on new roles and try out new perspectives on old ones and in doing so enhance our sense of life direction.

REFERENCES

Blair, K. 1983. Cubal Analysis: A Post-Sexist Model of the Psyche. Weston, CT: Magic Circle Press.

Csikszentmihalyi, M. 1975. Beyond Boredom and Anxiety: The Experience of Play in Work and Games. San Francisco: Jossey-Bass.

Lessing, D. 1975. A Small Personal Voice. New York: Random House.

Lichtenberg, B., Money, K. and Oman, C. 1984. Motion Sickness: Mechanisms, Prediction, Prevention and Treatment. Neuilly sur Seine: NATO Advisory Group for Aerospace Research and Development.

Social issues, game design, and research: An introduction

Cathy Stein Greenblat
Rutgers University, USA

Immersed in our own socio-cultural environments, we often lose sight of the universality of some of the social issues that face us in our daily lives, albeit with differences in salience. Yet issues such as prejudice and discrimination, the impact of technological change, the erosion of traditional principles and practices, rural development, population growth and its implications for everyday life, provision of meaningful participation in the decision-making process for persons in different social strata transcend national boundaries. In similar fashion, the search for better techniques to prepare students, whether they be youngsters in the classroom, villagers in rural environments, or corporate executives, to cope with these issues is being undertaken around the world. Teachers and trainers, recognizing the need for more effective modes of increasing information, enhancing analytic skills, improving praxis, and creating sensitivity to alternatives, have found that gaming and simulation offer advantages over more traditional techniques.

The papers in this section have been prepared by scholars in Australia, the Netherlands, Africa (Cameroon), the United States, Israel, Great Britain, and Canada. While each of the papers deals with a different specific topic, they share a concern with the potential of gaming to help cope with the social and pedagogical issues faced by all of us. Some focus on the use of gaming to deal with social issues; some offer insights to the game designer; and the last set deal with research issues.

Klaas Bruin's paper reports on the latest developments in his long-term project to develop tools for reducing prejudice and discrimination in the Netherlands. Readers in other parts of the world will surely learn from Klaas' work as they consider ways of dealing with similar problems in their own environments.

In similar fashion, Elizabeth Christopher, while reporting on the utilization of gaming and simulation for human resource development in Australia, offers lessons to all of us concerned with the impact of rapid social and technological change. Elizabeth discusses the ways in which such changes have eroded traditional principles and practices in management circles in Australia. As a result, managers require new approaches to human resource development, particularly approaches geared to coping with ambiguity. As new questions have arisen concerning job service, job rewards, criteria for status, work roles, timing of work, decentralization, the role of unions and the overall function of industry, organizational trainers face new problems and require new techniques.

Furthermore, the entry of increased numbers of women into managerial roles has changed the orientation of at least some portion of these cadres towards gaming techniques, previously undervalued as compared to more "intellectual" approaches. Her paper presents both a review of the social environment that has generated greater receptivity to gaming and practical examples of the effective utilization of gaming for exploration of organizational problems.

A somewhat different set of social issues is addressed by Jacob Ngwa and Philip Langley. Their concern is with the problems of rural development. While the descriptions of the problems of training both rural administrators and the rural citizenry have an African base, these problems beset teachers and trainers in less developed countries around the world. Jacob and Philip assess the limitations of conventional methods such as lectures and written assignments for bringing about change in the affective domain and change in praxis. They then describe the ways in which alternative pedagogical methods such as live-in field studies and field study seminars have been employed by the Pan African Institute for Development (PAID). With this base, and building upon traditional African simulation practices, PAID has recently turned to gaming-simulation techniques, notably the development of CAPJEFOS in collaboration with Cathy Greenblat, under sponsorship from UNESCO. This game simulates village level dynamics and the possibilities and problems of development in the areas of health and agriculture.

Ngwa and Langley summarize the issues dealt with in CAPJEFOS. In the paper that follows, Cathy Greenblat gives an overview of CAPJEFOS in operation, and comments briefly on the short run of the game at the ISAGA '86 meetings. It is hoped that this simulation will ultimately be utilized in teaching and training settings in many parts of the world. As this introduction is being written, CAPJEFOS has been run in the U.S., Canada, western Europe, Africa, and China, and Pier Corbeil has almost completed a translation of all the materials into French.

Rachel Mintz, R. Nachmias, and D. Chen take us to the second topic in this section: game design. Substantively, their concern, like that of the above authors, is with a worldwide problem: population dynamics. The need to understand the environment as a system, to comprehend the effect of specific variables on population dynamics, and to recognize the laws underlying such changes, is clear in our contemporary world. These authors, however, also have a concern with the dynamics of learning, particularly learning by discovery. The computer simulation they have developed, ECO-LIFE, is a fine example of the creative use of simulation techniques to guide students to comprehension of the interaction between system components. Game designers will find their creation of alternative output modes of particular interest. ECO-LIFE's designers have included pictorial, graphic and numerical representations of output, which make the simulation appropriate to learners at different levels and with different cognitive styles.

The design of any game relies on conceptual modelling and a theoretical framework. Israel Porat discusses these foundational elements in relation to media gaming, with particular attention being paid to news, reality, and feedback. Israel builds a theoretical framework which accounts for similarities and differences between real world media processes (e.g. news flow) and structures (e.g. media organization, media-consumer relations) and those to be found in media games. All these aspects shape the design and running of media gaming.

Finally, two papers offer insights to readers concerned with research. Rod Watson and Wes Sharrock, in an extremely innovative paper, subject the gaming enterprise itself to rigorous scrutiny. Starting from what has so often been assumed, but not demonstrated - that gaming-simulation interactions differ from interactions in the traditional classroom - the authors set out to assess whether and in what ways this is so. Imaginatively using ethnomethodology and conversation(al) analysis, they elucidate for us the dynamic of simulation-gaming. Their detailed presentation will hopefully inspire others to similar efforts to make gaming activities (and not simply their outcomes) the subject of research.

The next paper presents the "flip side" of this problem - the use of gaming to elucidate the research process. Florence Stevens' description of her ISAGA '86 workshop illustrates the potentials of simulations and "realists" for conveying a better comprehension of the nature of research investigations. The workshop activities are quite fully described, and insights are offered concerning how such exercises can lead to greater understanding of the major elements of a research project.

In sum, the papers in this section offer considerable enlightenment to many audiences - social scientists, game designers, researchers, and teacher trainers.

The application of simulation/ games in education when prejudices and discrimination are present

Klaas Bruin
Teacher Training Institute 'Ubbo Emmius',
The Netherlands

ABSTRACT: In the general literature on the nature of prejudices and discrimination it is often stated that, once they are formed, prejudiced attitudes are fairly resistent to change, whereas the public appearance of discriminative actions is largely dependent of the social norms that prevail. Consequently, in this paper educational measures that attempt to influence the social norms in a school as a community have a central place.

In simulation/games social learning is prominent and moreover, it takes place in an environment that is more or less under control. Over the years a range of simulation/games have been designed that tackle problems concerning prejudices and discrimination from different theoretical perspectives. When the social norms in a group openly approve of the public expression of prejudiced views, the risk is involved that with experimental learning methods such as simulation/ games these opinions are reinforced.

KEYWORDS: simulation, education, prejudice, discrimination.

ADDRESS: Klaas Bruin, Department of Sociology, Teacher Training Institute 'Ubbo Emmius', P.O. Box 1018, 8900 CA Leeuwarden, The Netherlands.

INTRODUCTION

In recent years in Western Europe the public expression of prejudices and the appearance of discriminative actions have unfortunately become part of the daily life of many people. Moreover, in many countries we have come to see again the rise of political parties that base their policies and public statements almost solely on ethnocentric themes. Over the last decade in many cases these parties have even gained some respectability and have furthermore shown their capacity to attract a considerable part of the youth. Consequently, education in the countries involved is faced with a major problem.

This paper can be qualified as an interim report after one year on a project on 'Prejudices, discrimination and education' that is financed by the Dutch government.

SOME LEADING PRINCIPLES

One of the starting points of our project was put well into words by Henri Tajfel (1981, p. 210-211), who has made an important, dilemma-like observation on the complexity of the subject at hand.

> 'Most teachers who attempt to provide children with knowledge about other countries would hope thus to provide a basis for tolerance and understanding in international relations. This hope appears to be based upon a common-sense assumption: that prejudice and emotion are based upon certain (possibly erroneous) 'facts' which the child may be said to possess.
>
> Thus, if the child believes certain 'good' things about a nation he will feel positively towards it, while if he believes 'bad' things about it he will be prejudiced against it. ....the study of racial attitudes in children suggest that the affective component of attitudes towards human groups may be acquired before the child possesses even rudimentary information about them ....'.

A second basic idea to start with in our project, was formulated by Pierre L. van den Berghe (1978, p. 20-21) in the sentences that follow:

> 'To arrive at a better understanding of the psychogenesis and psychodynamics of prejudice, we must relate these problems to the social context. ..... Racism for some people is a symptom of deeply rooted psychological problems, but for most people living in racist societies racial prejudice is merely a special kind of convenient rationalization for rewarding behavior. If this were not true, racial attitudes would not be so rapidly changeable as they are under changing social conditions.'

Which leads us to the third essential proposition of our project, formulated by Seymour Martin Lipset (1968, p. 6). For the study of prejudice and discrimination he suggested to abandon the focus directed solely at the individual, and proceed from the concept of the 'prejudiced community' instead.

ON PREJUDICES AND DISCRIMINATION

Prejudices are normally defined as a special sort of attitude. Its coginitive component consists of categorial perceptions and judgements, which include a tendency to neglect individual differences. These categorisations are mostly accompanied by negative, sometimes hostile feelings. Under specific circumstances prejudiced attitudes can lead to discriminative behavior. Research has shown (see: Allport, 1954; Katz, 1975; Kok, 1979; Milner, 1983) that prejudiced attitudes can have considerable distorting effects on perception and judgement. One of the obvious consequences, firstly, of the general process of categorisation is that differences between qualities of units within a certain category are likely to be deminished, whereas differences between qualities of units belonging to different categories tend to be increased. A second important characteristic of prejudices is, that they frequently result in a very typical reaction to information that contradicts the existing cognitive and emotional structure. It is quite normal that alternative facts when they are presented, are denied, distorted or called an exception, or that the reliability of the source of this information is questioned. This means that prejudices, once they are developed, are very hard to change under the influence of a direct informational approach. Because of the fact that prejudiced persons tend to seek and find the confirmation of their views, in this case attitude change is a complex and time-consuming process, which will normally transcend the possibilities of an educational setting.

The public expression of prejudices and its transformation into discriminative actions, finally, is largely dependent on the social climate of the relevant groups a person participates in. The social norms that are dominant in a group, mostly put into words by its informal leaders, will determine whether discrimination is disapproved, tolerated, possibly even demanded. Moreover, research has shown (VandenBerghe, 1979; Lipset, 1968) that forced change of social norms can be followed by change of the corresponding prejudiced attitudes in the long run.

SOME GENERAL CONCLUSIONS FOR EDUCATION

1. For the general subject of this paper the reasoning given above leads to two logical conclusions. If we want to teach meaningfully in a classroom-situation where prejudices, possibly discrimination are involved, we can basically move in two directions. We can either try to change the prejudiced attitudes, or make an attempt to influence the social norms in a non-discriminative direction. The first approach seems to be rather difficult (if not impossible) and time-consuming, given the long personal history of its development.

The effectiveness of the second approach will be dependent on the social composition of the formal and informal norms and values that prevail in the group concerned. This second approach will only be successful when the person in charge (the teacher, the trainer) possesses the necessary group skills.

2. Prejudiced views and discriminative behavior are transmitted and learned socially. It is socially important to involve students in learning activities that expose effectively the mistakes that are made in prejudiced opinions, and to take measures in the school as a community that contradict discrimination and its self-evidence.

3. It is quite normal that prejudiced views find their confirmation. Alternative, contradicting information will easily be denied, distorted or called an exception. In such a case even the teacher as a reliable source of information will be questioned as well. Consequently, it is doubtful whether one has to teach explicitly on subjects that are known to be 'risky', when prejudices and a discriminative atmosphere can be suspected. In this case one has to be warned in particular against the view, that the expression of one's feelings and opinions leads always to something good. Once the student's own prejudices are collected on the blackboard or uttered publicly in the context of a simulation/game, the risk will be that they remain in the classroom permanently, with even a gloss of respectability.

Educational Measures that Might have Influence:

1. Attempts to influence social norms:

   - measures on the level of the school as a community

     x appointment policy

     x the screening of the library

     x how to cope with cultural differences

     x meetings between teachers

   - adequate reactions to discriminative incidents

   - the teacher as a reliable source of information

2. The preparation of the teacher:

- correct estimation of the situation of the students

  x do they have prejudices

  x are social norms discriminative

  x of what age are they - there are stages in the development of racial attitudes

3. The subject matter and the ways of instruction:

- analysis and screening of all teaching material on stereotypes

- training in the logic of inference

- teaching on culture and cultural differences and the social consequences

  x the relation between nature and nurture

  x the cultural diverse history of all WE countries

  x factual information on different cultures

  x ingroup-outgroup differences

  x the relation between majority and minorities

- the reversal of roles

  x the person that discriminates is discriminated against.

THE SIMULATION/GAMING APPROACH

As has been summarised elsewhere (see Bredemeier and Greenblat, 1981), simulation/games can be an effective means to influence and change attitudes. In the case of prejudiced attitudes, however, the effectiveness of experiential learning methods will be largely dependent of the social composition of the groups involved. Given the 'openness' of the learning situation the risk of reinforcement of prejudices is a reality (see: Bruin, 1985; Milner, 1983). The exact relationship between the social learning that obviously takes place with most simulation/games, and the influence it can have on the social norms (both formal and informal) of a group and its social and cultural neighborhood is as yet rather unclear. Keeping these disturbing elements in mind, in the final section of this paper a list of concrete examples of simulation/games, aimed at different specific aspects of the phenomenon, is given.

1. On the level of the school as a community the application of simulation/games can result in social contact between individuals from different backgrounds, in a social situation that helps to reduce their mutual stereotyped views. Weissbach (Katz, 1975) has indicated that in simulation/games many of the facilitating conditions mentioned in the so-called 'contact hypothesis' are present. A second area on the school level where forms of gaming can be useful is in the training of teachers to react adequately in case of discriminative incidents.

2. A subject-area where simulation/games have shown to be very fruitful is teaching on culture and cultural differences. The main objective of games such as CULTURE CONTACT (Glazier and Isber, 1969), BAFA BAFA (Shirts, 1973) and SUMAH (Eden and Last, 1982) is to confront participants in a controlled learning environment with the consequences of culture shock, with a communication process that breaks down because of differences of culture.

3. Following the social psychological research started in the fifties (see: Sherif and Sherif, 1953; Sherif, 1966) a number of very simple and effective games have been designed that present to participants the dynamics of group norms, conformation and the distinction between different groups, which is likely to be intensified by competition between them.

4. In simulation/games rules that are played in real life can easily be reversed; the discriminator or the member of the majority group can be put in a situation in which he/she is discriminated against. Fruitful examples of this gaming approach are the A-B-Z GAME (Christiansen, 1977), ORANGE and GREEN (Cited by Weissbach in Katz, 1975) and MIJN MAAT IS MIJN MAAT (de Groot and van der Zwaag, 1984) in which participants are divided into majority-minority groups according to their shoe size. Another example in this category of simulation/ games is the well-known STARPOWER Game (Shirts, 1969) in which one political group is given the power to establish the rules of distribution in the game.

REFERENCES

Allport, G. W. 1954. The Nature of Prejudice. Cambridge: Cambridge University Press.

Benedict, R. 1982. Race and Racism (forword John Rex). London: Routledge and Kegan Paul.

Bredemeier, M. E. and Greenblat, C. S. 1981. "The educational effectiveness of simulation games: a synthesis of findings". Simulation and Games, 12:3.

Bruin, K. 1985. "Prejudices, discrimination and Simulation/Gaming: an Analysis". Simulation & Games, 16:2.

Bruin, K. 1986. "Prejudices, discrimination and education". School, November 1986. (Dutch only).

Christiansen, K. 1977. The A-B-Z-game. Defiance: Defiance College.

Eden, T. and Last, J. 1982. SUMAH. Brighton: Brighton Polytechnic.

Glazier, R. and Isber, C. 1969. CULTURE CONTACT. Cambridge: Abt Associates.

Greenblat, C. W. 1981. Group dynamics and game design: some reflections. In Greenblat, C. W. and Duke, R. D. Principles and Practices of Gaming-Simulation. Beverly Hills/London: Sage.

Groot, G. de and Zwaag, H. van der, 1985. MIJN MAAT IS MIJN MAAT. Leeuwarden: Teacher Training Institute Ubbo Emmius.

Katz, Ph. A. (ed). 1976. Towards the elimination of racism. New York: Pergamon Press.

Kok, G. J. 9ed). 1979. Vooroordeel en discriminatie. Alphen aan de Rijn: Samson.

Kok, G. J. 1981. Attitudes en energiebewust gedrag. In Ester, P en Leeuw, F. L. Energie als maatschappelijk probleem. Assen: van Gorcum.

Milner, D. 1983. Children and race: ten years on. London: Ward Lock Educational.

Sherif, M. and Sherif, C. 1953. Groups in harmony and tension: an integration of studies in intergroup relations. New York: Harper.

Sherif, M. 1966. In common Predicament: Social Psychology of Intergroup Conflict and Cooperation. Boston: Houghton Mifflin.

Shirts, G. 1969. STARPOWER. La Jolls: SIMILE II.

Shirts, G. 1973. BAFA BAFA. La Jolla: SIMILE II.

Taft, E. 1976. Coping with unfamiliar cultures. In Warren, N. (ed). Studies in Cross-cultural Psychology. London: Academic Press.

Tajfel, H. 1981. Human Groups and Social Categories. Cambridge: Cambridge University Press.

Tajfel, H. 1982. The Social Psychology of Minorities. London: Minority Rights Group.

VandenBerghe, P. L. 1978. Race and Racism. New York: John Wiley.

Training in an ambiguous world: new games for human resources development in Australian organizations

Elizabeth M. Christopher
Christopher Agency, Sydney, Australia

ABSTRACT: The paper suggests that some recent applications of games to human resources development in Australian organizations may be of interest to trainers internationally, because many organizations today find themselves in an ambiguous world where social, economic and technological changes are seriously eroding traditional principles and practices in the workplace. Managers are unclear about human needs in this stressfully changing environment. It is argued that the increasing presence of women in organizational training programs may in part account for an observed shift in emphasis from games as intellectual exercises to games for human resources development, and some examples are given.

KEYWORDS: Organizational ambiguity: changing work roles: games for human resources development: high tech and high touch.

ADDRESS: 19 Ryries Parade, Cremorne, NSW 2090, Australia. (02)909-8195.

INTRODUCTION: SETTING TODAY'S ORGANIZATIONAL SCENE

Most Australian organizations today are in a stressful state of change: many precepts and practices of management that were tried and found true in the past are now virtually irrelevant - and yet nobody seems to know what to put in their place. Once upon a time, for example, "the boss" was a person who made the final decisions and got paid more than anybody else for doing so. Nowadays high-tech staff are often better informed to make decisions than are their managers and they command higher salaries. There is ambiguity in job service - job security being a thing of the past - and ambiguity in job rewards; the traditional golden handshake has given way to the golden hello; the emphasis is on rich new blood, not length of loyal service. There is ambiguity in the working roles of young people, women, migrants and Aboriginals; in the timing of work now that much of it is no longer nine to five; in decentralization of jobs; in the role of unions; and in the overall function of industry. Organizational trainers now have to help management and staff to cope with this ambiguous world.

THE CHANGING ROLE OF GAMES IN TRAINING PROGRAMS

Two implications from today's organizational confusion are that training programs now more than ever need to focus on human relationships - and must be designed and directed as part of a company's or a department's overall strategic planning, in the light of rapid social, economic and technological change. For example, negotiation strategies now must be learned in an economic climate of large-scale industrial take-overs and mergers, in which unions jostle uneasily for new positions. Motivation skills and team building need to be acquired in a context of worker redundancy on the one hand and on the other hand a shortage of - and therefore a need to cherish - people with highly specialized and highly prized skills. Equal employment opportunities and affirmative action have to be provided against a wider social background of increasing resentment against married women in the workforce and a groundswell of antagonism about migrant - especially Asian - labor.

These are some of the circumstances in which games and simulations can really come into their own as learning methods in Australian organizational settings because their primary purpose is to study human responses to group situations - and it is becoming fairly obvious, not just in Australia but worldwide, that organizational ambiguity will only be clarified through more attention to interpersonal relationships in the workplace. If games become sufficiently widespread in training programs they may even cause a revolution in Australian organizational culture by demonstrating the validity of emotional experience as a measuring instrument for change, but this will require some flexible thinking by training managers.

The Historical Background of Games in Australia

Until very recently, where games and simulations were used in Australia in organizational training programs (which was seldom, if ever, done systematically) they were assumed to be for two purposes, one serious, the other frivolous. There was a polarization between two concepts, one of gaming as an intellectual exercise - which made games acceptable as learning strategies - and the other of games as light diversions, activities to warm up the participants before the serious business of a conference begins, but serving no other important purpose. Unfortunately this historical difference of opinion about the function of games has tended to cast doubts on the legitimacy of gaming as a comprehensive learning device. Thus a valuable training tool has too often gone unrecognized by Australian trainers.

The view that gaming is an intellectual exercise belongs mostly in traditionally male-dominated learning contexts that still exist, for instance, within the armed forces. For many years games have been used in Australia in the form of large-scale and complex simulations of problems such as proposed business mergers, organizational changes, potential natural disasters, drastic breakdowns in public amenities, and military tactics. These simulations nowadays are as realistic as computer calculations can make them and they are designed for management training and for the training of police, fire and other rescue agencies, safety personnel, and so on. The results are evaluated statistically and form the bases for operational and strategic planning. The fact that gaming is an ideal tool for exploring the *feelings* of the people concerned in such events has been almost completely overlooked.

Games are only just now beginning to be part of small-scale and scattered training programs that address such issues as workers' fears of technical innovation; stress management as part of occupational health and safety programs; uncertainty about the advantages of office automation; the threat of compulsory early retirement; prejudice against equal employment opportunities; sexual harrassment in the workplace, job discrimination; human risk management, and so on.

The Influence of Women on Games

Such programs represent a quite recent recognition of the relevance of feelings and relationships in organizational environments. This recognition may be increasing at least partly because more women are achieving management status and are becoming more involved in management training programs and women are culturally more inclined than men to emphasize the importance of human relationships - that is, emotions - in management contexts. Moreover, though the latest figures indicate that no more than three per cent of Australian managers are women, the percentage of women is higher in training departments than in most other organizational departments because teaching is traditionally an acceptable female professional role. Nevertheless, the great majority of senior managers who are ultimately responsible for training programs are men, and male managers have been brought up to discount emotions as indicators of organizational efficiency.

Thus male designers and users of 'serious' simulations are often inclined to regard the use of 'experimental' games in business and industry as unreliable, subjective and therefore impossible to evaluate, and unnecessarily time-consuming; they see players of these games as self-indulgent softies. Games as emotional experiences are all right for women and children - so goes the argument - but not for macho Aussies engaged in the serious business of management.

This is probably the result of cultural and historical factors. Culturally, Australian men tend to be pragmatic rather than idealistic, practical rather than abstract, inclined to action rather than reflection and oriented towards the present rather than the past or future. Historically, Australia's public education policy - designed primarily for boys and implemented by men - has stressed job-training rather than learning for its own sake, except for girls, for whom it has always been culturally permissible - even desirable - to study the arts and humanities. Furthermore in Australia, for thirty-odd years after World War II, there was an attitude that the country "belonged" in some fundamental way to white Australian males of Anglo-Celtic or Anglo-Saxon background. This was a viewpoint that seemed quite appropriate in that secure and vanished world where jobs could always be had for the asking and kept for life, salaries continued apparently for ever to rise regularly in annual increments, and Australia was a white man's lucky country.

Arguably these blinkered views of the Australian economic, social and industrial scenes in past decades have contributed directly to their ambiguous states today, and until very recently the form of Australian management training programs was profoundly affected by their male designers' underlying and mostly unexpressed sentiments that non-traditional teaching methods - such as games - could only be justified if they wre treated as systematic and methodical research tools.

Games as Change Agents

Now however the stress of organizational ambiguity has resulted in fresh insights to the relevance of human emotions because job redundancy is a daily commonplace even for those who used to feel most secure, and inflation and the rising cost of living have eroded traditional work patterns and salary bases. Migrants and women are not only firmly entrenched in the workforce but are actually acquiring management status and thus introducing a whole new dimension to traditional management thinking in Australia. In view of such rapid and fundamental changes in the Australian organizational scene, it is perhaps not altogether surprising that games and simulations should be looked at anew by trainers as a potential strategy for understanding these alarming and challenging phenomena.

Games are being used more and more, for example, as 'triggers' to fire players' imagination concerning positive - rather than negative - aspects of technological innovation; to evoke their experiential as well as intellectual responses to strategic planning; and to 'lock in' their learning about new work practices. Games are being used to tackle the 'real' reason for social and political problems, i.e. breakdown in human relationships.

This is in line with the arguments expressed in a large number of publications in recent years, from America and the U.K., whose highly qualified writers propose a new kind of emphasis within organizations if present ambiguity is not to become more dense. They argue that if organizations are to be economically successful in the future they must see themselves primarily as places in which people work together and only secondarily as operational settings in which to perform other functions such as manufacturing, buying and selling.

Thus there is in Australia the exciting possibility of a game revolution - in which games become strategies by which senior managers experience, for example, some of the frustrations of their subordinates and vice versa.

SOME PRACTICAL EXAMPLES

The Airplane Game

One such activity can be called "The Airplane Game". Teams of players compete for profits in making paper airplanes and selling them to an independent consortium. The consortium sells them the paper and offers them a choice of several models. The team managers have the responsibility of deciding how much paper to buy, what model to select, and how many airplanes their teams can reasonably make in the given time. Each finished airplane has to pass a quality control test by a member of the purchasing consortium (will the plane fly?). Team members learn the hard way that poor management decisions can cost them the equivalent of their jobs and in the second round of the game there is always considerable discussion about viable alternatives to traditional hierarchical organizational structures.

The game is also useful for comparing different forms of labor in terms of productivity and profitability. Some teams will set up a conveyor-belt system of manufacture, other team members will individually take specialized roles, and so on. Some team managers will provide their groups with relatively sophisticated equipment - rulers, set-squares and other such articles they may have in their briefcases. Resources are never evenly distributed in any given group of players, nor are they defined the same. It is worth noting and discussing whether there seem to be differences between the professional priorities of men and women as two distinct cultural groups. Which 'culture' pays more attention to function, to profit, to appearance of the product, or what?

It is also interesting to compare organizational profiles between teams in "The Airplane Game" - some teams will become more interested, as a group, in achieving a quality product, while other groups will cheat and lie in order to get as many planes as they can past quality control, to maximaize profit.

Money in the Middle

Another game, much shorter and simpler in form, though not in content, explores differences in value-systems between people whose real-life responsibility is the allocation of public funding (or of people who have to live on the dole or who are involved in minimum wage negotiations). Each member of the group has to contribute a sum (ranging from five cents to a couple of dollars, depending on the economic status of the players), which they know will not be refunded. The whole group then has to decide which individual is to receive all the money. The only rules are that the individual must be a member of the group; the money, in one lump sum, can only go to one person, whose absolute property it becomes; and the decision must be one of group consensus.

This game, which could be called "Money in the Middle", can evoke strong feelings, for example value-judgments about who "should" or "should not" receive a gift of money they have done nothing to earn. There is usually considerable discussion about criteria for selection - some people may suggest some kind of lottery and others may hotly dispute the idea as irresponsible; and so on. It can be debriefed as a negotiation exercise, as a study in personal and group priorities, and as an evocation of what individuals feel like when they hear people in a position of power (in this case the power of money) discussing their worthiness or otherwise to receive funding. It is a good game for exploring the prejudices of, say, senior managers in private sector employment in a real-life social climate of antagonism to "dole-bludging" (an Australian phrase referring to people, usally young people, who are suspected of living on unemployment or social security benefits without qualifying for them).

Catch as Can

Another negotiation game is "Catch as Can". There are two or more groups of players, and everybody starts with the same amount of money - how much is immaterial. The only rule is that at the end of a ten-minute round the winner will be the person in each group with the most money. Curiously enough, almost invariably there is a winner in every group at the end of round one. The winners have either persuaded one or more of the other players to give them money, or they have won it in an impromptu gamble, or they have stolen somebody's money when their back was turned. The winner in each team is allowed to set the rules for round two. Often the respective winners will become collaborators in order to fleece the other players more effectively, or cooperatives of some kind will form, depending on whether the leaders' temperaments are more or less competitive. Asian and Aboriginal players are likely to demonstrate cultural differences compared to the behaviour of the Australians, for example.

The discussion after this game, like all the others, is interesting on a number of levels. The debriefing can be used to discuss differences in strategies between male and female players as managers; or differences in leadership styles, persuasion skills, personal priorities, and so on.

Broken Squares

Another well-tested favorite is "Broken Squares". Players are in groups of five; each player is given a number of puzzle pieces and the object of the game is that each player shall end up with a square the same size as the other four. The only rules are no talking, no taking others' pieces, and no asking other people for their pieces. The only way the task can be completed is for all five members to share their pieces freely with each other. It can be hard, however, to persuade one player to break up an apparently good square even when it has become obvious to the others that an overuse of resources (pieces) is preventing other squares being made.

Concepts of group dynamics and ego-involvement in task accomplishment can be explored after this game, particularly in cross-cultural contexts. For example, Japanese players may have less difficulty with "Broken Squares" than Australians, perhaps because the game requires the players to behave as members of a group rather than as individual entrepreneurs.

CONCLUSION

The above are just a few examples of short, simple games that may well seem childish in description, yet are proving powerful tools to explore organizational problems. Australians see themselves under threat from many quarters - they are worried about inflation, the national debt, the weakness of the dollar, their falling international credit rating and escalating costs in government spending, but over and above these problems - about which as individuals they can do virtually nothing - Australian managers are now actively seeking innovative ways to bridge the gap between high tech and high touch.

Australia's saving graces historically have been an ironic humor, a sense of fun and a distrust of pretension. Moreover Australia is a young country and temperamentally optimistic. In a world of organizational ambiguity, games and simulations are fun to play, they appeal to participants' youthful optimism (however deep it may be buried), and to their sense of humor. Games as learning strategies are also (deceptively) simple and they offer fundamentally an optimistic view of human relationships. The Australian experience of games in human resources development may become a good example of games for growth in international training programs.

NOTE:

All the games referred to above are taken from Elizabeth M. Christopher and Larry E. Smith, 1987. Leadership Training Through Gaming: Power, People and Problem-Solving. Kogan Page, London.

Simulation-gaming and training for development: An African experience

Jacob N. Ngwa and Philip Langley
Pan-African Institute for Development, Cameroon

ABSTRACT: Those concerned with the promotion of local development initiatives in Africa must find ways to train development staff that include a focus on the affective domain and changes in praxis. Traditional methods of pedagogy, while useful for cognitive learning, have not proven effective in bringing about changes in attitudes and behavior. Training institutes in Africa have thus been exploring alternative pedagogical methods. At the Pan African Institute for Development the authors and their colleagues have utilized live-in field studies and field studies seminars. More recently, simulation games such as CAPJEFOS have been added to the repertoire of tools they employ in the training of development agents. Simulation-like enterprises developed by other African institutes are also discussed.

KEYWORDS: Africa; development; games; simulation; gaming-simulations; training.

ADDRESS: Pan African Institute for Development, PAID-WA, PO Box 133, Buea, Republic of Cameroon.

There is a saying in French that "le plus mal coiffé c'est le coiffeur". An English equivalent of this states that the worst shoed person is the shoemaker. It might also similarly be thought and stated that in many cases the major obstacle to the promotion of local development initiatives in Africa is the development staff themselves. Such a controversial statement needs to be further explained, especially coming as it does from persons directly concerned with the training of development staff working at various operational levels.

Such development staff include particularly the front line workers who are mainly extension service field staff and as such are in immediate contact with the mass of the population, both rural and urban. In general, extension workers in Africa are government employees, although over the last few years an increasing number of non-governmental organizations (NGO's), sometimes also referred to as Private Voluntary Associations (PVA's), have set up smaller localised extension services. Their major areas of activities are agriculture, including livestock production, primary and secondary education, health and related services such as environmental hygiene, water supply, family planning, and so on, and community development which includes the construction of feeder roads, village water supply construction, and adult education in general, notably in the form of literacy campaigns. More recently, support to women's income generating activities is coming from extension staff working for Women's Bureaus, Ministries of Social Affairs or the Departments of Community Development.

Field staff are usually supervised through a hierarchical structure of command beginning at the local level with the District and going through such intermediate levels as the Sub-Division and the Division (in French, "Sous Prefecture" and "Prefecture") where the respective administrative officers represent the central authority, and exercise overall authority over the respective development staff. At a higher level, generally known as the Province or Region, supervision is ensured by a governor and his close collaborators who are directly linked to the central ministries.

In training such development staff conventional pedagogical methods such as lectures and written assignments may be useful in conveying the cognitive aspects of relationships between development staff and the men and women farmers, or in more abstract terms, the relationship of the state to the peasantry. However, as concerns the affective domain and change in praxis, traditional pedagogy is much less effective. In fact, lectures on "correct" behaviour are likely to become mere sermons and a wide gap is created between the cognitive knowledge of relations and the actual practice of these relations. In a more advanced stage this gap is reified into institutional structures in which the state agents and the peasants live in different worlds and talk at or talk past each other without the needed developmental change ever taking place. It is such a gap which perhaps led the Secretary General of PAID, Professor Alfred Mondjanagni to observe: "Mais le constat est qu'il y a un décalage entre les discours et la pratique, entre les discours et les réalités quotidiennes parce-que les institutions et les structures en place ne sont pas toujours favorable aux changements necessaire a de telle pratique".[1]

Simulation gaming practitioners need no convincing that alternative pedagogical methods are necessary, and for some time now, training institutions in Africa have been exploring these alternatives.

The experience of the Pan African Institute for Development (PAID) in this area, includes:

a) Live-in Field Studies: The trainees are placed in the village where they live with the village people for a period of about 3½-4 weeks, in keeping with the adage "live with the people, work with the people, learn from the people". During this period, the trainees conduct a socio-economic study of the village so as to identify the development problems and potentials, and draw up a suggested development plan for the village, which can be implemented, based as much as possible on local resources and local self reliance. Preparatory work for the field studies already takes the trainees to the field where they get practice in field technqiues through such simulation exercises as mock census counts.

b) Field Studies Seminars: Field Studies are followed by a Field Studies Seminar held right in the same village. Here the trainees present their findings as to the developmental situation of the village as well as their practical proposals for improving the situation.

The findings and proposals are scrutinized by the villagers, and occasionally are completely "torn apart" during such scrutiny. This is a vital element for training in the effective domain, sensitising the trainees to the realities of the rural world. Field workers, their supervisors, as well as administrative authorities are invited to be present at the Seminar. A triangular dialogue which sometimes looks more like a verbal battle takes place, and this is also an important process for the actors to examine their consciences and hopefully then, change their attitudes and behaviours.

Simulation Games such as CAPJEFOS are seen by PAID as an important addition to these alternative methods.

Simulation, however, is not new to Africa as a pedagogical tool. Traditional story telling where animals often take on human attributes has always been an important tool for teaching social behaviour. Those who have read the Baole story as to why the hyena walks with its hind quarter near the ground will understand just how effective this method must be.[2] In Ruanda and Burundi but also in many other African societies, training in public speaking, negotiations, and use of wit, takes place through swearing matches for the younger children adn riddle competitions for older children. Elsewhere, praise exchanges during wedding ceremonies also serve the same training functions.

Without going into detail it should be noted that little has been done in terms of modern games to explore and exploit gaming and simulation as it exists within the African cultural context.

Modern gaming simulation has however been slow to take off in Africa as a teaching tool. This could partially be attributed to the fact that modern games are based on alien cultural experience which is valid in its own right but is difficult to transfer. The underlying models upon which most of the modern games are based are not built on an analysis of the African cultural and development situation.

One of the early attempts to use modern games as a pedagogical tool in Africa was in the mid 1970's. This was in a joint ENDA/UNEP/UNESCO programme of training seminars held in Ivory Coast, Ghana and Morocco, for management of human settlements. One of these seminars was led by Richard Duke, and Philip Langley was one of the authors. Games have also been used in Kenya and Zambia, and there are probably many other unreported examples.

PAID's work is to some extent an off shoot of these earlier attempts and was made possible through UNESCO funding which enabled Dr. Cathy Greenblat of Rutgers University to come twice to Cameroon and work on the spot with African based trainers.

The Pan African Institute for Development's continental spread, as concretized in its four regional institutes which recruit trainees from 42 African countries, puts it in a privileged position to spread innovative pedagogical methods, including simulation gaming, across Sub-Saharan Africa.

CAPJEFOS, the game developed as PAID-WA, seeks to capture in the form of a gaming simulation the Institute's 15 years' experience in training rural development workers using the other active methods already described above. Essentially the game seeks to simulate the factors which hinder or promote rural development, and the role(s) of the development workers in this interaction. The action is situated mainly at the village level, and special attention is given to self-help programmes in the areas of health and agriculture, and the interaction between the village and the higher levels of national organization such as the division, the province, and the central national authorities. At the end of the game, the participants should be able to:

a) develop a better understanding of the factors that are at work in a development situation and the nature of their interconnections;

b) develop greater empathy for the rural population (villagers) and better knowledge and appreciation of their rationales;

c) develop a better grasp of what the development agent's role should be.

In the game, the players assume the roles of villagers and development agents and go through activiteis and experiences similar to what their counterparts do in real life. Among the issues that have been built into the game or that could be expected to emerge from the players' interactions are the following:

Increasing differentiation of village life; Socio-economic structures, class solidarity, class conflict; Health problems and their sources; Agricultural Problems; Protein shortages, nutritional levels, problems of food preservation; The interrelation of agricultural and health problems; Gender relations in village communities, including the sexual divsion of labour; Marketing relations and their impact; Food self-sufficiency at the household and community level; Economic conflicts at the household and community levels; Problems of education; The relationship between education and development; The role of religion in the village; The relationship between religion and development; The relationship between members of the village community and the outside world; Problems of the development of "rapport" between villagers and outsiders; Conflicts between needs and aspirations in village communities and "top down" development programmes and projects; Differing interests of villagers and development staff; Problems of setting priorities for development projects; Problems of implementation of development projects; Intersectorial relations; Needs assesement: "bottoms-up" vs. "top-down" approaches; The necessity of interrelated projects for benefits to be reaped; Social indicators; Problems of the relationships between the development agents themselves which raise the question of integration in development planning and implementation.

These and other issues are the themes that should form the focus of the post game discussion. For maximum pedagogic effect, the discussion is an absolute and important component of the game, and should be led skillfully. Adequate time should also be allocated for it. Themes may be selected for special emphasis depending on the lesson objectives and the nature of trainees as well as their environment. Discussion need not be limited to the observable behaviour during the game only, but beginning from such behaviour, one can probe the deeper issues underlying social behaviour.

For example, one can use a conversation or interaction between a farmer and an extension worker to explore the ideological beliefs and "class" interests underlying that behaviour and the implications of such beliefs and interests for the state in particular and the society as a whole.

The game has one other relatively important characteristic in relation to training of staff in self awareness and expression: the role play instructions are specific, but leave adequate room for individual creativity and expression. The game manager is always surprised by the new ideas that come up each time the game is played.

Our experience with the game so far suggests that the game is particularly well adapted as a training instrument for middle level development staff where it can constitute an excellent introduction to field work, but not an adequate substitute for it.

However as it is somewhat difficult to get higher level staff to undertake a reasonable amount of field work, the game can help to introduce them to the issues and realities that occur in the field. Where field work is possible, the game can serve as an introduction to such field work. In this case, the game will, among other things, serve to soften the "cultural shock" between the high level bureacrat and rural people in their own setting expresing their own minds about the inadequacies or even the incompetences of the state.

However, for the training of men and women farmers and other rural people in the African setting there is a need to explore mdoels more akin to the traditional forms of simulation already mentioned earlier. Such work is being attempted in an initial stage in the form of Rural Drama by Professor Eyo of the University of Yaounde; by Gog, a Cooperative Theatre Company from Britain which performed in rural areas of South West Cameroon in late 1985 in collaboration with the

Cameroon based Musinga Drama Group of Buea; and also by IPAR Buea, in its Environmental Approach to Curriculum Design for Primary Schools in Cameroon. These are pioneering pointers in the right direction for future work and collaboration opportunities by modern gamers in Africa and abroad.

When one considers that about 80-90% of the African population live in the rural areas and that it is upon them that the burden for any meaningful self reliant development rests, the need to develop adequate and appropriate pedagogical materials to train these people to understand properly the developmental issues they face, and to acquire the necessary competencies to solve these problems, is clear. The immensity of the task may be frightening, but the challenges and opportunities for simulation gamers interested in training for development work are obvious.

NOTES

1. Alfred C. Mondjanagni "La Participation Communauture au Soins de Sante Primaire". Douala: Institut Panafricaine Pour le Developpement, Secretariat General. (June, 1986), p.2.

2. "La Vache de Dieu" in Dadie B., La Pagne Noire Paris: Presence Africaine, 1953.

APPENDIX

WHAT IS PAID?

The Pan African Institute for Development (PAID), is a Non-Governmental International Association composed of individuals and statutory bodies. It has consultative status with ECOSOC, ECA, FAO, ILO, UNICEF and UNESCO. It also has observer status with the OAU.

The Association was formed in 1964 through the solicitude of the Cameroonian Government and grants from European Governments. Later on non governmental organizations and governments from Europe and elsewhere followed with their support and grants.

The first Institute was the Ecole des Cadres established in Douala, Cameroon in 1964. Today as a result of growth and decentralization, PAID has four regional Institutes as follows:

1) Kabwe, Zambia, (PAID-ESA) which focusses specially on the needs of the Eastern and Southern AFrican Region;

2) Ouagadougou, Burkina Faso (PAID-OAS) which focusses on the problems of Francophone West Africa;

3) Douala, Cameroon (PAID-AC) which takes care of Francophone and Lusophone Central Africa; and

4) Buea, Cameroon (PAID-WA) whose main focus is the Anglophone West African countries.

PAID through these regional institutes carries out activities specializing in Rural Development.

These activities include:

- Training in the Planning and Management of Projects and Administration of Development Programmes, Training of Trainers and Management of Small Business Enterprises.

- Applied Research directed towards solving concrete development problems identified in the field.

- Support to national and regional training institutes to further enhance their efficacy, and also to local development projects to improve upon their goal attainment capabilities.

- Consultancy services to private and public, national and international organizations.

PAID provides these activities within the framework of the following guiding principles:

1. PAID serves exclusively the African region.

2. PAID seeks to promote integrated and participatory development of people by seeking out for implementation innovative and original methods.

3. PAID believes that there are many roads to development and therefore rejects the importation of foreign development models having little or no applicability to African realities.

4. PAID considers the field as the permanent reference for its activities, and the local or regional milieu as the privileged framework within which the needs and realities of development as well as the demands of national planning are expressed.

CAPJEFOS: The Village Development Game

Cathy Stein Greenblat
Rutgers University, USA

ABSTRACT: On the second day of the ISAGA '86 meeting, approximately 30 participants including both English and French speaking persons, participated in an abbreviated demonstration of CAPJEFOS, run by this author and Jacob Ngwa. Although time constraints permitted play of only two rounds, all roles - villagers and development agents - were included. Both play and the discussion that followed were animated, as participants engaged in analysis of the game design, the substantive issues of development, and the parallels and divergences between the rural development problems currently facing African and other nations. The following pages offer a brief overview of the design process, parameters, and dynamics of CAPJEFOS.

KEYWORDS: development, games, simulation, gaming-simulation, Africa, training, game design.

ADDRESS: Dept. of Sociology, Rutgers University, New Brunswick, NJ 08903, USA.

HISTORY

Design of CAPJEFOS began at a workshop led by Dr. Cathy Greenblat, consultant to UNESCO and Professor of Sociology at Rutgers University (USA), held in Buea, Cameroon in December 1982. The workshop was attended by 25 participants from 7 African countries, all of whom were professionally engaged in formal or informal training activities or in educational planning. The workshop included one week of training of participants in the principles of gaming-simulation design, including a hands-on experience. Teams were constituted and guided in the design of the basic components of a village development game, tentatively entitled "SOMNAS".

In January 1985, following much interim correspondence, a second workshop was held, this time with the express purpose of having a smaller team work with Dr. Greenblat to complete design of the village development game, and to field test it with development staff trainees and trainers. The current version of CAPJEFOS is the result of the work of this second group, which consisted of Cathy Stein Greenblat (USA), Philip Langley (Great Britain), Jacob Ngwa (Cameroon), Saul Luyumba (Uganda), Ernest Mangesho (Tanzania), and Foday MacBailey (Sierra Leone).

CAPJEFOS simulates on an abstract level the factors which hinder and/or promote rural development at the scale of a village. Particular attention is paid to self-help programmes for health and agriculture, and the manner in which village actions are influenced by regional/national planning objectives and policies. Through their experience with CAPJEFOS, players should:

a) develop a better understanding of factors in development and their interaction;

b) develop greater empathy for villagers and better knowledge of their rationales;

c) find a "safe" means for exploration of what the development agent's role could or should be.

While CAPJEFOS is modelled after problems typical of African villages, the designers believe that the development problems simulated are similar in many other areas of the world.

GAME PARAMETERS

CAPJEFOS can be played by a number of different types of players. It was initially designed as a training tool for development agents in in-training or in-service programmes. First priority is for village level workers, second priority for division/sub-division/micro-region workers, and third priority for provincial and national level agents. The game is also useful as an educational tool for students of development, African studies, rural agriculture, rural economics, etc. at the University level. Bright secondary school students could also play CAPJEFOS in the context of courses dealing with development issues.

The minimum number of players for the game is 20. With this number, only the village is played (i.e. there are no development agents). A minimum of 23 players are needed to play the basic form of the game - 20 villagers and 3 development agents; the optimal number of players is 30 - 25 to play villagers and 5 to play development agents. A few additional roles could be added by an Operator familiar with the game if the numbers were between 31 and 35.

Two operators are required to run the game. Both should be familiar with it from a thorough reading of the manual. If possible, a third person should be recruited to serve as Operator 1's Assistant. This person should help set up the room, and needs only to be familiar with the basic character of the game, not with all the details. Where Operators have first had the experience of being players, they will find running the game the first time is much easier.

There are several ways to run CAPJEFOS, depending upon the time available, the teaching/training objectives, and the experience of the Operators. In the SIMPLE VERSION, only the village roles are played, and two to three rounds (years) of play are followed by a discussion of village life and structures and general development issues. Five hours should be allocated to play of this version. Where a full day (7 hours + a lunch break) can be scheduled for the game, the BASIC VERSION, involving both village and development staff roles, can be run. Finally, a FULL VERSION requires a day and a half and permits the full complexity of the game to emerge.

Two rooms - one large and one small - are required for play. The first is set up as the village of Capjefos, and the second, as the regional market town and headquarters of the Development Agency. Ideally, introduction of the game should take place in a third room, as this permits the Operators to set up both rooms in advance and to preserve the low level of familiarity with the "other place". In both rooms, furniture must be movable. Approximately 30 chairs and 6 tables are required in the large room and 10 chairs and 5 tables are needed for the smaller room.

THE GAME IN OPERATION

When players arrive, they are greeted and introduced to the general character of the exercise by one of the two Operators. Within a few minutes, they are divided into two groups: villagers (20-25) and development agents (3-5). The latter are taken by Operator 2 to another room which is thereafter known as "Somnas" - the regional center of the Pawafra Division. There they are informed that they are the recently hired staff of the newly formed Rural Development and Industrial Agency of Pawafra. They are given role cards and instructions about the activities they are to perform each round. A document about the region and the general characteristics of the rural villages within it is given them to read, and specific role assignments are made (1 Divisional Officer, 2 Health Field Officers, 2 Agricultural Officers).

At the same time, villagers are given their briefing by Operator 1. All are residents of the village of Capjefos (named for its founding ancestors.

CAthy, **P**hilip, **J**acob, **E**rnest, **FO**day, and **S**aul

They are told of the demographic, economic, social, and cultural characteristics of the village and are informed that the Development Agents will soon be coming to visit and work there. As the development agents are not informed of the specific cultural traditions and needs of the villagers, so too, the villagers are not told the specifics of the development agents' aims. They are informed that these people intend to introduce some projects, but they must, through interaction with them, learn about the nature, costs, and rewards of these projects. Specific role assignments (Chief, Traditional Doctor, Religious Leader, Teacher, Town Crier, 10 Men and 10 Women Farmers) are made. Four of the farmers are told that they have found it more profitable to engage in trading than in farming, and are given special instructions about buying food and cash crops in Capjefos and selling them in Somnas, or about buying commodities in Somnas and selling them to the villagers. All these players then move to a large room set up as "Capjefos" and play begins.

Play takes place in 2-5 rounds which last approximately 1 hour, each representing one year. The game begins part way through 1980 (Round 0); the first part of this preliminary year has been "played" by the designers, so participants are given record sheets indicating what they have done so far. These records can then serve as guidelines for the next year's decisions and for completion of the next year's forms, though players may make quite different decisions.

The rhythm of life in the two locations differs. In Capjefos, a round is divided into 5 steps: Work Time, Market Time, Group Meetings, Community Meeting, and Social Life. In Somnas, there are three major periods: Divisional Development Committee Meeting, Office and Field Work Time, and Reporting Period. Field agents may go to Capjefos beginning in 1981, and will discover that it is difficult to find a convenient time to talk with villagers, especially during the work and market times when they are heavily occupied with the tasks of production and meeting subsistence needs.

The village economy is set up so that it operates just above subsistence. Little surplus is available to individual villagers, although there is some differentiation in economic resources. Likewise, on a community level there is little capital available for development. Sickness is rampant, partly due to the drinking of contaminated water, partly due to poor diets, and partly due to regular outbreaks of disease that are not properly treated. Villagers suffer mild, moderate, or severe cases of measles, malaria, worms, and diarrhea, on a regular basis; no clinic is available, and treatment is via herbs or care from the traditional doctor. Because of the need for labor in agriculture to meet subsistence needs, and the drain on available time due to sickness, little surplus time is available for villagers to invest in development projects.

"Event cards" are tailored to each role player. They can be given out at random or selected by the Operators to introduce particular issues (e.g. demands from Regional Authorities for the agents, the religious leader and the teacher; requests to farmers for funds from children who are elsewhere in school; visits from relatives which drain already low time and money resources; etc.)

The task of the development agents, then, is made difficult by their lack of familiarity with some of the village customs and by the severe limitations of village resources. In addition, they have the job of obtaining considerable statistical data to file their annual reports in the office and must try to learn what type of projects the villagers would like and would support. By Round 3, (in the basic version of the game) they must try to implement those projects they have devised. If 4 or 5 rounds are being played, in Round 3 they are given regional/national mandates for programs which are likely to conflict with the projects they have developed alone or in conjunction with the Capjefos villagers.

Three types of projects exist in the game. The third type is only used in the full version:

Village Projects:

> These are the type of activity which villagers might well propose themselves and undertake on their own initiative with little external assistance. Some have direct effects on production; others do not. Some of them tend to be stereotyped ways of making Capjefos into a "modern" village. The Chief receives information on these during Round 1 (1981).

Local Projects:

> These are suggested to the Field Agents at the start of Round 1 (1981). In most cases they will require some external assistance to the village in the form of technical advice from the field staff and/or money from the D.O.'s discretionary fund. These projects are the type which development staff are more likely than villagers to propose. In this case, the projects have mostly been chosen to be simple, labour-intensive, and reasonably close to villagers' needs.

Small Scale Projects:

> These are proposed by the Minister of Rural Development. In spite of their title, they tend to be capital intensive and to increase economic and technical dependency without necessarily being well-tuned to the specific needs of Capjefos and/or of the smaller farmer. The D.O. will be informed of these projects at the start of Round 3 (1983) before the field staff leave for Capjefos (in full version only).

The rationale of these projects in terms of game play is to introduce a series of events in which "village" and "local" projects will be complementary IF those playing the roles of field staff realise that this is a good tactic, IF they attempt to graft the "local" projects onto already existing initiatives in the village, and IF they explain them to the villagers. However, whether the field staff do this or not, the "small scale" projects will tend to conflict with both the "village" and the "local" projects. This will create the contradictory situation in which the field staff often find themselves, squeezed between their knowledge of the village setting and the villager's problems on the one hand, and the instructions received from higher authorities on the other. However, the field staff see themselves as the carriers of modernity, which is often epitomised to them through the projects proposed by the State. Even though they may at times have a critical position in regard to the higher echelons of their own technical service, this may be because the State does not bring them the benefits and the status they feel they should receive, as their aim is often to climb up the hierarchy or at least to acquire the symbols of status and authority. Their position with regard to the three types of project will be influenced by this, and the discussion after the game can bring out these contradictions.

Computerized simulation of ecological systems: A multidimensional dynamic presentation

R. Mintz
R. Nachmias
D. Chen
Tel-Aviv University, Israel

ABSTRACT: An ecological system is characterized by the multiplicity of variables affecting any biological population. In order to demonstrate the effect of specific variables on population dynamics and to teach the laws underlying these changes, a computerized simulation called ECO-LIFE was designed.

Population growth depends on the characteristics of the species and the properties of their environment. The user is asked to define the population by the number of offspring, life span, the ability to endure starvation, etc. The environment is specified by the initial food quantity and its time related additions.

The population dynamics element is presented in three different complementary forms. One is the dynamic picture representing actual changes in the population (birth, death, increase in density, food consumption, etc.). This detailed information is required in order to simulate a real life situation. However, the overload of details with regard to individuals interferes with generalization concerning the population dynamics. The second level of presentation is therefore a graphic summary serving to arrive at a generalization of the population dynamics. Another form of output is a table presenting a quantitative correlate of the pictorial and graphic outputs.

The above mentioned versatility of output presentation enables the adaptation of the simulation to a broad range of students. It is suitable for students relying on an inductive learning process, who prefer the detailed picture before turning to the summarizing graphs and tables. Students who progress through deductive learning may reverse the order of output presentation. The preference of either the verbal or the spatial mode made by any student is reinforced by the pictorial and graphic presentation, on the one hand, and the tabulated presentation, on the other.

KEYWORDS: Computerized simulation, simulation, ecological system, symbol system, system approach.

ADDRESS: Tel-Aviv University, Faculty of Humanities, School of Education, 69978 Tel Aviv, PO Box 39040, Israel. Telephone: 03-420763.

A systems approach is, nowadays, one of the common ways of studying the environment and organisms living in it. According to this view, a living environment can be seen as an ecological system; within this sytem, living units (organisms) and environment interact toward a dynamic equilibrium (Forrester 1971, Krebs 1972).

The aims of ecology as a science are to find out the laws governing the system. Until recently, this could only be taught at the university level in biology courses, because the use of elaborate mathematical models was involved. The use of mathematics enables simplification of the system, description of the components of the system as variables, and the assignment of different values to parameters which affect the whole system.

Attempts to teach about the environment as a system at lower educational levels (elementary and high school) have encountered many difficulties, due to the complexity of the processes, the large number of variables involved, and the length of time needed for useful observation. As a result, students are unable to discover the consistent and coherent nature of the ecological system on their own. The principles have to be taught in an abstract and verbal way, leading to difficulties in comprehension and application. Under school conditions technical difficulties make it very hard for the students to deduce any scientific rules by experimentation.

The computerized simulation ECO-LIFE (1986) was developed in an attempt to overcome these difficulties. This simulation is intended to equip both teacher and learner with a tool for studying an ecological system in which animal populations grow in varying environments. The simulation enables teaching through discovery, suggesting hypotheses which may be verified rapidly by running experiments through the computer, examining the results, and if necessary, re-running the experiment with modified values of the variables. The simulation provides the results in three modes: pictorial, graphic and tabulated data.

ECO-LIFE SIMULATION: THE MAIN POINTS FOR DEVELOPING THE MODEL

We focused on three major aspects while designing the ECO-LIFE simulation:

I The first aspect was an attempt to define the part of reality to be examined and to simplify it. Our goal was that the learner should face a simple system with clearly defined components and variables, relevant to the issues under study. ECO-LIFE simulates a sub-system of the ecosystem, a single population interacting with its environment. Eight variables affect the system.

1. Average number of offspring of an individual.

2. Average life span of an individual.

3. Age of sexual maturation: that is, the number of time units required for the organism to reach the age of reproduction.

4. Endurance of food shortage by the individual, i.e. the number of time units which an individual might survive without food.

5. Sociability, i.e. the number of close neighbours an individual tolerates without dying.

6. Initial number of individuals.

7. Initial quantity of food.

8. Quantity of food added in each time unit.

II The second crucial point was to bring the students to comprehend the interaction between the system components. This meant that the students should understand cause and effect in the system, feedback relations and direct or indirect relations in the system (Roberts, 1984).

In the ECO-LIFE simulation, the population and the environment interact according to these rules:

1. Each individual moves randomly to the right or to the left, up or down, provided the space it moves into is not occupied by another individual.

2. If two adult individuals meet, multiplication occurs, with the resulting appearance of the average number of offspring around the couple.

3. If the individual collides with a food unit, the food is consumed and disappears from the screen.

4. A young individual becomes an adult if it reaches the age of sexual maturation.

5. An individual may die of old age, starvation or crowding.

The system which behaves according to these rules produces a long term process which is affected by the initial values given to the variables by the student at the beginning of the running of the simulation. Through examination of the relations between the variables, the student learns that the state of the population in the system is determined by the combination of a number of processes - birth, growth and death - which take place over a period of time. Direct relations, such as the effect of the environment's carrying capacity, as well as indirect relations, such as the influence of population variables on the environment, may be seen in the system.

III The third important point was to bring the student to understand a real system through the use of the model. This raised questions about ways of designing the computer screen in order to achieve young learners' maximal understanding.

The simulation is expressed through dynamic output presented in three forms: pictorial screen, graph screen and numerical data table.

Pictorial screen. The screen can show animals and food units. The pictorial screen shows the movement of animals, meeting of adults, appearance of offspring and disappearance of adults.

This form of output is suitable for pupils used to an inductive pattern of learning. Before they are able to understand graphs they need a detailed pictorial representation. The picture gives the learner the feeling of a real-life living environment, which creates the link between model and real system.

Graph screen. This shows x-y coordinates and a curve describing the number of individuals vs. time. The graph is updated at each time-unit. The graph provides the learner with a comprehensive view of the state of the system over time. It can provide information about cyclic or linear changes which appear in the system after a long period.

Numerical data table. The table can explain every point on the graph. It includes the following information: number of individuals, number of food units, number of individuals born, number of individuals dead, and is then later subdivided into cause of death (starvation, over-crowding and old age). The table is updated at each time-unit during the running of the simulation. The table adds information about cause of death, the total number of births and the total number of deaths. It enables the learner to analyze the graph thoroughly.

The three modes complement each other. The pictures provide the feeling of an environment, the graph gives a general view and the table provides full and detailed information on each time unit in the process. The three symbol system allows the program's adaptation for a variety of students at different levels.

SUMMARY

The computerized ECO-LIFE simulation is a tool for teaching about an ecological system. Its main goals are: simplification through a dynamic model; variable manipulation by the learner and data processing by the computer. Identical output is expressed through three different symbol systems: pictorial, graphic and numerical tables.

REFERENCES

Forrester, J. W. 1971. World Dynamics. Wought-Allen Press, Inc.
Krebs, E. J. 1972. Ecology: The Experimental Analysis of Distribution and Abundance. Harper and Row Publishers.
Roberts, N., Andersen, D., Deal, R., Garet, M. and Shaffer, E. 1984. Introduction to Computer Simulation: The System Dynamics Approach. Addison-Wesley Publishing Company.

SIMULATION REFERENCES

ECO-LIFE. Mintz, R., Nachmias, R., Naaman, N. and Rotary, N. 1986. The Computer in Education Research Lab. Tel Aviv University. Ramot Pub.

Toward a theoretical framework of media gaming

Israel M. Porat
Hebrew University of Jerusalem, Israel

ABSTRACT: The intention of this paper is to suggest a theoretical framework in order to organize and clarify various elements of media simulation games. Such a framework will be useful to both game designers and operators/ users. The characteristics comprising the theoretical framework include reality as input, the subsequent processes of selection and interpretation by newspersons, and the media product as output. A final element of the framework is the feedback of the consumers. As we regard the simulation game as a reflection of the real world, we recommend that the game characteristics be examined against their actual representation in the real world.

KEYWORDS: media gaming, reality, real news, feedback.

ADDRESS: School of Education, Faculty of Social Sciences, The Hebrew University of Jerusalem, Israel.

INTRODUCTION

The intention of this paper is to organize and clarify the elements of similarity and difference in the field of media games, and to enable evaluation of them. Such a framework will make it easier for game users to choose a game suitable to their needs and might be helpful to game designers to define the profile of the planned game.

A theoretical framework can be useful to the game user in helping him/her learn about the system which the game simulates. The game user's interest in a media simulation can stem from several factors: he/she may be interested in it for instruction on the media, for learning purposes such as in professional training, for developing critical consumption of media products, or for research purposes.

It should be noted that over-lapping can exist among the noted objectives. As noted by Crookall (1985), there exist two broad overlapping types of simulation: substantive and performative. The former concentrate on the study of issues or content areas, the latter on the development and practice of skills. Games incorporating these two elements will demonstrate overlapping, for elements of both substance and performance necessarily will enter into the same game. The difference lies in which element is emphasized more.

Despite the recognition that experience-based games and simulations have various strengths standing alone, a number of concerns remain as to their use. Ruben (1977) has noted that many of these criticisms have been directed at the trainer, instructor or teacher, who often lacks the necessary expertise for operation of the games.

Therefore, the game user should be aware of the various differences in games, such as in the kind of audience they are designed for, the appropriate age of participants, and whether they are aimed at special programs, before he/she operates a particular game. The theoretical model makes it easier to analyze the game according to the game user's personal criteria and enables him/her to understand the media according to certain objective, profesional standards. As a sort of checklist, it guides him/her toward acceptance or rejection of a given simulation game.

For the game designer, the framework aids in the effort to suit the planned game to a previously-defined audience. After having received the data on the target audience, the designer is compelled to research in order to get a total picture of how the game will look. But the theoretical framework defines the "look" of the game for him. This saves time and gives a sense of direction.

Consideration of the audience is of ultimate importance to the game designer, and interaction with it lies in construction of the rules. Ruben (1977) points out that rules may enumerate actions which are permitted of players or they may list only those activities which are precluded. In some instances, the designer utilizes the behavioral rules and behavior patterns players bring to the situation as a primary source of rules for the simulation. This manipulation of game participants is what creates the special character of a particular game.

It may be asked, why is there a need at all for new games and improvement of existing games in media? Despite the existence of varied media games, there is a lack of coverage of important aspects of the media, particularly of reader feedback. In addition, Crookall (1985) points out that although news institutions have inspired a number of published media simulations, they are based on 'news' which is inevitably old or second-hand. A theoretical framework, therefore, insures that all the pertinent components of the typical media model will receive attention.

Furthermore, even though there is a need for different games representing each individual medium, only a uniform framework can allow comparisons between games to be made. This uniformity is also important for the creation of common criteria for the game designer and the game user.

SYSTEMS ANALYSIS

The theoretical framework takes the form of a systems analysis approach, which allows observation of the complete media product, such as a newspaper, or concentration on smaller components, such as an article. It permits both the user and the designer to approach the model through a triangular comparison between the medium, its reality, and the simulation. The game represents a reflection of the reality.

Ruben and Kim (1975) note that the systems approach to communication provides a means for dealing with a large number of communication links simultaneously, with the interaction and the complex circular flows of media systems. Once it is established that a particular system is a model of a portion of the real world, the system can be analyzed and tested.

In order to understand how the game uses the system to reflect the reality - and thus provides the user and designer with the theoretical framework - first it is essential to understand the system's relation to reality.

TABLE 1 Media and Media Games as Systems

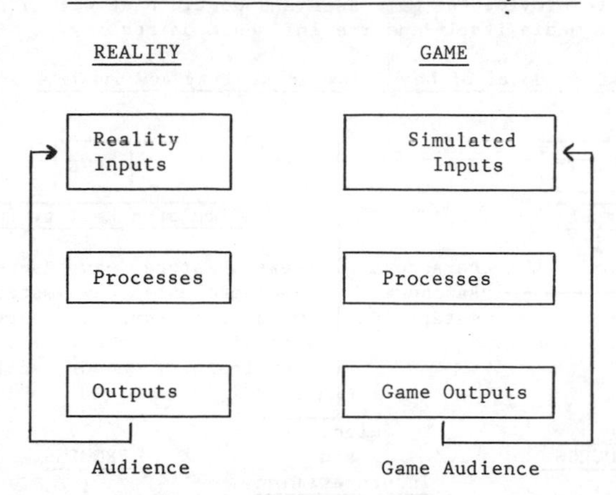

Defleur and Ball-Rokeach (1966) describe the analysis of social systems, which concerns itself with the patterns of action exhibited by individuals or subgroups who relate themselves to one another within such systems. According to this analysis, the first major component is the audience, the main interest of the designer. Some of the major variables that play a part in determining how this component will oeprate within the system are the major needs and interests of audience members, the various social categories represented in an audience and the nature of the social relationships between audience members. It should be noted, however, that the audience is on the receiving end of the "media as systems" model.

We are dealing with the reconstruction of the media reality in order to fully involve the all-important audience, since this is the main concern of game designers and operators. Just as Defleur and Ball-Rokeach speak of "social systems", Adoni and Mane (1984) discuss "social reality" - "social" because it can be carried out only through social interaction, either real or symbolic. They note three types of reality: objective social reality that is experienced as the objective world existing outside the individual; symbolic social reality, which consists of any form of symbolic expression of objective reality, such as media contents (here the individual's ability to perceive different spheres of symbolic reality comes into play); and the subjective social reality, where both the objective and the symbolic realities serve as inputs for the construction of the individual's own subjective reality.

This brings us to the kind of reality required in a simulation. This social reality is perceived along a continuum based on the distance of its elements from the individual's everyday life experiences (Adoni and Mane, 1984). According to the media-dependency hypothesis (Defleur and Ball-Rokeach, 1976), the degree of media contribution to the individual's construction of subjective reality is both a function of one's direct experience with various phenomena and consequent dependence on the media itself for information about these phenomena.

There is a complementary relationship between direct experience and media presentation for creating a clearer picture of what is going on. It is the designer's task to go beyond a mere symbolic representation of reality, which is based on selection and editing of material derived from reality, thus depicting only a certain part of reality and portraying it from a specific point of view (Adoni, Cohen and Mane, 1984). This is approximately the media's task. Instead, the designer has to fill in what the media misses, leaves out or hides, in order to provide the game user and participant with a full understanding of the media itself and its influence on reality.

TABLE 2 Model of News Flow in Reality and in Game

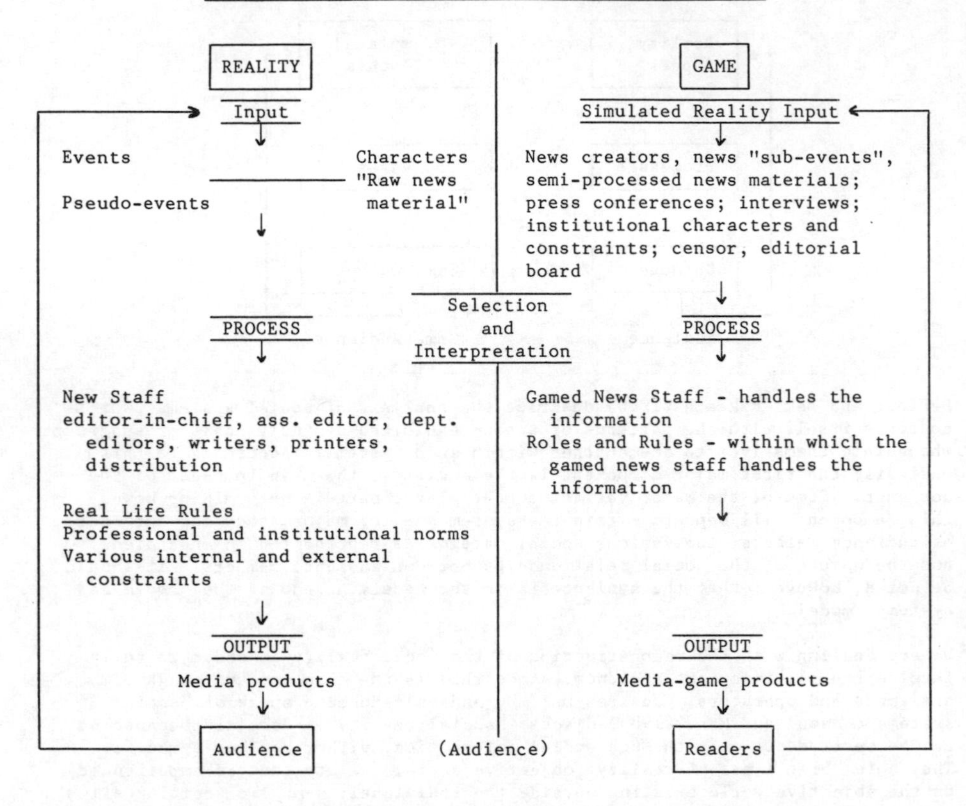

The variables that comprise the theoretical framework are our main concern. The first group of variables refers to the reality which the media portrays. As shown in Table 2, the media reality, which serves as the input to the media system, consists of events and characters. The characters include both people involved in the events and those who generate them.

INPUT

In order to understand the input variables, it is essential to understand what the characters do. it is desirable to attain a level of "reflection" in which the media reality and the game reality are very similar. Of course, the game is by nature a relatively closed system, as opposed to the reality, which is a series of interacting subsystems which exert control on each other (Defleur and Ball-Rokeach, 1966). The more the game exhibits verisimulitude, the more "open" - and thus "real" - it will be.

It follows that our examination of media simulation should be a comparison of components in the reality with those of the game. The elements of the reality must be present in the game for it to bear the characteristics of a simulation, and this fact is taken into account in the design process. This element creates the need, first of all, for "real news" input (Crookall 1985). Pre-prepared, semi-processed news items limit the excitement and experience of being editor. The more a media simulation concentrates on performance the greater the benefit of "real news".

The game input is made up more of simulated occurrences and written instructions and less of characters because it is easier to present in a game framework. The roles within the events are determined in relation to reality. They carry great influence on the input of the game and organize game processes according to the rules. McCombs (1979) categorizes the roles according to their place in the media process: collectors, information processors, news managers, and technicians.

There are three main activities in journalism (Johnstone, 1976) for the designer and user to keep in mind when dealing with a media game: newsgathering, reporting and other types of writing; editing and news processing; and the supervision and management of editorial operations.

There are at least three types of news writers: the reporter, who writes what he sees; the interpretative reporter, who writes what he construes to be the meaning of what he sees; and the "expert", who interprets things he has not seen at all (Johnstone, 1976). A majority of journalists also perform behind-the-scenes editorial duties.

Journalists perform the "gatekeeping" function in mass media, which includes all forms of information control that may arise in decisions about message encoding, such as selection, shaping, display, etc. "Gatekeeping" presents one of the characteristics of the media process. From a system perspective, attitudes of mass communicators might be hypothesized to be reflected in message control or in the way messages are encoded for mass audiences. Donohue notes that communication accuracy is a function of the controls in the process. Editor assignment is a control factor with motivational aspects.

THE PROCESS

The process in which the journalists function is whatever occurs in the media system. On the one hand information is received and on the other hand a communications product is created. It represents information flow by way of selection and interpretation of media material. The main decisions concerning information flow involve content, and these decisions are made by editors. Part of the selection is the decision-making process which is performed according to professional and institutional norms. The interpretation component deals with decisions of space and location. Interpretation gives context to information selection.

In the media reality there are events in which characters and written details are integrated in random fashion, which occur with great frequency. On the other hand, the game contains fewer events and is more subject to the control of the designer and/or operator.

OUTPUT

The role characters bring the information through the process to the output stage. The output is the product which has been influenced by all the events since the input/reality stage. This product includes a part of the initial events in altered form, after design. Only part of the initial information arrives in the output stage. By comparing the final product - whether newspaper, or TV or radio program - with the input, it is possible for the designer and user to see the selection, interpretation and bias. The differences provide a justification for game design and use.

The significance of the output product can be looked at in several ways (Tuchman, 1978). Either news presents to a society a mirror of its concerns and interest, or it helps to constitute society as a shared social phenomenon, for in the process of describing an event, news defines and shapes that event. In any case, news readers work to find meaning in the product.

If the designer is to replicate the media reality, he/she needs to not only take the individual roles of the media player's into account, not only to construct thoughtful occurrence, but also to consider the reaction he/she expects from the consumer.

FEEDBACK

The product reaches the consumers, who are liable to react by purchasing it, by answering reader and viewer surveys, or in other ways. These reactions represent the feedback and the ways in which the consumers can influence the media product. In our opinion, without feedback the media products do not really reflect reality, in that they do not answer the needs of the consumers to complete their picture of what's going on around them. In the same way the designer must reflect the reality in his simulation, so that the user who intends to direct it, as well as the participants he will use, will be able to complete and correct their picture of the media reality.

Feedback is defined as the way in which system adjustments are made in response to information about system performance (Donohue, 1972). Media systems themselves may provide a feedback function. In order to get the most accurate and helpful feedback from the consumer, it should be understood that the consumer's level of comprehension imposes the major constraint on the journalist's use of words.

There are situations in which every journalist receives a one-sided picture of the reality, and this is what he/she deals with because he/she doesn't know any better, or has no choice. At other times he/she sees the reality from several angles and chooses what to address. This relates to the multiple-reality aspect of the game, as described by Greenblat (1982).

The multiple-reality aspect is important in that it adds variation and verisimilitude to the game. Obviously, if the same event was used uniformly for all the players, there would be a similar reaction from them. If we want to demonstrate the richness of a multiple reality, only varied events within the game will do. It is generally a question of how close the game approaches the reality.

SUMMARY

In this article we attempted to suggest a theoretical framework for mass media games through concepts and a basic model from systems theory. It seems to us that this model clearly demonstrates the parallel relationship between the game as a system and the mass media system. We hope that this view will aid users and designers in mass media games.

BIBLIOGRAPHY

Adoni, H., Cohen, A. A. and Mane, S. 1984. Social reality and television news. The Journal of Broadcasting, Vol. 2801.

Adoni, H. and Mane, S. 1984. Media and the social construction of reality, Communication Research, vol. 11, No.3, pp. 323-340.

Crookall, D. 1985. Media gaming and NEWSIM In Crookall, D. 1985. Simulation Applications in L2 Education and Training. Oxford: Pergamon Press. Special issue of System, Vol. 13, No.3.

Defleur, M. L. and Ball-Rokeach, S. 1966. Theories of Mass Communications. David McKay Company, Inc., New York.

Greenblat, C. S. and Duke, R. D. 1982. Principles and Practices of Gaming Simulation. Sage, Halstead, New York.

Johnstone, J. 1976. The News People, University of Illinois Press, Urbana.

McCombs, M. E. 1979. Using Mass Communication Theory. Prentice-Hall, Inc., Englewood Cliffs.

Ruben, B. 1977. Toward a theory of experience-based instruction Simulation and Games, vol. 8, 2: 211-231.

Ruben, B. and Kim, J. Y. 1975. General Systems Theory and Human Communication. Hayden Co., Rochelle Park, N.J., 1975.

Tuchman, C. 1978. Making News The Free Press, New York.

Some social-interactional aspects of a business game for special purposes in the (L2) teaching of English

D. R. Watson and W. W. Sharrock
University of Manchester, England

ABSTRACT: The authors recommend that simulations and games be subjected to rigorous scrutiny as research objects in their own right, and claim that the cognate sociological perspectives of ethnomethodology and conversation(al) analysis contain the analytic resources required for this task. These approaches are used to render researchable the pedagogic notion of 'de-classrooming the classroom'. Using transcribed excerpts of video recordings of a business management game for teacher training, a comparison between games and traditional educational settings is offered. In addition, some data-based observations are made concerning the distribution of game roles and of speaking rights in the interactional organization of the game.

KEYWORDS Game, simulation, business, ethnomethodology, conversation(al) analysis, speech exchange systems, interaction, categories, roles, category-bound activities.

ADDRESS: Department of Sociology, University of Manchester, Manchester M13 9PL, England. Phone: 061-273-7121.

INTRODUCTORY COMMENTS: TOPIC AND APPROACH

Our intentions in this paper are twofold. First, we wish to elucidate, from a sociological point of view, how simulations and games are collaboratively 'brought off' by participants through their communicative interaction with each other. Secondly, we wish to show that the analytic resources of the cognate sociological approaches of ethnomethodology and conversation(al) analysis may be applied in detailed and rigorous ways to simulations and games as social phenomena. In line with these approaches we shall introduce two transcribed data excerpts, though the brevity of the paper will only allow us to use the data illustratively, rather than in the close and thoroughgoing manner in which ethnomethodologists and conversational analysts usually try to approach their detailed interactional data. [1]

Of course, these two aims are inextricably interrelated. We hope to render visible various social-interactional features of the game - features which participants may usually take for granted - through the use of these approaches. Consequently, we should begin by outlining the major aspects of these approaches so that readers unfamiliar with them can assess the kind of light they cast on the data. [2]

Ethnomethodology and conversation(al) analysis both focus upon society members' practical uses of the corpus of procedural knowledge in the culture (termed 'members' methods') and upon the ways in which they use this knowledge in order to make sense of everyday actions, interactions and setting, including the ways in which they build this sense into the production of their own actions. Included in this focus would be the ways in which participants, through their communicative practices, display, align and share the sense they make.

A major topic of study, both for ethnomethodology and conversation analysis, is members' mastery of ordinary language. This mastery is a pivotal resource in members' collaborations in deploying their practical sense-making procedures. Conversation analysis has, in particular, sought to subject to close analytic scrutiny the interactional organization of the exchange of speech. This mode of analysis has sought to show how, for instance, a second speaker, in his or her design and production of utterances, displays a practical analysis or treatment of other speakers' prior utterances. In this sense, conversation analysis shares ethnomethodology's generic concern with the social-organizational relations of actions conceived as recognizable states of affairs (for the society members themselves in the first instance and not just for the analyst.) From this description of these approaches it is clear that the frequently heard orthodox sociological criticism that they involve 'subjectivism', reducing social-organizational phenomena to events happing 'in the heads' of individuals is entirely misconceived. In their focus on practices or (inter)actions ethnomethodology and conversation analysis might better be termed 'praxiological.'

By and large, these modes of analysis insist on using what is sometimes termed 'retrievable data', mainly audio and/or video recorded materials and transcripts thereof. [3] The transcript conventions are designed to highlight the social-interactional nature of speech exchange. This type of data is regarded as forestalling the analyst's reliance upon his or her tacit, unexplicated, common sense and *a priori* definitions of what is purportedly 'typical' or 'possible' in this or that kind of social setting. Instead, the analyst is constrained to deal with the details of the naturally occurring and naturally situated actions and interactions. This technique, therefore, replaces the analytic characterization of 'stock' actions and settings cast in terms of commonsense preconceptions or the professional analyst's biases; these *aprioristic* conceptions would, for the ethnomethodologist and conversational analyst, place unacceptable 'extrinsic' constraints not only on what counts as data in the first place but also on the analytic characterization of the data. In this regard, too, ethnomethodology and conversation analysis, in their methodological insistence on studying 'observables', cannot properly be defined as 'subjectivist' or 'reductionist' in focusing on what are sometimes called 'mental events', particularly in view of what many orthodox sociologists rely on, namely that which is to them given by intuition in their characterization of actions, interactions and settings.

RENDERING 'DE-CLASSROOMING' RESEARCHABLE

In another paper in this volume, we have outlined the educational ideologies which made for the 'de-classrooming' of the school classroom. These ideologies hold that the traditional organization of the classroom, including its spatial configuration, casts teacher-pupil relations in a formal, didactic and authoritarian mould, individuating students, making it difficult for them to co-operate in learning, rendering them passive recipients of instructions and authoritative statements from the teacher, and thereby sapping their motivation to actively involve themselves in learning. These traditional classrooms were seen as discontinuous with real-world situations, encouraging rote-learning and conformity, rather than practical adaptiveness. Games and simulations as pedagogic devices were seen as overcoming, and indeed reversing, these features of traditional classroom situations, by motivating pupils to learn together, relatively independently of the teacher and by 'learning elliptically', so to speak - learning about one thing (e.g. a second language) by doing another (simulating a household situation or whatever) where the second provided a putatively 'real-world' context for the first.

The question is, how do we render these issues not just matters of educational orthodoxy, but researchable in a thorough, rigorous and detailed manner? Here, we feel, ethnomethodology and conversation analysis, with their emphasis on retrievable data can show us the (or a) way to do this. Moreover, we already have several studies of educational situations which have been conducted from this perspective, notably one by Alec McHoul (1978).

McHoul provides an analysis based on transcribed data, of the 'feeling of formality' which co-participants typically have in the more traditional school classrooms. He provides analytically for this orientation in terms of the procedural rules or conventions for the speech exchange systems prevailing in such classrooms, and which depart from or transform the procedures of ordinary 'informal' conversation. [4]

McHoul notes that the speech exchange system prevalent in formal classroom situations involves a differentiated distribution of speaking rights as between teacher and pupils. It is this differential distribution which generates what McHoul terms a 'social identity contrast' between teacher and pupil. Following the conversation analysts Sacks, Schegloff and Jefferson (1974), McHoul terms this differentiated distribution of speaking rights the 'pre-allocation' of turns and turn-types as between participants. The upshot of these differentials is that the range of options which one finds in ordinary conversation is restricted in classroom talk. This is especially so for the pupil who only has the choice between continuing to speak and passing the floor back to the teacher. The teacher has a significantly greater range of options and is largely the one who effects the 'local management' of classroom talk.

The rules or conventional procedures McHoul finds in the formal classroom are, briefly, as follows: first, there exists a 'teacher's rule', where the teacher can select the next speaker (unlike ordinary conversation where any participant can select any other). The pupil thus selected has the right and obligation to speak. If no pupil is nominated to speak, the teacher must be allowed to continue by the pupils.

Second, the student's/pupil's rule. If a pupil is selected to speak then he/she does not have the right or obligation to select themselves, nor may they select other pupils to speak next. Thus, as McHoul puts it, only the teacher can 'creatively' direct speakership in a traditional classroom, and the more formal the classroom situation, the more likely this is to be the case. The teacher is in the pivotal position. For instance, the teacher may self-select him/herself to speak at an appropriate juncture (a speaker turn transition point) in the pupil's utterance.

McHoul goes on to specify what he takes to be the procedural rules for the exchange of speech in the classroom, and outlines some of the outcomes of participants' operation of this speech exchange system - outcomes such as the minimization of gaps and overlaps in the large, multi-party settings which classrooms comprise, also the generating and reproduction of the teacher-pupil social identity contrast and a systematic asymmetry built along the lines of the contrast. The more formal the classrooms, the greater their asymmetry.

Now, if games and simulations are claimed to 'de-classroom' the classroom, then one way in which we can test that claim is to inspect the systematics of speech in actual simulation and gaming simulations. For instance, we can examine, using retrievable data, the way in which the game/simulation frame of reference is established, sustained and transformed in and through participants' (including game directors' and their assistants') speech practices. How, for instance, are game instructions issued, how are written-textual rules turned into talk and put into interactional effect, how are games started up and closed down, how are transformations effected from one phase of the game to the next, how are post-game 'de-briefings' done, and so on? Our initial inspection of video-taped data suggests that there are, in our data, major switches in the features of the speech exchange system which prevails in the game/simulation setting, depending on whether the pre-game instructions are being issued, or whether the game is at the intra-phase juncture, for instance. It is from such an intra-phase juncture that the following three short data excerpts are derived. The recordings are of a simulation game designed as a teacher training device for teachers of English for Special Purposes, the purposes in this case being 'Business English' education. The game designer's intention was to show the teachers how a game simulating a relatively complex business situation could be used for educational purposes (see Coote _et al_., 1985, for a fuller description of this computerized game, which is titled MANEDES - 'Management Negotiated Decision Simulation.')

Data Excerpt 1

(A game of business decision making where participants are divided up into departments. The aim is to fill out a sheet itemizing the allocation of company resources.)

1. A: What about the two million revenue reserves? (1.5)
2. B: Hold on where have you got that?
3. A: In the balance sheet
   ]
4. D in the balance sheet
   ]
5. E in the balance sheet
   ]
6. left
7. hand side of the (sheet)
8. B: two million reserve
9. Yet to find out what it means
10. A: That means (background noise) two million (.) got two million
11. in the bank (yes)
12. D: Those are probably in the form of loans since it on the(.)er
13. liability side of the balance sheet (.) it's not cash
14. (.) it's not cash that's available
15. C: Oh, i-if it's my job to get this sheet filled out I'd like to
16. inform you that thus far I have nothing filled (out)
17. (laughter)

18. E: Yes but if you look on the asset side (.) cash in bank

19. two million three three five

20. D: That's right that's what I'm saying we have two million

21. dollars in cash that has to get us through the next

22. period or we have to er take out loans

Note: C = Chairman.

<u>Data Excerpt 2</u>

(The same game).

1. Sales: Could I ask what the basic objective of this meeting is?

2. Chair. To get this sheet filled out.

3. ....

4. Sales: Where's the information we need for the sheet?

<u>Data Excerpt 3</u>

(The same game)

1. Chair: You have to bear in mind I have no idea what's going on.

2. (Laughter)

3. Chair: raw materials(.) number of kits purchased the

4. <u>first thing</u> on this list is (.) raw materials number

5. of kits purchased supplier number one, supplier number

6. two, supplier number three now if somebody can tell me

7. what <u>that</u> means we'll (.)

8. Production: These are production decisions now in other words

9. Chair: this is the

10. production area

11. Production: right so we have to decide how much (.) how

12. many kits we need to purchase for the next quarter

13. alright? but we can only decide this in function of

14. how many the marketing department estimates are

15. gonna be sold (.) and I think everybody yeah do we

16. all agree? that we will function basically in terms of what

17. marketing decisions have been made (.) so

18. I suggest if we get this basic information from

19. there from the marketing department then we

20. can say that we can produce (.) you can say whether

21. you've got the money for investments that may be

22. necessary to produce it(.) you can say what

23. implications there are for personnel yeah?

24. Personnel: yeah.

25. Chair: Would you like to be chairman of the board?

26. Production: No.

ANALYSIS OF THE DATA

Perusing these data excerpts we can see that the talk in the game situation approximates more closely to ordinary conversation; that is, it shows a diminished degree of pre-allocation of turns when compared with the classroom situations studied by McHoul. In these excerpts there is no 'incoming' by the game director (who is present). Certainly, the conventional three-turn sequence (with the teacher producing, say, a question in first turn, the pupil answering in second turn and the teacher evaluating that answer in third turn) so frequently found in classroom discourse is, consequently, not found here. Game participants self-select, or are selected by prior speakers, to take a turn at talking in ways that would by no means always be possible in more formal, traditional classroom settings. Moreover, it is more the case that anyone can produce any turn type, any type of conversational action (e.g., the act of questioning is not restricted to the teacher/game director) in the game setting.

However, as we shall see, the social organization of the game talk is by no means fully 'returned' to ordinary conversation and there is what Sacks, Schegloff and Jefferson (1974) term a 'formula' operating within the game, whereby incumbents of some game roles have special rights over particular turn types, and/or utterances incorporating certain domains of knowledge/information (e.g. marketing information). Thus, whilst there is no pre-allocated provision for the incoming of the teacher/game director, there does exist an element of pre-allocation as between players' roles within the *mise en scene* of the game. This attests to what, in another paper in this volume, David Francis has termed the 'double settinged' nature of games and simulations. It is the activities of participants within the game upon which we shall concentrate in the analysis which follows.

An initial inspection of the transcript also reveals a central characteristic of simulation and gaming, and one that will indeed be apparent in a *prima facie* way to anyone who has participated in a game/simulation. This characteristic is that no rubric, no finite list of instructions, no amount of pre-game briefing can be automatically or self evidently 'sufficient in itself' for participants. As with all sets of social rules, game rubrics must be actively interpreted by players; game rules do not provide for their own application but must be applied through members' practices. Indeed, the very sense in which the rules or items of game information apply has to be established *in situ* by participants where the particular game context itself has unrelievedly to be consulted in order to establish that sense. The specific sense of the rule/item of information has to be established for participants from within the 'here and now' context of this particular time of playing the game.

This establishing of game sense requires participants' joint use of their commonsense cultural competences, their ordinary sense-making procedures, derived from everyday life outside the game, upon which they must rely within the game setting to make 'game sense'. These sense making practices and procedures comkprise the corpus of procedural cultural knowledge - 'knowledge how' - which participants share, hold and use in common. Sense making, then, is a matter of public knowledge, and this provides for its character as a collaborative activity. In this game it is a task shared amongst participants and, accordingly, one that is interactionally managed. These aspects of making sense, then, are not a matter for psychologists; they are generically social or collective phenomena.

Moreover, the tasks of making sense of a game/simulation are not exclusively analysts' concerns. They are, in the first instance, members' or participants' concerns and are eminently practical, not theoretical, matters for those participants. For instance, Transcript 1 exhibits participants' concern with establishing the sense of an item of information concerning revenue reserves amounting to two million; the sense is to be established in a manifestly practical way, i.e. what kind of resources, if any, do they comprise? What do these resources mean for the playing of the game? The sequence transcribed manifests an array of cultural/communicative devices of practical reasoning and enquiry.

These devices by no means comprise straightforward interrogative sequences (that is, question-answer pairs); note the complex set of insertions between the initial question 'What about the two million reserves?' and an utterance that might stand as an answer on line 10, the answer which is also potentiated by the immediately prior utterance 'Yet to find out what that means.' The reasoning procedures involve consulting game features, e.g. the placement of the 'two million' on the balance sheet and getting participants' co-orientation to and agreed interpretation of that placement.

Note, too, that as in ordinary conversation, diffusely operating through this data is a normative preference for agreement and confirmation, which is often made manifest in their solicitation. Observe, for instance, that where some disagreement or objection seems to be forthcoming (excerpt 1, line 18) the format adopted is 'agreement token + disagreement') i.e. 'yes, but ....' and the following utterance endorses the agreement and minimizes the discrepancy without the speaker having to abdicate her position/interpretation.

Similarly, in excerpt 3, lines 15-17, we again have an invocation of an agreement, and here we also have a built-in tentativeness in the form of an agreement-solicitation ('Yeah, do we all agree?'). Particles such as 'I suggest..', interrogative tags such as 'yeah?' at the ends of utterances also attest to the preference for the solicitation of agreement and confirmation, as in line 23 of excerpt 3, where agreement is issued as an immediate next action on line 24 by the speaker from the Personnel Department.

One might suggest, then, that one of the 'seen-but-unnoticed' tasks for game/ simulation participants is to establish a set of resources (i.e. a shared domain of game-relevant working knowledge) in common which thereafter may be assumed to be held and known in common and invoked and used in practical ways without further questioning, where the game rubrics as such cannot in themselves be treated as furnishing these resources in a necessary and sufficient way. One might also add to this observation that the basis of agreed resources is a condition for the advancement of the game, e.g. in participants' formulation of a plan of action, or 'what to do next' (the latter being a central, pervasive concern.)

One major focus for the ethnomethodological and conversation analytic approach to games/simulations is, then, the examination of what we might, following

Pollner (1979), term the 'explicative transactions' which are involved in gaming and simulation - that is, the organized linguistic transactions which make visible for all parties the basis of (in this case) the game, the features of the game, what has been achieved so far, and the like.

Again, trading on Pollner's terminology, these transactions involve the (interactional) making and management of meaning in the game/simulation setting. These transactions may be seen not only to establish shared meanings in the game but also to exhibit their here-and-now situated relevance to given events during the course of the game. Note that the meaning of a given game feature is not simply or primarily problematic for the game analyst but for the game participant, as excerpt 1, line 9, excerpt 2, line 1 and excerpt 3, line 1, all indicate.

The first two of these instances occasion some specification of the sense in the next turn at talk. The third, the Chairman's utterance, prefaces a 'recitation' of part of the game rubric, which in turn gets a specification from the Production representative, duly endorsed by the Chairman in next turn. No rule or rubric can, in and by itself, provide for its own application, whether in gaming/simulation or in 'ordinary' settings. This is a nice example of how game rules do not 'automatically' generate game conduct. The playing of the game involves participants in jointly conducting explicative transactions, i.e. in collaboratively working out a situated ad hoc interpretation of some item in the game rubric. To say this is not to point out some flaw in the game design, it is simply to point out a feature of any rule or body of rules. Nor are we saying that the presence of explicative transactions is some kind of defect in the playing of the games; on the contrary, these are a constituent feature of the playing and are crucial for developing game order; they build in the 'visibility arrangements of the game'.

Similarly, game players have to work out what conduct is appropriate to their game role. Again, game roles cannot be 'scripted' to the extent to which they become self evident. As Cicourel (1972) aptly observes, what counts as conduct relevant to a given game (or any other) role might well be subject to a great deal of interpretive work and negotiation over time - and the temporal ordering of games and simulation is itself an absolutely central feature. What counts as conduct-relevant-to-a-role always involved judgemental work and 'ad hocery' on the part of interactants.

Participants' orientations to role, or to what counts as role-relevant conduct, can be found at several points in the transcripts. On lines 15-17 of excerpt 1, for instance, we have the Chairman (C) displaying such an orientation in order to make a joke - namely, that if the meeting's job is to make decisions on the deployment of resources he has not so far been able to record any. Similarly, in excerpt 3, when the Production Department representative proposes a modus operandi, that is, establishing some basic marketing information and distributing tasks to different departments on the basis of that information, the Chairman, again by making a joke, ties the prior speaker's activities to the role of chairperson.

Less explicitly, the Chairman, again after producing a joke based on what we might commonsensically expect of someone in that role (i.e., that chairpersons might be expected and required to know what is going on) then begins to read from a list provided in the game - again, of course, something we might expect of a Chairperson.

In conversation analytic terms, participants are displaying an orientation to category-bound activities (Sacks, 1974) - that is, those activities which may be treated as conventionally tied to a category or limited set of membership categories such that, for example, the distribution of tasks, the reading of a schedule and the rest may be seen as properly performed by someone occupying the position 'Chairman'. Similarly, in excerpt 3, the Production Department representative organizes his proposed distribution of tasks on the basis of category-bound predicates of the Marketing Department's and Personnel Department's representatives respectively. In short, membership categories in conversation serve as a locus for the distribution of rights and obligations in the game. thus generating the game's 'division of labour'. Category-bound activities, then, can comprise a central resource in making sense of a game, as indeed David Francis shows in his contribution to this volume; making game sense is, for participants, an eminently practical matter, a matter of establishing (say) the object of a meeting, as in excerpt 2, but more specifically, of establishing what to do as an immediate next action, how to organize matters in a coherent manner from moment to moment.

Participants' orientations to category-bound activities may also be seen to be implicated in the allocation of speakership, such that the Chairman has one set of speaking tasks (keeping the meeting aligned to the task, summing up and the rest) and representatives of the Production, Marketing and Personnel departments have other tasks, i.e. speaking on behalf of their departments, etc. To a certain extent, then, we can begin to speak of what Sacks, Schegloff and Jefferson (1974) refer to as a 'formula' for the allocation or pre-allocation of speakership or the taking of particular turns and turn types/conversational actions, and the recognition of appropriate and/or 'authoritative' speakership, where, for example, Marketing Department representatives may pronounce authoritatively on - for the practical purposes of the game - marketing matters, etc. In this regard we may examine the allocation of speakership as being based on what (within the terms of the game) is 'owned knowledge' (Sharrock, 1974), where ownership of knowledge is premised on the basis of membership of categories and names or labels attached to those categories. The allocation of speakership is not independent of the distribution of responsibilities on the basis of the categories of game membership - including that of game director.

There is also the issue of participants' orientation to the duplicative organization (Sacks, 1974, pp.220-1) of category-collections within the game, where sets of membership categories are treatable as organized into or team-like units (e.g. the Marketing Department's 'team' may have categories such as 'Marketing Director', 'Marketing Assistant' etc.) these categories often working in a 'Chinese box' fashion (department fitting into firms, etc.). In this respect, of course, the game may be designed so as to putatively map out 'actually existent' business organizations. Some predicates may 'travel across' all categories in a duplicatively-organized collection, i.e. some activities may be bound to all the categories in a given unit.

One may examine in excerpt 3 speakers' orientations to these levels of duplicative organization - again as part of their making sense of the game (including the overall organization of the game as a frame of reference). Categories, the organized relation of categories, and their bound activities also provide for game participants a set of organizationally given built-in motivations to ongoingly and closely listen, speak and participate; e.g. if a prior speaker refers to an activity which is treatable as tied to a particular category or unit, one or more participants may thereby be selected to speak next or be subject to pressure to self-select, such that if he/she declines to do so, some kind of excusing or justifying account may be seen as relevant by fellow participants.

Participants, then, must incessantly monitor the ongoing conversation for such occasions, and this is one way in which continuing co-orientation and active participation is secured - for listening, too, is an active, not a passive, matter, and has its socially organized features.

We have only been able to chart the contours of these data excerpts in the most general manner. However, we hope to have done enough to indicate that at least one way of analyzing games and simulations is to subject to close and careful scrutiny the culturally-bound reasoning procedures participants bring to the game and which they deploy in making situated, practical sense of game features.

NOTES

1. We wish to thank David Crookall of the Laboratoire de Simulation, Universite de Toulon et du Var, for furnishing the video-recorded data derived from a business game which he conducted.

2. For an elementary introduction to ethnomethodology and conversation analysis, see David Francis, Chapter 5 in Cuff and Payne, eds. (2nd edition, 1985); for considerably more advanced treatments see Heritage (1984) and Sharrock and Anderson (1986).

3. For the transcribing conventions used particularly in conversation analysis, see Atkinson and Heritage (1984).

4. For other examples of the ethnomethodological and conversation analytic approaches to classroom talk see Cuff and Payne (1983).

REFERENCES

Cicourel, A. V. 1972. Basic and normative rules in the negotiation of status and role. In Sudnow (1972), pp.229-58.
Coote, A., Crookall, D. and Saunders, D. 1985. Some human and machine aspects of computerized simulation. In Van Ments and Hendon (1985).
Cuff, E. C. and Payne, G. C. F. eds. 1985. Perspectives in Sociology (Second Edition). London, George Allen and Unwin.
Heritage, J. 1985. Garfinkel and Ethnomethodology. Oxford, Polity Press.
McHoul, A. W. 1978. The organization of turns at formal talk in the classroom. Language in Society, Vol. 7, (August), pp.183-213.
Pollner, M. 1979. Explicative transactions: making and managing meanings in traffic court. In Psathas (1979), pp.227-55.
Psathas, G. ed. 1979. Everyday Language; Studies in Ethnomethodology. New York, Irvington Publishers.
Sacks, H. 1974. On the analysability of stories by children. In Turner (1974), pp. 216-32.
Sacks, H., Schegloff, E. and Jefferson, G. 1974. A simplest systematics for the organization of turn taking for conversation. Langauge, 50:4 (December), pp.696-735.
Sharrock, W. and Anderson, R. 1986. The Ethnomethodologists. London, Tavistock/Horwood.
Sharrock, W. 1974. On owning knowledge. In Turner (1974).
Sudnow, D. ed., 1972 Studies in Social Interaction. New York, The Free Press.
Turner, R. ed. 1974. Ethnomethodology. Harmondsworth, Penguin.
Van Ments, M. and Hendon, K., eds. 1985. Effective Use of Games and Simulations, Loughborough SAGSET.

The use of simulations for purposes of research

Florence Stevens
Concordia University, Canada

ABSTRACT: A participation workshop provided the opportunity for a hands-on approach to learning about the use of simulations and realsits in research. During the session which took place in both English and French, participants collaborated in a realsit, half of them collecting data on verbal and non-verbal communication strategies used by their colleagues who were involved in a simple activity. Their observations were reported and examined. The fundamentals of undertaking simulation-based research were reviewed and examples of work already done in the field were presented. It was concluded that simulations or realsits were valuable additions to the variety of techniques available to researchers.

KEYWORDS: communication, realsits, research, simulations, strategies

ADDRESS: Concordia University, Dept. of Education, 1455 de Maisonneuve West, Montreal, Quebec, Canada H3G 1M8

THE USE OF SIMULATIONS FOR PURPOSES OF RESEARCH

The purpose of this workshop was to involve the participants in a practical session aimed at demonstrating how simulations or realsits could be used in research.

Recent research (Stevens 1985) has shown that a simulation can provide an open-ended situation in which the objectives of the researcher are not apparent to the participants, thereby overcoming what Labov (1969) has called the "Observer's Paradox", i.e. observing behavior which is calculated to please the experimenter. A simulation therefore permits collecting "clean" data, providing that its parameters have been carefully thought out.

What exactly is a simulation of realsit?

Garvey and Seiley (1968) consider simulation to include activities which "produce artificial environments or which provide artificial experiences for the participants". Guetzkow et al. (1963) suggest that a "simulation is an operational representation of the central features of reality". the latter's definition is close to that of Jones (1980) who nevertheless considers "simulation" to be inappropriate terminology and coined the term "realsit" to convey the data of a real not a simulated situation, one in which the participants are engaged in an interactive context. In other words, real people are behaving in a real way in an artificial situation which has been set up to provide a context for their participation.

During the session in Toulon, experience was to go hand-in-hand with theory, and participants were therefore invited to engage in a simulated research task. After they had taken part in the action, principles and guidelines for setting up realsits or simulations were discussed and research using these modalities was reviewed.

THE PARTICIPATION WORKSHOP AS REALSIT

Participants

Seventeen people took part in the workshop; approximately half were English-speaking and half were French-speaking. They came from a variety of backgrounds ranging from engineering to history, and included researchers in such diverse fields as communication, play, literacy, organizational theory and group dynamics. Because of their linguistic background, the participants formed two groups (one English, the other French), which were further sub-divided into two sections (observers and players), the functions of which were not divulged to all the participants.

Materials

Three sets of materials had been prepared:

- a "training kit" for the observers, which contained a sheet of instructions to be followed while the workshop leader was engaged with the player group; a one-page glossary of communication strategies; several copies of a form on which communication strategies could be recorded; a brief review of principles and guidelines for setting up realsits or simulations for purposes of research;
- an article (Williams, 1985) for the player group;
- sufficient "Go-Bots" (plastic toys which can be transformed from a vehicle to a robot) to provide one per player.

Preparation

Before the group arrived, tables and chairs were set up to provide appropriate playing space and observational vantage points.

One group, (the players, who were kept unaware of this designation) was given 15-20 minutes to read a short article (Williams 1985) and to prepare a simulation based on it for the other group. The article concerned miscommunication in employment interviews between people of different cultures and lent itself well to the construction of new situations based on the author's findings.

The second group had also been given some reading material (which identified them as observers) and had gathered together out of hearing of the first group. The workshop leader then worked with this group, making clear the object of the exercise, and explaining the difficulty of obtaining spontaneous behavior from people who knew they were being observed - accurate results being obtainable only when the observed are not quite sure what the researcher or observer is looking for.

In this instance, the observers were to identify communication strategies demonstrated by the players while they were absorbed in manipulating the Go-Bot toys. By using a simple system of numbers and arrows, each observer was to note the behavior of one player of the same langauge group. The verbal and non-verbal communication strategies to be identified were reviewed, and the group was instructed on how to observe individual and interactive events and how to record them on the prepared form. As soon as play had stopped, they were asked to total the number of occurrences of each type of behavior, so as to be able to report on these to the group as a whole; they were also asked to describe in a general way the behavior of the player, e.g. more verbal than non-verbal, etc.

Action

The observers then joined the participants who had been reading the article. After it was explained that the suggestions of this group would be discussed following an activity in which both groups would participate, bags containing the Go-Bots were distributed, one to each player sitting in the two groups of four which had been formed according to language preference. They were told: "Here's something a little unusual for you to explore. Let's see what you can do with them and with each other"; "Voilà quelque chose qui pourrait vous intéresser. Amusez-vous donc avec entre vous."

The players then opened the bags, removed the Go-Bots and proceeded to examine them, transform them, exchange them and discuss them. The observers noted the verbal and non-verbal strategies used by the players to communicate their reactions, intentions, requests and responses. The activity was stopped at the end of 10-12 minutes.

Debriefing

It was at this point that the objective of the exercise was revealed to the players, many of whom of course had been trying to figure out what was going on while they were playing with the Go-Bots.

As has been indicated previously, the session was conducted in two languages, information and discussion occurring in either one as required. Generally speaking, the members of each group had sufficient knowledge of the other language to understand the gist of what was said, but were unable to express themselves as well as they might like to in the other language. It was therefore unnecessary to translate everything that was said, but to ensure that each participant could feel comfortable and become involved in the experience, French and English were used to permit full communication.

It was first outlined that in research, there lies an inherent difficulty in obtaining spontaneous behavior from people who are being observed: the workshop was used as an example. When queried as to their perception of the purpose of the activity in which they had just engaged, the players offered suggestions which ranged from their ability to effect transformations on the Go-Bots and to figure out what the transformations might produce, to creating something with them, e.g. a game or dramatic representation. Some volunteered "communication strategies", because of having read the article by Williams (1985) but were unsure as to what that might mean.

They were told that they had just participated in an experiment which was a realsit in which each member of the observer group monitored the verbal and non-verbal communication strategies of one of the players.

Communication strategies were loosely defined as means used to get a message across. The message might be the expression of a thought apparently to one's self, but intended to be overheard; or one which was directed at one or more persons with the intention of communicating a message. Verbal strategies were construed to be those which made use of words, noises, laughter, etc. and non-verbal strategies were construed to be those which got messages across without the use of vocal sound. In the latter category, only gestures and the actions associated with play had been identified and recorded; the verbal strategies category included speech related to the immediate situation as well as speech related to a secondary or imaginary situation set up by one or more players. Each of these verbal categories contained three sub-divisions: dialogue, monologue, or noises.

The members of the group who had observed the French-speaking players first reported their observations, followed by those who had observed the English-speaking group; all of the results were noted on a large chart in full view of everyone.

The results showed that the French-speakers engaged in a considerable amount of dialogue, two of them talked out loud to themselves, and one tended to laugh a lot; most manipulated the Go-Bots more or less constantly, while one person frequently helped the others to figure out details; they all showed the Go-Bots to each other, expressing satisfaction at having accomplished a transformation; one person grabbed the others' Go-Bots three times; and two people exchanged their Go-Bots to have a try at working with a new one.

Most of the English-speakers talked a great deal among themselves, and one person used a wide range of tactics to communicate with all members of the group, e.g. questions, comments, exclamations, requests, suggestions, and encouragement. Several announcements to the group as a whole were classified as monologues because they were not apparently intended to initiate conversation. One person in this group rapidly assumed the role of leader and directed some of the interaction. Three of the four players entered into a play mode, i.e. assumed the role of a Go-Bot, and either made it speak or make noises or interact with other Go-Bots in a created situation. Their non-verbal strategies included frequent manipulations of the Go-Bots, a considerable number of instances of showing off the results of their transformations, and a few exchanges; no one grabbed anyone else's Go-Bot.

It soon became evident that individuals did not always behave according to the categories which had been specified, and that it was not always easy to classify some behaviors as belonging exclusively to one category and not another. It was therefore important to capture everything that was going on, and analyse it later. Theories must be based on all the facts, and not facts made to fit theories, represented in this case by predetermined categories.

However, in order to run through the various stages of a research project with the group, a simple analysis of the results was attempted, which itself could serve as an example of how to go about determining essential points in the qualitative analysis of data. It was decided that comparisons could be made between the behaviors of the two groups, first noting their similarities and then the differences between them, and questions could be asked, e.g. generally speaking, did one group's behavior differ markedly from the other? In what way? Why? Next, substantive questions could be investigated, e.g. was one type of behavior associated with a particular verbal or non-verbal strategy? What were the functions of the different strategies? It was suggested that if clarification of the data were required, it was always possible to check back with the participants, by asking them how they had obtained certain results, or why they had used certain strategies, even though they might not have been conscious of their motives at the time.

This very schematic way of demonstrating how one does research using a realsit led to an examination of the principles involved and a review of research previously undertaken using this approach.

Synthesis

The procedures which had been experienced by the participants were then reviewed and conceptualized. The latter were reminded that in setting up the research, its purpose must be clear, with specific questions in mind or testable hypotheses, and that its objectives must not be known to the participants. The realsit or simulation must be carefully planned so that its parameters provided a rich situation for data collection; it should frame the problem under investigation in such a way that analysis of the corpus would yield data which could be analysed. An authentic situation must be created to provide for the spontaneous involvement of the participants. The researcher must be ready to accept and analyse what actually occurred in the situation, truly testing original hypotheses and framing new hypotheses as needed by analysis of the data. It was mentioned that the realsit or simulation and the debriefing might be concurrent or might follow one another.

Some operational guidelines for research were proposed. They included the selection of an appropriate environment or location for the realsit or simulation, with consideration given to the availability of facilities (electrical and other). The subjects themselves must be selected according to accepted criteria for research. Permission must be obtained for the use of the space required from autorities connected with the site of the research, and for the participation of subjects (either from the subjects themselves or from appropriate authorities, if the latter were not adults). The importance of following the ethical guidelines published by fund-granting associations with regard to use of human subjects in experiments was emphasized, as was the benefit to be derived by the researcher of maintaining excellent relations with all those connected with the research. The procedures for the realsit or simulation must be determined and the personnel engaged in the research must be thoroughly trained ahead of time. Training should include being informed of the purpose of the research, a dry-run of the activity, as well as training in the use of any equipment required or observation techniques to be used. Data collection materials should be prepared beforehand and be available for use; equipment and supplies should be checked and in readiness for the realsit or simulation. Transcription procedures of the observations to be made (whether by video, audio or personal observation) during the realsit or simulation should be specified; decisions should be made on the kinds of data analysis to be undertaken; the workload distribution for assistants and a timetable for completion of each phase of the work should be set up.

Research projects which followed these guidelines were briefly cited: a Christmas simulation and a farm realsit (Stevens, 1985); the Go-Bot realsit used in the participation workshop which had been used earlier in the year in Paris with Portugese and North African immigrant children learning French as a scond language; and a recent realsit which took place at a location in Canada where maple syrup was being processed. On this occasion, after a tour of the sugaring-off facilities, a ride on a horse-drawn haywagon, and an excellent lunch with Quebecois specialities, a debriefing session took place: each group of 4 children was asked to talk about their morning's activities. A tray of objects related to the sugaring-off activities served as a reminder and a point of departure for conversation. The French native-speakers and the anglophones learning French as a second language who participated in the event were able to test their communicative skills in a non-threatening environment, one in which the researcher could identify their communication strategies and analyse their linguistic ability.

At the end of the participation workshop, a full set of materials was distributed to the player group who had not received them previously, as they had been kept in the dark concerning their part in the session. Because of time constraints, it was not possible for the player group to construct a realsit for the observer group and acquire some practice in implementing the guidelines which had been discussed. Perhaps, however, this was too much to expect from a 1-1/2 hour session.

CONCLUSIONS

A simulation or realsit can provide a valid research situation if it incorporates an appropriate research design. The research design is the component which transforms a learning environment to an experimental situation for purposes of data analysis. The way in which the experiment is set up, the elements reported, the detail with which they are reported, must reflect features of good research design.

The simulation and realsits described are all simple, do not take too long, and require little preparation before the action can begin. Authentic situations provide a means of getting away from a familiar environment and circumscribed conditions, in which the participant could try to supply what he/she thought the researcher was looking for (Labov's "Observer's Paradox", 1969) instead of participating spontaneously in the activity.

The use of informal situations as research settings has many advantages and is a particularly rich potential source of information in understanding how communication takes place. In this area, and in others still to be explored, I believe that realsits or simulations have a considerable contribution to make by enlarging the variety of techniques available to researchers.

REFERENCES

Crookall, D. (ed). 1985. Simulation Applications in L2 Application and Research. Oxford: Pergamon Press. Special issue of System, Vol. 13, No. 3.
Garvey, D. M. and Seiley, W. H. 1968. On simulation teaching. Phi Delta Kappan, 69: 473-80.
Guetzkow, H. et al. 1963. The simulation of international relations. New Jersey: Prentice Hall.
Jones, K. 1980. Communication, language and realsits. In Race and Brook (1980).
Labov, W. 1969. The study of language in its social context. Stadium Generale 23: 30-87.
Pride, J. B. (ed). 1985. Cross-Cultural Encounters: Communication and Mis-communication. Melbourne: River Seine Publications.
Race, P. and Brook, D. (eds). 1980. Perspectives on Academic Gaming and Simulation 5: Simulation and Gaming for the 1980s. Loughborough: SAGSET.
Stevens, F. 1985. Simulations as research instruments. In Crookall, D. (1985).
Williams, T. 1985. The nature of miscommunication in the cross-cultural employment interview. In Pride (1985).

Management simulation and business gaming: An introduction

Alan Coote
The Polytechnic of Wales

Managers at work spend much of their time weighing the chances (and extent) of success against the risks of failure (in economic as well as psychological terms.) Come to think of it perhaps that is what we all spend much of our time doing. As time passes many of us will get better at this 'game', particularly if each time we 'play' we can transfer previously learned principles and behaviours to new situations. Thus we need to recognize (whether consciously or subconsciously) that a new situation in a different context warrants the application of appropriate risk/success balancing skills. Presumably there are some situations (or contexts) which are quicker and more effective at providing us with the experience we need to react most favourably in subsequent situations. There is nothing like 'having onces fingers burned' to concentrate the mind wonderfully next time a potentially dangerous situation arises. Some managers, of course, minimize the risks of failure by reacting to situations in tried and trusted ways. 'Safe' decisions are taken, sometimes to the detriment of the organization and people working in it.

The business of taking risks is difficult to talk about (or lecture about) in a way which effectively modifies risk taking behaviour. Even simulating the feeling of risk taking and its consequences is not easy. Recreating the sensation of risk taking through experiential training requires the careful choice of simulation game context and format. And the problem is complicated since, paradoxically, we may sometimes be more successful in inducing risk taking behaviour in the working environment if we take out or reduce risk taking behaviour in the simulation game. In this way perhaps we can decontaminate the simulated environment so as to allow thorough exploration of techniques and behaviours which will be required in the risk taking environment of the 'real' world.

All of this leaves simulation game designers and users with the need to choose carefully the context in which the simulation game will be set. As someone once said: "A simulation game is like a kiss, interesting to read about but much more interesting to participate in. And those that do tend to repeat the experience."

But it does depend on who one is kissing - and why!

Should participants be asked to be explorers lost in a desert, members of the Board of a multi-national corporation, middle managers in a medium sized company? Should they be urged to be themselves and make decisions about production, manpower and sales in a manufacturing and retail company? Or should they not be provided with any context, being encouraged to create their own?

Such choices will be governed by the types of participant and their expectations, what the trainer hopes to acheive, the desire of the trainer to appear credible and, of course, personal foibles. Amongst the host of other factors which might govern the choice of simulation game; context perhaps one of the most important is the degree to which the context provides participants with the motivation to proceed (or 'simulate', or 'play') with interest and enthusiasm. 'Business' contexts tend to be successful vehicles for generating such interest.

All of the simulation games discussed in this section utilise business contexts of one sort or another. These business contexts are, however, used to create somewhat different sensations of risk taking for the participants. They also aim to achieve different objectives. The simulation games described range from those in which there is an overt effort to simulate risk taking to those which create a 'climate' in which risk taking (during the activity) is deliberately minimised. Thus participants may be placed in contexts which allow risk-taking-tension to be created and sometimes in contexts which enable processes, skills and behaviours to be explored outside the contamination of a 'real life' risk taking environment. The selection of articles also illustrates a range of uses to which simulation games based on business contexts can be put. Some of the simulation games described aim to assist (or speed up) the acquisition of analytical, planning and functional skills (see, for example, Robinson, Kelly and Ibrahim, Brand and Keith) while others are used primarily to affect understandings and attitudes and, presumably, subsequent behaviours (or example, Sawyer-Laucanno, Gernert, McMahon and Coote, Freeman). One article describes a simulation game set in a business context which is used as a vehicle for exploring the acquisition of game competence (Francis), and another discusses participants' assistance in the building of a complex, computer generated simulation (Teach). Each simulation game discussed manipulates, to varying degrees, the risk/success balance of the everyday working or recreational environment, using the models created to suit a variety of participant groups and resulting in a facinating cross section of applications.

Intercultural simulation

Christopher Sawyer-Lauçanno
Massachusettes Institute of Technology, USA

ABSTRACT: Simulations used in management training, if they are well constructed and played, can contribute to better intercultural understanding. The goal, however, of most of these simulations is solely to enable the participants to make effective decisions and devise strategies in relation to business practice. As a result, cultural dimensions are not reflected or examined in the simulation. In order to show how simulation can contribute to a better understanding of intercultural communication we played THE MERCURY SHOES GAME (at the ISAGA '86 Conference in Toulon), a simulation constructed around an intercultural business issue.

KEYWORDS: Intercultural, management, simulation, training

ADDRESS: 14N-236, MIT, Cambridge, MA 02139, USA. Tel: (617)253-4743.

INTRODUCTION

A major, but often overlooked factor in the success or failure of an international business operation is the intercultural relationship between the host and guest. In most cases the role of the guest is played by a multinational corporation; the host, by the native employees of the company in the country in which the multinational is doing business. Despite the importance of intercultural understanding in this type of international business, few management games take this issue into consideration. In order to acquaint the ISAGA '86 conference-goers with the way in which cultural issues can be integrated into the core of a business game, we played THE MERCURY SHOE GAME, an interactive in-basket management game, originally developed by the author as a segment of THE TIME-LIFE BUSINESS GAME.

THE GAME

The game structure itself is fairly simple. It consists of three variations on one basic scenario with accompanying documents: statistical and financial information, intercompany memos, letters and some general background information on doing business in Japan. The three scenario versions are as follows:

NY Head Office Scenario

Your firm, Mercury Shoes Corporation, is an American manufacturer of sports shoes. Since 1960 you have had a marketing subsidiary in Japan which has been completely Japanese-managed since 1972. Although sales were brisk in the first 15 years of operation, over the last five years there has been only a negligible increase. This trend is in contrast to the other international branches, all of which have had constant growth in sales during this period.

Since the subsidiary's founding, the Japanese marketing has relied on three wholesalers for product distribution. In the early years, these wholesalers were instrumental to Mercury's success, but it appears that recently they have been considerably less aggressive in their sales efforts. At the same time, both domestic and foreign-import competition has significantly increased.

At present the wholesalers' contracts are up for renewal. The Japanese office has recommended that they be extended, arguing that it would be a severe violation of trust to terminate the contracts at this time. A detailed memo has been issued by the subsidiary president (see Docs.) on this matter.

Your task is to go to Tokyo and investigate what can be done about increasing sales. The head office has further informed you that another Japanese company, Freetime Sportswear, has indicated interest in acting as a distributor for Mercury's products. Your New York managers feel that a new distribution arrangement might be highly desirable. They are aware of the difficulties involved in overruling the local management, but feel it important to make some changes. They have informed you that you might consider changing personnel at the subsidiary level, setting up a new distribution arrangement and/or investigating changes in product lines being sold in Japan.

Japanese Office Scenario

Your firm, Mercury Shoes Corporation - Japan, is a nearly wholly-owned marketing subsidiary of the American Mercury Shoes Corporation, a manufacturer of sports shoes. Your subsidiary was founded in 1960; since 1972 it has been completely Japanese-managed. Although sales were brisk in the first 15 years of operation, over the last five years there has been only a negligible increase. This trend is in contrast to the other international branches, all of which have had constant growth in sales during this period.

Since the subsidiary's founding, the Japanese marketing has relied on three wholesalers for product distribution. In the early years, these wholesalers were instrumental to Mercury's success, but it appears that recently they have been considerably less aggressive in their sales efforts. At the same time, both domestic and foreign-import competition has significantly increased.

At present the wholesalers' contracts are up for renewal. Your subsidiary president, Mr. Matsumoto, has recommended that they be extended, arguing that it would be a severe violation of trust to terminate the contracts at this time. A detailed memo has been isued by Matsumoto (see Docs.) on this matter. This does not seem to have impressed the head office, as they are sending over a marketing specialist from New York to "look over" the situation.

You are aware that the effectiveness of your office is in question and that it is probable that the head office will institute some rather drastic changes, possibly not only confined to distribution. Your task, then, is to try to convince the parent company that the Japanese office is capable of making the necessary changes to increase sales because you have a superior understanding of Japan, Japanese business and Japanese consumers.

Freetime Sportswear Scenario

Your company, Freetime Sportswear, is a relatively new and highly progressive Japanese manufacturer of sportswear. Although you entered the market comparatively late, you have managed to more than double your sales in the last five years. You have been able to do this through an aggressive marketing strategy, high-quality products and a wholly-owned distribution system with outlets in more than 1,100 Japanese cities. This network includes small and large speciality stores, department stores and general merchandisers.

Your major competition in Japan is New Life Sports, a company which sells not only sportswear, but also sports equipment and acts as an exclusive distributor for West Germany's Wunder Sports Shoes. You have learned recently that the American Mercury Shoes Corporation, which has a marketing subsidiary in Japan, is thinking of making a change in its distribution system. You have contacted the head office of Mercury in New York, after being told by the Tokyo office that no change was being proposed. You have learned from the parent company that, in fact, a marketing specialist is coming to Japan to investigate new distribution possibilities. You feel you are in an ideal position to negotiate a distribution agreement, although you realize that it is a somewhat sensitive issue locally.

You have a great deal to gain through becoming a distributor. Mercury Shoes had gross sales last year of nearly 900 million yen (200 yen = 1 US dollar). A recognized leader in the sports shoe business, they make high quality shoes, popular with Japanese consumers. With little effort, you feel you could sell their shoes, as the same retailers currently buying your sportswear also stock sports shoes. Your task, then, is to meet with the American marketing representative and attempt to work out an exclusive distribution arrangement.

THE GAMING SESSION

About 18 participants attended the session at the ISAGA '86 Conference in Toulon. After a few opening remarks, I formed three groups of about six each, and gave each group its scenario with accompanying documents. The groups were told to keep the information in their packets confidential. During the first 20 minutes, in separate parts of the room, the groups discussed the scenarios and documents. Quickly assuming their respective roles, they began actively debating the advantages and disadvantages of various strategies. Interestingly, a leader seemed to emerge in each group.

Within a few minutes, three distinct assessments, and hence game plans, emerged. The Japanese Office group immediately realized that they were on the defensive and that it was in their best interests to attempt to persuade the reprsentative from the head office that they were doing (and would continue to do) all they could to increase sales with the same distribution system. The New York head office group was initially unsure of what the best strategy was, but decided to arrange meetings with both Freetime and the subsidiary office. Freetime adopted an aggressive strategy, sending a representative to New York at once to contact the Mercury head office.

During the actual play a number of issues came to light, but predicatably the cultural factors seemed to be the least of the concerns. Almost all the participants, as in real life, focused on the bottom line. The New York group used pressure tactics on their subsidiary office to get them to do something about the distribution; Freetime eventually sent a whole delegation to New York to persuade Mercury to appoint them as distributors; even the subsidiary office failed to overtly bring up the potential flap that could result from up-ending the well-established relationship between the wholesalers and the Japnese office. The only mention of this concern was that it would probably be in the best interests of the Americans to allow the Japanese office to handle things the 'Japanese way'; hardly a strong defense.

Unfortunately there was not sufficient time to play the game to completion. When I stopped the action, the situation was this: The Japanese subsidiary believed that the American Head Office was sympathetic to its plight; in reality, the New York group was actively negotiating distribution arrangements with Freetime.

DEBRIEFING AND DISCUSSION

In the brief debriefing a number of important points were raised. The first was that although the cultural issue was in everyone's mind, almost all the participants focused on the business problem. Some remarked that for them, this was much easier to deal with than the cultural issue. One gamer, from the New York group, had become convinced that the concern about sensitivity to the host was a red herring and that the main problem was simply the subsidiary's financial performance. A couple of members from the Japanese office remarked that they felt enormously victimized, and to some extent this contributed to their inability to devise an effective defense of their position. 'All we had was the ethical issue - you just don't come in from outside and totally change the entire system we had built, and that had worked - but that didn't seem to matter,' one member remarked. Related to this issue of feeling powerless was the awareness on the part of the Japanese office members that somehow their countrymen at Freetime were making inroads on their territory.

In a long discussion with several of the participants in the afternoon following the session, other facts emerged. A couple of them remarked to me that they had wanted to bring up the issue of cultural sensitivity, but that as the dominant mood was so business-oriented they didn't feel they could really speak about this matter. Only towards the end of the session, did the perceived issue of unfairness on the part of the Americans begin to emerge as a potential issue.

CONCLUSION

The primary lesson to be drawn from the game session is that in the short term, at least, the pragmatics of turning a profit took precedence over any other consideration. Therefore, the debriefing is vital in creating awareness of the importance (or non-importance) of intercultural considerations, for while these issues emerge during the playing of the game, the profit-and-loss problem seems at the time far more important.

Having played this game with several other groups I have found that after a couple of hours, though, the cultural issues begin to become far more important, mainly because the Japanese office usually becomes quite hostile towards both the Head Office and Freetime. Occasionally, members of the New York group will respond to this and back off on making abrupt changes. In most cases, though, this does not happen. At times, the subsidiary office has taken extreme measures. Some of the responses have included: 'leaking' the news to the local press that Mercury, New York, is attempting to violate Japanese law; taking precipitous actions (firing the wholesalers, presenting new business plans) and in one game, hiring gangsters to kidnap Freetime's president.

The longer the session goes on, the greater the emergence of the cultural dimension. Even if time limits the play, as with our session, a debriefing can often highlight the importance of intercultural understanding, as well as the difficulties inherent in making this issue the central concern of the game.

SIMULATION REFERENCES

THE MERCURY SHOES GAME is available from the author. It was originally one segment of THE TIME-LIFE MANAGEMENT GAME, currently out of circulation.

The competent player: Some observations on game learning

D. W. Francis
Manchester Polytechnic, UK

ABSTRACT: This paper proposes a 'naturalistic', observational approach to the phenomenon of game learning. Taking as a case study the play of a business negotiation game, it examines what game compeence consists in and how it is acquired and displayed by players. In particular, it shows how 'game adequate reality' is interactionally constructed and sustained. This approach constitutes an alternative to the psychological conceptions of learning which predominate in the literature on gaming/simulation.

KEYWORDS: Learning, interaction, game competence.

ADDRESS: Dept. of Social Science, Manchester Polytechnic, Cavendish Street, Manchester, England. Tel: 061-228-6171.

INTRODUCTION

What do persons learn in a game or simulation? How do they learn it? Is gaming distinctive as a learning situation? These and similar questions have been around for as long as the gaming movement itself. They are continually debated. Yet clear-cut answers seem as remote as they ever were. The conceptual and methodological difficulties which surround the evaluation of gaming as a teaching/learning method are notorious. (See Garvey, 1971 and Barnett, 1984). I have no intention of entering the fray on these issues. Instead, I would like to propose an entirely different way of approaching the phenomenon of game learning. Most studies of game learning set out to apply or test a (psychological) theory of learning (e.g. Rogers, 1980). But what if one's aim was to describe the learning that happens in a game, rather than explain or evaluate it? Arguably, the best strategy might consist in looking closely at what actually happens, behaviourally, in the conduct of a game. In this way one may follow Husserl's advice to return 'to the things themselves' (Husserl, 1970).

The problem of game learning is addressed, implicitly if not explicitly, every time a game is constructed or run for educational or professional purposes. But it is addressed from a certain point of view. The game designer or organiser necessarily has an 'instrumental' orientation. The game is an instrument for the achievement of some educational or professional objectives. The concerns of the designer or organiser centre around the relationship between these objectives and the structuring and organisation of play. The critical questions are things like: 'What do we intend that players will get out of playing our game' and 'How can the game be constructed so as to maximise the players learning experience?'. In asking such questions of their game, designers and organisers necessarily make assumptions about the nature of the task or activity and the knowledge involved in performing it. Understandably, they are concerned to make these assumptions explicit and subject them to scrutiny. But there are inevitable, built-in limits to the extent to which the foundations of gaming behaviour are available to instrumental scrutiny. The instrumental orientation constitutes a framework in terms of which much that happens in the course of play appears simply irrelevant. Much that players do and say in the course of playing a game does not concern the topic of the game, the tasks set in the game instructions or the achievement of game objectives. Even behaviour which does appear to bear, directly or indirectly, upon the 'business' of the game often does so in ways tangential to the aims of those running the game. The 'problem' of player compliance arises out of the fact that actual players, unlike the ideal players conceived of by the game designer, can and frequently do operate with different relevancies from those of the designer or organiser. Players have _their_ reasons for playing the game. From the perspective of the designer or organiser, getting players to play the game _in the way required_ can be a recurrent problem. It sometimes seems as though games would work well if it wasn't for the players!

Therefore what a player learns in the course of a game may relate not so much to its topic or objectives as to what it takes to play the game in approved ways. Game learning involves coming to understand, in the course of playing a game, what it takes to be a player. Even in circumstances where players are keen to comply and play the game 'seriously', it may not be clear to them what 'playing the game seriously' should involve. And even where instructions and game rules are clearly spelt out, players have to discover for themselves what these mean _in practice_. What does 'proper play' of this particular game consist in? These matters are problematic in ways which the literature on game learning and gaming behaviour fails to recognise. (For example, Schild, 1968.) Much that players learn about a game is 'invisible' from the instrumental point of view. The reason is not 'lack of awareness' or any other 'failing' on the part of game designers and organisers. It has to do with the _essential incompleteness_ of rules and instructions and the taken-for-grantedness of everyday knowledge. (See Garfinkel, 1967). The 'seen but unnoticed' character of much game activity is the product of the fact that players are assumed, by each other as well as by game designers and organisers, to possess everyday interactional competencies. Most of the time players do make adequate sense of game instructions; they do take appropriate account of game information; they do comply with the requirements for proper play. Therefore the problem of 'game adequate behaviour' is not the subject of overt investigation or inquiry. From an instrumental point of view it does not matter _how_ players do these things, only that they _can_ do them. The behavioural competencies and everyday knowledge that a game presupposes are hidden beneath its overt form and content. To discover what such competence consists in, how players acquire and display it, it is no use looking to designers or organisers intentions. Nor is it much use relying on players reminiscences of what they did and how they did it. Only by examining data which 'captures' the game play in its course, such as audio- or video-tapes, can the character of implicit game competence be brought out.

PLAYING A BUSINESS NEGOTIATION GAME

The data on which this paper is based consists in a video-recording of a business negotiation game. Since my aim is exploratory, I will not attempt, on the basis of this data, to produce a 'general' or 'comprehensive' analysis of game learning. If one eschews this kind of _rush to generalisation_, it may be possible to learn something about the fundamental properties of game learning from a particular case. A game is a social situation. It involves action and interaction, typically both among players and between players and organisers. How does learning occur in and through such interaction? How is grasp of the game displayed by players in the course of play? What does it mean behaviourally for a game to be played 'intelligently' or 'thoughtfully' as opposed to 'unimaginatively' or 'uncomprehendingly'? These questions could be pursued via a set of stipulative definitions and theoretical idealisations (e.g. 'types of learning', 'forms of interaction'). By contrast, the 'naturalistic' strategy adopted here involves suspending all such a priori definitions in favour of close and detailed examination of interactional data. (For a more extensive discussion of the rationale behind this investigative strategy, see Schwartz and Jacobs, 1979).

The game I recorded is A. W. Gottschalk's CHEMICAL CONSTRUCTION COMPANY. It was played as part of a three day course in business negotiation, organised by the Business Studies Department of a large institution of higher education in the U.K. the participants in the course were all sales personnel in the advertising department of a large national newspaper. The course was run by two tutors in business studies and the training officer of the newspaper. The game was one of a variety of activities that made up the programme of the course. Playing it took up the morning and most of the afternoon of the second day.

How did the players of CHEMICAL CONSTRUCTION COMPANY learn how to play the game? As with most games, they learn how in the course of playing it. They were not taught in advance, nor did they acquire knowledge by repeated plays. The game was played only once. At no point was the game halted or suspended by the organisers to give advice or instruction to the players. Players did on one or two occasions seek guidance from the organisers, either on matters of information or on how to conduct themselves as players. But such guidance as was given was kept to a bare minimum, seemingly as a conscious strategy on the part of the organisers. None of the participants had ever played the game before, and for most of them it was their first experience of a role-playing game. So how did they know what to do?

Let us consider the information and instructions the players were given prior to the commencement of play. CHEMICAL CONSTRUCTION COMPANY consists of three teams of players. These represent, first, executives of a multi-national petrochemical corporation, 'Ajax Petrochemicals', secondly, executives of a large civil engineering construction company, 'Chemical construction', and thirdly, officials of a trade union, 'The Civil and Cosntructional Engineering Union'. The course participants had already been divided into three teams for a previous activity and those teams were retained for the negotiation game. All were given one set of notes describing the game situation, the titles of the members of each team and other 'general information'. In addition, each team was given a separate, shorter set of notes containing information about its own circumstances and other details unknown to the members of the other two teams. After these documents had been distributed the course leader gave a brief introductory talk about the game. He outlined the game situation and its 'background'. He then allocated each team a room as its 'headquarters'. It was emphasised that teams could and should communicate with one another in a variety of ways: by telephone and letter as well as face-to-face meetings. Also, he indicated a timetable for the game. The course programme required that the game be completed by mid-afternoon, since other activities were scheduled for the late afternoon and early evening sessions. The course leader suggested that each half hour, by clock time, should be taken to represent one day of negotiations within the game.

The framework of the game, as described by the course leader, is as follows. Chemical Construction ('C.C.') is engaged in the construction of a refinery and associated facilities for Ajax Petrochemicals in a rapidly developing third world country ('Nigeria'). The estimate of costs originally submitted by C.C. was U.S. $48 million, which was eight millions more than the Ajax management had forecast. Despite this difference in estimates, Ajax's pressing need for the facility and the fact that other cost estimates were even higher led the Ajax board to request C.C. to commence construction work immediately. C.C was asked to submit a more detailed cost breakdown on the basis of which a final, fixed price contract would be agreed. 'At this moment' work has been in progress for eight months, but Ajax and C.C. have been unable to agree a final price for the project. Therefore C.C. have suspended all work at the site pending the outcome of negotiations. This has created a number of problems between C.C. and the Civil and Constructional Engineering Union ('the union'), which represents the entire workforce on the site, both the expatriate workers ('mainly engineers and technicians') and locally recruited employees ('mainly skilled manual workers'). The union have a wage claim outstanding. In addition, a competitor of C.C., 'Global Construction', are beginning work on another large civil engineering proejct nearby and are seeking to recruit workers. The course leader stressed that while the interests of the parties were not identical, all three had much to gain in reaching agreement with one another. He then asked if there were any questions. After one or two factual clarifications the teams were told to retire to their respective rooms to discuss the game information and consider their courses of action.

What was this 'setting up' phase of the game designed to accomplish? The organiser's introductory remarks clearly were intended to orient the players to the 'game setting'. The information they were given, both orally and in writing, amounted to more than just a list of 'facts'. It served to define a setting of a kind which the players should have found broadly recognisable. An unstated, but omni-relevant condition of game play is that it should reflect the 'kind of situation' indicated by the game instructions. Neither the written instructions nor the introductory talk spelt out what was necessary behaviourally to realise that setting in and as the course of play 'here and now'. But it would be wrong to see this as a failing, or thereby to regard them as 'incomplete'. The organiser could have said more. Having run the game numerous times before, he presumably could have said more about the internal organisation of the teams, the management of communication between the teams or the possible scenarios which could lead to agreement. He might also have said more about the international petrochemical industry (real or invented) or the problems of industrial development projects in third world countries. But however much the organiser had said and however extensive the game instructions, players would still have to decide in the course of play itself what was required behaviourally to produce and maintain 'adequate realism'.

The participants in CHEMICAL CONSTRUCTION COMPANY demonstrated awareness of the requirement for 'adequate realism' in a variety of ways. They interpreted and applied the game instructions on the basis of their commonsense understanding of the kind of situation the game was intended to represent. They did not treat the instructions as self-evident rules of conduct, to be followed mechanically or uncritically. Nor did they treat the information they had been given as the only source of relevant knowledge. Realising the game setting in and as the course of play involved filling out this information on the basis of 'what we know' about situations 'of this kind'. They were thus able to improvise courses of action and introduce 'new' elements into the game in an ad hoc fashion. In the game the setting which players invoked to contextualise their activities was 'business life'. As a known-in-common context, 'business life' constituted an open set of relevancies in terms of which the meaningfulness, appropriateness and effectiveness of game actions could be interpreted and judged. By treating business life as a commonsense interactional order, players were able to manage game activities and relationships in ways which 'brought off' their roles as putative members of business organisations. They used resources available to hand to create an appearance of business life that was adequate for the purposes of the game. They did so not just individually but concertedly and collaboratively. Furthermore, the commonsense features of business life were improvised in the course of play itself, not in addition to or outside of it. Players took it for granted that others would recognise what they were doing and see its point or relevance. Explanations were only given in the event of another's failure to understand. This is not to say that there were no discussions among team members about 'what to do next'. Such discussions went on continually. But they took place against the background of actions whose sense and meaning were 'obvious'. By virtue of this commonsense obviousness the course of play flowed, without constant interruption due to failures of understanding. But it is important to appreciate that the recognisable sense and meaning of game actions was an accomplishment on the part of participants. Bringing off the business setting of the game involved 'work' of reasoning and interpretation.

What were the components of the game setting which players produced in and through their actions? First, there were 'business identities', such as 'chairman', 'accountant' and 'secretary'. These identities were created for the sake of adequate play. They were not simply taken from the game instructions. For example, there was no mention of a 'chairman's secretary' in the instructions but one team felt that this role was necessary for verisimilitude. Without this role, certain activities would lack 'adequate realism'. As the members of the team put it, 'chairmen don't make their own 'phone calls'. In this way participants drew on their commonsense knowledge of action-identity ties to simulate 'business life'. Sacks speaks of activities which are commonsensically tied with identity categories as 'category bound activities' (Sacks, 1974). Social actors treat these ties as adequate grounds for inference and action. Thus the perceived performance of certain actions can serve to identify who or what the persons concerned 'must be'. Identities routinely can be read off from activities, while knowing 'who' someone is can establish 'what' is being done. Therefore business identities often provided game participants with the key to the second component of the setting, 'business activities'. Participants selectively invoked 'archetypal' business activities and situations to spontaneously structure the ongoing play. Events such as 'the board meeting', 'the working lunch' and 'the shareholders meeting' were treated as features of the setting as and when the course of play was seen to require them. Activities such as 'reporting back to my board', 'consulting with my union members' and 'discussing the financial implications with our accountant' were used by players to manage their dealing with members of other teams. The distribution of game identities was an occasioned matter, related to the 'local' interactional circumstances that obtained when a particular game action needed to be taken.

Thus participants did not treat 'business life' as a rigid set of conditions which had to be adhered to regardless of the state of the game. If the 'local' circumstances of play dictated the need for actions that would be abnormal or out of place in a business context, such as a 'company chairman' delivering a letter by hand, this would be rationalised by a remark such as 'We're a bit short staffed, old boy, our messengers are all on strike'. The implausibility of such actions was not challenged by other players, though it might be marked non-disruptively by smiles or laughter. In this way players collaborated to maintain a 'game adequate' definition of reality. Similarly, the simulation of 'business life' was produced in ways sensitive to the physical and organisational context in which the game was being played, such that these 'real' circumstances were not allowed to undermine the 'game reality'. Players managed the practical constraints created by the 'real' context in three ways. The preferred strategy was to incorporate aspects of the physical and organisational context into the play (e.g. the hotel bar was treated as a location for 'informal' meetings between executives). Alternatively, where such incorporation was not possible, players would intentionally disregard the 'real' circumstances or 'suspend play' until non-game intrusions had been dealt with. The first alternative had the effect of making players actions appear bizarre to anyone not operating with the interpretive relevancies of the game. For example, when Tim and Alistair treated their lunch in the hotel restaurant as a 'working lunch between company chairmen' they received some puzzled looks from other guests. In these ways players collaboratd with one another to 'make the game work'. Precisely what was necessary to acheive this was something that they discovered as they went along.

CONCLUSION

Discussions of game learning invariably have been cast in psychological terms. Most often, learning is conceived as a process in which the structure and content of a game serves to transform the state of knowledge of the individual participant. One effect of this is to make learning a mysterious process, one which can only be elucidated via a theory of the mind. Studies of game learning repeatedly have sought explanation (and justification) of how game players learn in one or other psychological learning theory. (Bruner's is the current favourite - see Rogers, 1980). Another effect is to focus attention wholly upon what the game does for the player and away from what the player does for the game. Despite the claim that gaming presupposes an 'active' learner in an 'open-ended' learning environment (Dudley, 1980), little is known about how players create and sustain the 'reality' of the game situation. Yet, arguably, it is this task of 'reality construction' which, from the point of view of the player, is gaming's characteristic requirement. The philosophy of gaming and simulation suggests that, if only designers can achieve 'isomorphism' between their game and the segment of reality it models, then players will attain the real-world knowledge which is the game's objective as they acquire competence in the game. But a game is first and foremost just that: a game. Therefore the competence players develop and display is first and foremost the behavioural ability to sustain a 'game adequate' reality. How they learn what this involves is the question I have tried to open up.

REFERENCES

Barnett, T. 1984. Evaluations of simulations and games: a clarification. Simulation/Games for Learning, vol. 14, no. 4.

Dudley, J. 1980. Discovery learning in simulation/gaming. Simulation/Games for Learning, vol. 10, no. 3.

Garfinkel, H. 1967. Studies in Ethnomethodology. Englewood Cliffs, Prentice Hall.

Garvey, D. M. 1971. Simulation: a catalogue of judgements, findings and hunches. In P. J. Tansey (ed.) Educational Aspects of Simulation. London, McGraw-Hill.

Husserl, E. 1970. Cartesian Meditations: An Introduction to Phenomenology. The Hague, Nijhoff.

Rogers, P. 1980. Some games with primary school children: an application of Bruner's theory. Simulation/Games for Learning, vol. 10, no. 3.

Sacks, H. 1974. On the analysability of stories by children. in R. Turner (ed.) Simulation Games in Learning. Beverly Hills, Sage.

Schild, E. O. 1968. Interaction in games in S. Boocock and E. O. Schild (eds.) Simulation Games in Learning. Beverly Hills, Sage.

Schwartz, H. and Jacobs, J. 1979. Qualitative Sociology. New York, The Free Press.

SIMULATION REFERENCE

CHEMICAL CONSTRUCTION COMPANY. A. W. Gottschalk, 1976, London Business School.

Management simulation games and international co-operation

Hans R. Gernert
Humboldt-Universität zu Berlin — GDR

ABSTRACT: Many more or less successful attempts to transfer games between different institutions within the GDR and between socialist countries during the last decade established the idea to develop a complex multilevel management simulation game that provides insight into the mechanisms and the necessary negotiations of international trade between the GDR and the UK.

The paper describes the scenario of the game TRIPOD briefly, stressing its modular structure so that parts of the simulation can be used as games separately.

Advantages and problems of international co-operation occur in this project. The authors live in countries with different social systems. They had relatively rare direct discussions but were very engaged in designing and the use of simulation games in higher and further education at both ends.

The topic of the game allows the demonstration of mutual advantages of international trade with socialist and capitalist countries.

The game TRIPOD is used during the regular exchange of students between the two institutions involved. It provides broad knowledge of foreign social-economic and cultural environments and proved to be an excellent opportunity to stimulate foreign language learning.

KEYWORDS: Management simulation games, transfer of gaming methodology, international co-operation, multi-level games, international trade games, games in higher and further education.

ADDRESS: Humboldt-Universität zu Berlin, Sektion Wirtschaftswissenschaften, Bereich Wirtschaftsinformatik, Spandauer Str. 1, DDR 1020 Berlin, Tel: 21 68 340.

THE MANAGEMENT SIMULATION GAME BES 1

The dissemination of management simulation games (MSG) within one country and especially the international transfer of complex MSG is as challenging and interesting as difficult.

There are several objectives of this promising task including: to demonstrate and compare different economic mechanisms, to teach various concepts of management, to study gaming methods used by experienced authors, to support the development of games in new areas, and to save time and money if the adaptation of a game is easier than individual developments.

One example of the successful dissemination of games in higher education in the GDR is the complex computerized MSG BES 1. This management game was developed at the Department of Statistics at Humboldt-University Berlin - GDR. It aims at simulating a manufacturing process in order to support studies in the fields of business administration, finance, general management, and economic theory. The industrial enterprise simulated by this model is described in terms of its product line, marketing conditions, fixed and current assets, manpower, and the necessary financial framework. In each simulation period decisions are taken on manpower, working conditions, investments in fixed assets, research and development, production, turnover, and distribution of profits.

The development of BES 1 was completed at the end of 1973. Since 1974 it has been applied in training full-time students and high level executives of enterprises in several branches of industry. A recent version of this game has been used in about 20 universities, polytechnics and institutions of further education in the GDR. It is now the most widely used large scale rigid rule simulation game in this country.

There are several reasons for this dissemination. The new users followed up similar educational aims based on the theory of the socialist economy. Many scientists became aware of the educational power of games in higher and further education. There have been few problems transferring the computer programs because the hardware and software facilities were overwhelmingly compatible. Furthermore it proved important that the game was extensively described and all algorithms were given to the users to provide the opportunity to change parameters by themselves.

INTERNATIONAL TRANSFER OF GAMES

The international transfer of games and gaming methodology provides a very important opportunity to disseminate gaming experience. Several attempts to transfer games between socialist countries have been made (Gernert _et al_. 1983). As far as the GDR is concerned the co-operation with the Kasakh Polytechnic Institute in Alma-Ata, USSR, Hungarian universities and institutions for further education, and the Institute for Social Management in Sofia, Bulgaria, has been exceptionally fruitful.

The main objective of the co-operation with the Alma-Ata Polytechnic has been to exchange ideas and experience in running games on several topics in order to improve education especially for postgraduate students. While the Humboldt-University and other educational institutions in the GDR were interested to use several Alma-Ata games in the area of inventory control, transport, and quality control, the Kasakh Polytechnic planned to develop a general management simulation for an industrial enterprise. During the co-operation the model BES 1 was transferred to Alma-Ata, implemented on the computer system of that university and is now available to other universities and polytechnics in the USSR. Due to this fruitful co-operation and the excellent contacts between the scientists involved at both ends the joint development of a branch-level game is planned.

The Hungarian game DÖNTS and the BES 1 model have been exchanged to the mutual advantage of both sides. The main incentive for this research task has been the understanding that these models complement one another. Both the models simulate the behaviour of industrial enterprises. While BES 1 sees the company in a more generalized way and aims at developing different strategic concepts in managing the socialist company, the model DÖNTS is more concerned with simulating the production process of the enterprise in detail.

In using the translated documentation (briefing papers, users manual, program description, etc.) an outline of the proposed usage of the game under GDR conditions was prepared. Experience from those who had played the game previously in its original version was very useful in this respect. The outline included recommendations of the most appropriate game variants to use. It was discussed with interested scientists of several GDR universities and polytechnics and with the working group Management Simulation Games under the Ministry of Higher Education. We believe that as a result of these efforts, a clear-cut concept of the proposed use of this game in education was the main reason for the successful transfer of this model.

Since 1982 research work has been done to transfer an agricultural game from Karl-Marx-University, Budapest, to Humboldt-University Berlin and to combine the Bulgarian Game IM-1 (Institute for Social Management, Sofia) with the game BES 1 in Berlin.

THE GAME TRIPOD - JOINT VENTURE IN INTERNATIONAL CO-OPERATION

Structure

The development of complex computerized simulation models for use within the framework of games is a long lasting and expensive process. Discussions with experienced scientists will help shorten this time. Another possibility to increase the efficiency in developing MSGs can be seen in creating a set of games relying on a very flexible computer software package. In using the experience from developing and running the simulation model BES 1 for many years the following system of simulation models and games has been set up:

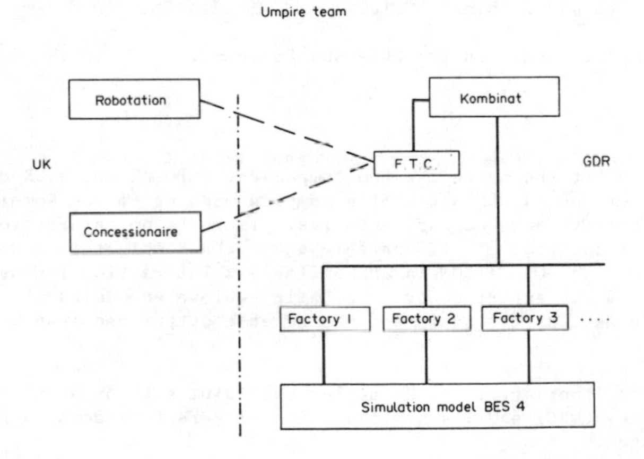

Fig. 1 TRIPOD, UK-GDR

Participants of the game are responsible for managing the GDR factories and represent the lowest level of management within this multilevel set of games. Their decisions are the input data to the simulation model BES 4 that will simulate the behaviour of each of these factories for a simulation period of six months.

On the level of the GDR-Kombinat the player team is in charge of running the part of an industrial branch consisting of several factories. Special responsibility goes with the Foreign Trade Company (F.T.C.) attached to the Kombinat management. This team has to analyse the international markets, to negotiate terms of trade for imports and exports, and to carry through all import/export business on behalf of the Kombinat and the respective factories.

It is possible to extend the game described into the international trade exercise TRIPOD. In close co-operation with the School of Management, Buckinghamshire College of Higher Education, High Wycombe, Great Britain, the interface to two English companies based on independent simulation models has been developed. The two English firms are a company called ROBOTATION, producing robot-type machinery and a UK-Concessionaire (UKC), set up during the game as a joint venture of the Kombinat and ROBOTATION for selling products from Kombinat in the UK market (see Figure).

The umpire team is responsible for controlling the game, elaborating the strategic aims of the simulation, and evaluating the results. They have to provide information on national economic developments, trends, and special additional events. In the role of the GDR Ministry of Industry they have to issue the plan targets and the allocation of resources to the Kombinat.

The interaction between the teams of participants is of special concern in this complex game. Therefore the authors always have been stressing the negotiation interfaces. The interaction is not built into the computer model but takes place through negotiations between and within the player teams. These areas of discussions are:

- within the factory-, the Kombinat-, and both the UK-teams,
- between factories and Kombinat, F.T.C., and ROBOTATION,
- between the umpires and both the GDR- and UK-teams.

TRIPOD - Scenario

At the starting point there are the two companies, ROBOTATION, a UK capital goods manufacturer, and a GDR automobile complex made up of the Kombinat, its Foreign Trade Company, and the car factories. There is no interaction between the companies but the position is unstable since the Kombinat is required to expand production and ROBOTATION is unprofitable. Interaction is ensured because the latter needs a GDR export contract to help achieve economies of scale, and the former needs new capital equipment to increase output and launch its new design.

Since the deal is important to both parties the joint setting up of the importer-distributor agency (UKC) and its staffing with players from both so far existing teams is envisaged.

The function of the Kombinat model is largely to indicate the necessity for expansion, and, since there is full employment in the GDR, there is the need for this to take the form of increased productivity. The task is, therefore, to optimise production through the variables of research, training, workstudy, and automation and the effective allocation of resources. Financial penalties for holding surplus stock mean that precision in forward projections is required and the quintessential problem is one of planning.

The team activities in this part of the game therefore aim at showing how:

- the planning system is carried through in a socialist economy,
- the macroeconomic interdependence of various factors must be taken into consideration,
- import and export transactions are initiated and handled,
- import-export business on the basic principle of mutual advantage can stimulate the production of high quality goods on very efficient production lines.

The market model of ROBOTATION is designed to represent the problems arising from a wide range of differing markets for a technically sophisticated product requiring a high degree of backup service. The production model simulates, in a simplified manner, problems of an assembly industry dependent on the regular supply of components with order lead-times in excess of delivery commitments with the added complication of stock degradation and obsolescence. The work-rate is affected by local wage levels, bonus schemes, overtime and the state of the order book. Expansion requires labour recruitment and training and has to be funded. The key to an effective strategy lies in accurate forecasting.

In the car industry sales depend on a large and effective dealership and dealers are attracted only by the prospect, based on historical data, of large sales. The way to achieve both would seem to be to provide a product which is competitive in design, quality and price and to offer immediate delivery, skilled and widespread servicing and plentiful cheap parts. The UK-Concessionnaire is responsible for national advertising and to create the right image. This is the situation constructed in the distributor model and feedback can be obtained by the company through consumer advisory publications. Success depends on assessment and application of the marketing mix.

The game is interactive in the sense that communication and negotiation is required between the groups of players, but not in the widely-used but restricted sense of resting on competition between the teams: on the contrary the ultimate lesson is the need for co-operation to obtain mutual benefit.

Educational Possibilities

The authors of Exercise TRIPOD have produced three games each of which is capable of being run on its own with interface data supplied by the umpires.

The specific teaching objectives of the GDR-model are concerned with the planning process in a socialist economy, the influence of macro-economic factors, the techniques required to optimise production in an enterprise within the framework of socialist planning, and the multi-level relationships involved.

ROBOTATION represents primarily the production problems of an assembly industry and the marketing problems involved in creating and meeting demand from five different markets under heavy competition.

The distribution model of UKC is concerned with the entrepreneurial problems of breaking into an established market and the cash-flow problems resulting from keeping high stocks and offering credit.

All three parts of TRIPOD involve the application of a number of management techniques from financial analysis to production planning and market appraisal.

Thus from each exercise derive three teaching objectives:

(i) Promoting understanding of the environment in which business decisions are made and of the methods used.

(ii) Helping potential managers cope with specific situations by the application of scientific techniques rather than guesswork.

(iii) Helping potential managers cope with complexity. Lots of decisions have to be made, many of them simple, all in some way inter-related.

To these objectives one may add a fourth, though the emphasis is different in the differing economies:

(iv) To cope with uncertainty. Managers repeatedly have to make decisions before they know all the facts.

What dimensions have been added by putting the three games together? As they stand the GDR producer is hierarchical, the UK producer, though autonomous, has at least two management levels. To these a distributor is added, probably a subsidiary to both organisations, though with possible areas of conflict. There has also been introduced the necessity of conducting international negotiations, preparing contracts and monitoring them, and several other potential areas of conflict.

Therefore another skill area has been opened: Learning to cope with other people.

To sum up, using the TRIPOD exercise extends the area to promote understanding to encompass international trade and the appreciation of how each others economic system operates.

The complex game TRIPOD has been run successfully in Berlin and High Wycombe with GDR and UK students. These participants not only enjoyed the game but appreciated the wide range of decisions to be made and the interesting discussions they were able to create during negotiations. Furthermore the game has proved to be an excellent tool to stimulate foreign language learning and practice.

REFERENCES

Assa, I. Gevrenov, S. Kolarov, N. and Petrov, S. 1976. The Management Game "Production Enterprise Economic Mechanism". In: Proceedings of the 3rd International seminar on Simulation Games. Sofia.

Doman, A. 1975. Bericht über ein komplexes Unternehmensplanspiel, gennant DÖNTS. In: Proceedings of the 2nd International Seminar on Simulation Games, Berlin.

Gernert, H. Conlan, J. and Pope, A. 1983. A GDR-UK Trade Game. In: Stahl (1983).

Stahl, I. (ed). 1983. Operational Gaming: An International Approach. Oxford: Pergamon Press.

SIMULATION REFERENCES

DÖNTS. Cf Doman, A. (1975).

IM-1. Cf Assa et al. (1976).

BES 1. Gernert, H. and Kölzow, W. 1973. Ein betriebliches. Simulationsmodell (BES 1). Humboldt-Universität zu Berlin, Sektion Wirtschaftswissenschaften.

BES 3. Gallenmüller, O. and Wagner, W. 1979. Anleitung zur Durchführung des Planspiels BES-3. Bergakademie Freiberg.

BES 4. Gernert, H. and Messerschmidt, K. 1984. Humboldt-Universität zu Berlin, Sektion Wirtschaftswissenschaften.

TRIPOD. Gernert, H., Conlan, J. and Pope, A. 1984. Humboldt-Universität zu Berlin (GDR), Sektion Wirtschaftswissenschaften and Buckinghamshire College of Higher Education, School of Management. High Wycombe (UK).

QUALSIM: an approach to help managers establish quality assurance mechanisms

Laurie McMahon
King's Fund College, London, UK

Alan Coote
The Polytechnic of Wales

ABSTRACT: This paper describes a simulation (QUALSIM) that is simple in structure and that was developed specifically to enable management developers to help senior managers from the health service in the U.K. to establish formal quality assurance systems. The background to the development problem is discussed, together with a description of the simulation itself and the context in which it can best be used. Consideration is also given to applications of the simulation for managers who work in settings other than health care.

KEYWORDS: Game, health service, management, quality assurance, simulation.

ADDRESS: LMM: King's Fund College, 2 Palace Court, London W2 4HS. Tel: 01-229-9361; AC: Department of Management and Legal Studies, The Polytechnic of Wales, Pontypridd, Mid Glamorgan CR37 1DL, Wales Tel: (0443) 405133.

INTRODUCTION

The national health service (NHS) in the UK is currently undergoing a fairly radical change in the way that it is managed. A report on the management of the service was commissioned by the government and undertaken by Roy Griffiths (the managing director of Sainsbury's - a large national food retail chain). The report's (Griffiths, 1983) recommendations were accepted without qualification by the government, and as a result the largest organisation in western Europe set about the task of changing itself.

The most noticable change has been the introduction of general managers at hospitals (and at each higher level of administration) to replace the consensus team of administrator, doctor, nurse manager and finance director that previously was held jointly responsible for the delivery of health services at the local level. However, the fuss created over these stuctural changes has tended to mask even more significant elements of the Griffiths initiatives. These relate to the substitution of the administrative and professional cultures that are dominant within the NHS with a more distinctly mangerial one. This more entrepreneurial approach to managing means that there needs to be greater autonomy for local managers. Equally as important, those that provide the services locally need to be more responsive to the needs and expectations of the population that they serve. It is this latter issue, that of helping local managers develop systems that will make them more responsive to the expectations of the consumers, that produce very particular development difficulties.

THE DEVELOPER'S PROBLEM

The nub of the problem is that NHS managers have never been encouraged to think about the consumer before. Perhaps we should clarify this rather sweeping statement. We would suggest that there is difference between being concerned about the quality of the technical or clinical aspects of treatment, and being concerned about the environment in which the treatment takes place and the impact or impression that that environment has on the consumers (patients and their relatives). So the health service may have been providing excellent clinical treatment, whilst at the same time producing unhappy and dissaffected consumers whose expectations about the manner in which they should be treated were not being met.

In most other service industries, this failure to meet expectations would result in the consumers taking their custom elsewhere. But in the UK, we have a national, state funded service, so the possibility of 'shopping around' is extremely limited. The result is that the natural control of the market (that of falling demand for the product) does not alert service managers to the problem. We should point out that we are both great supporters of the NHS, and that this comment should not be interpreted as a plea for more private medicine - the country can ill afford the increase in costs and the decrease in standards that we are convinced would result.

In any case we would argue that even in countries where private health care flourishes the power of the market is somewhat illusory. This is because most health care consumers lack perfect knowledge about what constitutes good treatment, and more particularly, because of the dependent nature of the health care consumer on the provider. Consumers are disinclined to complain whilst in a vulnerable position.

All of this means that health care consumers in general, and those in the UK in particular, are unlikely to be a powerful force in raising and maintaining levels of service quality. It follows that, from a management point of view, the need for managerial systems to assess or monitor the quality of what we can call 'service delivery' is of prime importance. Griffiths intended that something be done about it, and that in turn generated a need for those involved in management development to consdier how they could support managers in developing such systems.

THE LOGIC OF QUALITY MONITORING

The logic of management systems designed to provide a continuing assessment of the quality of service delivery can be described in a deceptively simple way. It involves establishing the criteria by which service quality should be judged, operationalising those criteria so that they are capable of being quantified, some more or less objective process of assessment or measurement, and finally (and this is too often forgotten) some mechanism by which the organisation can improve its performance.

A key difficulty is whose criteria does one use? Equally as important, how do you make staff sensitive to the needs of the consumers? As we have already suggested, the managers and professionals within the NHS are not used to thinking about service delivery in this way because they have become used to seeing services from the institution's point of view. We would argue that people who have been socialised into this way of viewing things find it almost impossible to strip off the desensitising filter, and to view what they do from a consumers' perspective. This seems to be the case even where managers <u>recognise</u> the need for such a perspective.

This problem is not an easy one for the management developer to overcome. We suggest three ways in which the providers can be assisted. The first is to help managers and professionals become more aware of the need to get into quality monitoring; second to help them understand the process of establishing the systems to do it; and thirdly to skin the institutionalised eye and let them see what they do from the consumers' perspective.

THE SIMULATION

When faced with this challenge, we felt that using traditional teaching techniques, and telling people to view a familiar environment from an unfamiliar viewpoint was likely to be particularly inappropriate and unproductive. From our experience, the problem seemed an ideal candidate for simulation/gaming. Allowing people to experience the process of building up a system of quality monitoring and enabling them to take a consumer's view of a familiar environment seemed the only way that any significant learning could be achieved. But what sort of simulation could we design that could achieve this objective?

We found the answer in a bar! Not only did they provide a complex customer service with which most people, even those who were teetotal, would be familiar, but they were also environments which no matter how often used, would not have been assessed for quality in a systematic way. They could therefore be used to demonstrate the difference between an institutionalised view (the unthinking, uncritical pub experience) and the view sharpened by a more structured approach. It was also the case that a range of pubs and bars are usually available locally, and that they gave a similar enough service to allow comparison between them. Finally, but by no means the least important, we knew that a simulation involving assessing the service quality provided by pubs and bars would be enjoyable.

We have found that the best time to schedule the activity on residential programmes is towards the end of the afternoon, through the early evening to the following morning. On non-residential programmes, running the simulation over lunch time using the sessions before and after lunch is also possible.

STAGE 1: Preparation

Having provided the group of managers with an introduction about quality assessment and it's importance to health care mangers (not that dissimilar in content to the points that we made at the beginning of this paper) we divide them into groups of three or four. The reason for not having bigger groups, is that publicans may become less than cooperative if large numbers of people troop through a bar taking notes!

Having explained the basic structure of the simulation to them, we then ask the small groups to develop the criteria by which they would judge the quality of service provided by a bar or pub. This stage usually results in about 10 criteria being established, which are fairly general in nature. For example, 'ambiance', 'cleanliness', and 'range of drinks' are commonly identified. The exercise generates a great deal of discussion and we do not structure the groups' deliberations, nor do we try and synthesise the different lists to provide a common starting point. This is because we feel that a strong learning point can be made from the range of criteria and the differences in value systems that produced them.

Stage 2: Operationalising Criteria

The next stage is for the groups to operationalise their criteria. In other words we ask them to define a little more closely what 'ambiance' might mean, for example, and then to consider what, in reality, would make them decide whether or not the ambiance in a bar was a good or bad one. We also ask them to think about how they would record their judgements (i.e. whether they need to use yes/no boxes or a more refined scale). Finally at this stage, we ask them to consider whether or not each of their criteria is as important to them as the other. Thus we introduce the idea of weighting the criteria to balance out their relative value to the group.

We have been tempted to assist the groups at this stage with a standardised form with appropriate headings but our experience suggests that this problem is best handled by the groups themselves. This is partly because the groups then tend to feel greater ownership of their criteria and methods of operationalisation, and also because we have found that the diversity between the groups provides a rich seam for learning. However, we do provide photocopying facilities so that the groups can reproduce their measuring instruments if they wish.

The groups, of course, develop a group subjective view about appropriate criteria and their relative importance, and we have been careful not to remove this subjectivity by asking groups to play a role (such as agents of a consumer magazine) lest they then move towards criteria which others might use to assess quality. The views of the participants themselves are all-important.

STAGE 3: The Quality Assessment

The next phase of the simulation is more structured. We have previously identified (prior to the simulation) a number of local bars and pubs that are within easy walking distance of the facility being used for the programme. We prefer to have visited the bars involved, so that we can ensure that a wide range of environments are surveyed, and also so that we are able to understand, or at least to challenge, some of the judgements made by the groups.

It is obviously important that the evident 'fun' element of the simulation does not take over from the 'learning' element. We therefore stress that only two bars should be surveyed using the measuring instruments developed by the group, but groups may like to visit bars being surveyed by other groups to help them understand the judgements being made. Whilst on this 'task orientated' theme, we also make it clear that this exercise is not just for drinkers, pointing out that how pubs cater for those who prefer non-alcoholic drinks or coffee must be an important element when considering the overall quality of services. Without making too much fuss about it we ensure that each group has a mixture of regular and infrequent bar users. Even those who rarely visit bars are usually enthusiastic in their approach to the exercise.

We have found that the best way of ensuring that people do not go astray (or at least do not get lost!), is to provide an enlarged section of the local map with the chosen bars and pubs marked clearly on them. Two bars are allocated to each group and then, armed with their maps and their measuring instruments, each group sets off on the survey. We tend to pay fleeting visits to a number of the closest bars, to maintain contact with the groups and make them feel that it is 'work', and also to provide us with a few anecdotes to help us get into the review later on. For example, we have seen a seasoned pub-user, who during the preparation stage claimed that the only thing he cared about in bars was the quality of the draught beer, walk into one of the premises he was surveying and, without casting an eye towards the beer pumps, go straight to a shelf and run his finger along it to check for dust! In the feedback sessions we were able to use this to demonstrate that in structuring one's assessment, awareness of other factors becomes heightened.

STAGE 4: Reporting Back

Following the survey we allow the groups some time to put together the reports they are going to make to the whole group. At the report stage the atmosphere is usually relaxed and humourous since every group will have anecdotes about bar staff, the clientele or the bar's facilities with which to amuse the others. However if handled sensitively, a great deal of serious work can be done.

STAGE 5: Evaluation of Criteria

The reports over, we then ask the groups to evaluate the criteria and their operationalisation, the quality of their measuring instruments, and their performance as surveyors. We have found that the groups tend to be harsh critics of their performance and are able to think of many ways in which they could improve their quality assessment system. We also encourage the groups to consider how they could convert this first tentative exercise into something that could be marketed to local bar and pub owners as an indicator of their customers' satisfaction.

It is vital at this stage that the difficulties in doing the job of service quality assessment are highlighted. We consider it essential that the generalisable principles about developing a similar system for any consumer service are clearly established, thereby starting to build the bridge between the bars and the participants own world of work.

STAGE 6: Translating the Experience

The final stage of the simulation depends to some extent on the circumstances in which it is being used. It is often appropriate to get small groups to begin the process again, this time using a range of facilities where the quality of service delivery is particularly important in health or other relevant settings. For example, we might have participant groups concerned with an outpatients department, the waiting area in accident and emergency, the anti-natal clinics in a health centre or the enquiry desk in a large hospital. This exercise could either be 'for real' if we were working to establish quality assessment mechanisms in a hospital facility, or would provide participants with the 'starter kit' to begin the process once they return from an off the job programme.

CONCLUSION

Getting managers and profesionals onto the path of developing quality assessment systems is both important and, given the values of the health service in the past, difficult. We believe that using a simulation provides an excellent way of not only helping managers to better understand the process, but also to enable them to see the worth of a structured approach. We have found that this simulation in particular provides a high level of intrinsic satisfaction and is enjoyable for participants.

ACKNOWLEDGEMENT

We are grateful to Michael Nightingale of PQCS Management Consultants, 11 Palace Court, Palace Road, London SW2, for the original idea behind this simulation, and for helping us translate the idea to a gamed reality.

REFERENCES

National Health Service Management Inquiry (Griffiths Report), Department of Health and Social Security, London 1983.

Simulation gaming and the induction of new business students in a large university department

James M. Freeman
University of Manchester Institute of Science and Technology, UK

ABSTRACT: The increased use of simulation gaming in business education is widely acknowledged. A particularly important - but little studied - application of games has been in the area of student induction.

In this paper we summarise the key findings from a survey of new students before and after a simulation game exercise conducted during their first week at University. Empirical results are found to be in close agreement with those reported elsewhere in the literature - where games have been used for quite different purposes. Notwithstanding this agreement, there are some unexpected outcomes from the survey which appear to shed new light on the effectiveness of gaming.

KEYWORDS: induction, business students, management game, regression analysis, discriminant analysis.

ADDRESS: UMIST, PO Box 88, Manchester M60 1QD, England. Tel: (061) 236 3311.

INTRODUCTION

Many universities and colleges run brief induction programmes for new students, before the beginning of each academic year. In the UK, these 'fresher's week' programmes typically last between five and seven days and are scheduled during the week immediately preceding the formal start of term. It is expected all new students participate in the fresher's week programme and attendance rates, in practice, are often close to a hundred percent.

Fresher's week has been described as 'a shoehorn to ease students into their new community as swiftly and effectively as possible' (Kingsbury 1974). The impact of this kind of event on the social and academic integration of new students should not be under-rated: though its effects may not be felt directly, exposure to voluntary orientation programmes in the USA has been found to significantly influence the numbers of students who persist with their first year studies (Pascarelli et al., 1986).

Viewed in terms of meeting specific induction goals, the fresher's week programme operates at many different levels: it allows new students formal and informal opportunities for meeting teaching and support staff at the admitting institution. It provides a framework for interactions between fresher students and acts as a gateway to student organisations, activities and services. It gives students the chance to acquaint themselves with the administrative regulations of the institution and finalise their course registration and (where necessary) accommodation arrangements. At a more basic level, fresher's week provides new students with the breathing space needed to familiarise themselves with their new surroundings.

Seen in the longer-term context of orientation (Warren 1975), the contribution of fresher's week can be judged in relation to the four objectives: (1) maintaining the motivation of new students, or their desire to succeed at a level the same as or higher than they were inducted into the institution; (2) developing in new students a feeling of belonging together and a sense of satisfaction in being students at the institution; (3) bringing about a commitment by new students to the educational standards of the institution; and (4) developing in new students an acceptable conformity to the institution's formal and informal rules of behaviour.

Above all, the time, money and enthusiasm put into the fresher's week activities by an educational institution represent a significant investment on the part of that institution. This is not made lightly and is evidence of the belief in 'the more or less proven fact that a good start means a good year' (Kingsbury 1974).

Department of Management Sciences, UMIST and the Fresher's Business GAme

An integral part of the annual fresher's week proceedings at the Department of Management Sciences, University of Manchester Institute of Science and Technology (UMIST) is the business game event. The game is slotted into four consecutive mornings of the full week's programme, and for the 1985 intake of new undergraduate students (who account for some 20% of the total undergraduate entry to UMIST) an attitude survey was undertaken to help monitor the effectiveness of the business game in supporting the Department's induction programme.

METHOD

Design and Variables

The design of the study was longitudinal, with fresher students, being surveyed immediately before and after the business game was played. (A copy of the survey document used prior to the commencement of the game - the pre-test questionnaire - is shown in the Appendix. The corresponding post-test questionnaire circulated to students at the end of the game exercise, is identical to the pre-test form except items have been changed from present to past tense). At the beginning of fresher's week (in late September 1985), details for a total of 176 new entrants to the Department's BSc in Management Sciences course were confirmed to staff involved in game preparations. By this time, entrants had been divided up into tutor groups of between six and eight students, (each having access to a nominated member of staff who acted as personal tutor to all tutees in the group). The 24 tutor groups obtained from this arrangement were complemented by a further three groups, drawn from the 16 new students who had enrolled for a course run jointly with UMIST's Chemistry Department, the BSc in Management and Chemical Sciences (MACs) course. The resulting 27 tutor groups formed the basis of teams allocated to three parallel sessions (nine teams to each session) of a competitive supermarket management game, SUPERTRAIN.

The briefing for the business game was given on 1st October 1985 and lasted 60 minutes. Game play took place on the mornings of 2nd and 3rd October and a debriefing session held on 4th October. Students, based for the most part in the offices of their personal tutors, were involved, in all, in four decision inputs.

Of the 192 students targetted to take part in the simulation exercise, there were 26 (Group A) for whom there were no recognisable responses to either pre- or post-test questionnaires. In addition, 31 students (Group B) who completed the pre-test questionnaire did not identify themselves in the post-test survey and 28 students (Group C) who answered the post-test questionnaire could not be linked to pre-test respondents. After making allowance for partial or full non-response to the survey, 107 students (Group D) were left for whom data were available both pre- and post-test. The latter group represents a response rate of 56% and is the basis of the major part of the statistical analysis that follows.

Student participation in teh business game must, of necessity, be seen against a backdrop of many other activities taking place during the fresher's week period, including registration and, for example, social occasions such as the Department's Wine & Cheese party not to mention events organised centrally by the Institute. This creates difficulties for isolating specific effects of the game from fresher's week schedulings as a whole. It also strongly affected the choice of variables for the study.

The variables finally selected for analysis centre on the bank of nine items used in the pre-and post-test questionnaires. The bank is broadly split into items concerned directly with student induction (items one to four) and those (items five to nine) measuring attitudes towards different aspects of the game. (It was thought possible that items five and eight on Motivation and Contribution might represent vicariously the degree to which new students are oriented towards the Department).

Geography: Much of the time spent by students in the Department was a natural outcome of their participation in the game exercise. It seemed reasonable therefore to test the influence of the game on student perception of the Department's geography.

Support Staff: Many of the Department's support staff, encountered later in the students' first year, were prominently caught up in game proceedings, (for example by inputting team decisions to the computer) and it was hypothesised the game assisted in the process of introducing support staff to students.

Teaching staff: Beyond the game organiser and his assistant, a number of the Department's teaching staff helped in the administration of the game (for example by graphing key statistics in order that students could judge their team's performance relative to that of other teams). A source of further interaction between students and teaching staff stemmed from the siting of students in their tutor's offices. The inevitable visits by staff to collect documents, leave papers behind etc. produced a variety of gratuitous exchanges. In summary, it was felt the game might have facilitated early relationships between teaching staff and students.

Other new students: By their active involvement in team decision-making, it was considered participants would develop a strong kinship with some or all of their fellow team members as well as members of opposing sides.

The motivating qualities of business games have long been appreciated (Orbach 1979). The link between Motivation, Interest in the area covered by the game, and prior knowledge of that area appears less well understood, however. Also the connection between motivation and the Contribution by students to team decision-making or their willingness for Risk-taking. It was thought these items might provide possible insights into the effectiveness of gaming as an aid to induction - as well as some measure of the efficacy of the game package adopted.

Other variables considered in the analysis against each student were: (1) the game Session (coded 1 - 3) and Team (coded 1 - 9); (2) the Rank of the team's performance across all four decision rounds. (This was based on the team's average 'profit' in the game, by round - though profit was only one criterion of team outcome, it was seized on by students as being the prime indicator.); (3) the number of rounds a team was profitable; (4) the Rank of the team's performance in the last (fourth) round of the game; (5) whether the team was composed of X type students (having Advanced level Mathematics qualifications) or Y type students - whose Mathematics qualifications were only at Ordinary level. (It was thought this variable could be important given the essentially quantitative character of the SUPERTRAIN game. Note that MACs students were treated like X students and coded 1 where Y students were coded 0); and (6) whether students had taken part in the Department's 'Schools Visits' programme held for applicants the previous March. (It is not normal Departmental practice to invite University applicants for interview). The Schools Visits programme offers the prime opportunity for would-be Management Science undergraduates to formally see round the Department before entry. Students who attended the Schools Visit event were coded 1, and those who had not, 0).

Of the resultant 25 variables considered in the analysis, those corresponding to responses to the nine items on the post-test questionnaire were treated as endogenous (i.e. determined by other variables within a causal model) and the rest exogenous (i.e. determined from outside the model).

Statistical Analysis

Assuming questionnaire responses to be binomially distributed and relying on the large sample normal approximation to the binomial distribution (Patel et al 1976), the data have been analysed from a number of different standpoints, but primarily using multiple regression. Wherever statistical significance is referred to, it is assumed to be at the five percent level - unless other stated.

Results

Table 1 displays the mean responses to the survey - also their standard errors - for the three groups of respondents, B, C and D.

T-tests conducted between Groups B and D respondents pre-test, reveal no significant differences between the groups by item except for motivation, where Group B students can be shown to be significantly less motivated than their Group D peers.

TABLE 1 Mean Responses to the Pre-Game and Post-Game Questionnaires (Standard Errors in Brackets)

	Group B respondents	Group D respondents		Group C respondents
Introduction to the Department	pre-Game	pre-Game	post-Game	post-Game
1 Geography	2.387 (.221)	2.224 (.098)	2.850 (.089)	3.000 (.171)
2 Support staff	2.226 (.221)	2.028 (.107)	2.514 (.095)	2.679 (.200)
3 Teaching staff	2.968 (.205)	2.822 (.111)	3.178 (.094)	3.500 (.182)
4 Other new students	3.226 (.226)	3.402 (.100)	3.907 (.089)	4.071 (.145)
The Business Game				
5 Motivation	3.355 (.183)	3.701 (.100)	3.645 (.090)	3.643 (.128)
6 Interest	2.484 (.222)	2.467 (.098)	2.364 (.099)	2.286 (.177)
7 Knowledge	1.742 (.160)	1.748 (.095)	2.056 (.090)	1.857 (.160)
8 Contribution	3.968 (.118)	3.935 (.078)	3.720 (.090)	3.893 (.195)
9 Risk-taking	4.258 (.139)	4.084 (.087)	3.748 (.085)	4.000 (.206)

An analysis of Group D students, based on paired tests of pre- and post-test responses, shows significant differences against all items except motivation and interest. In the case of the induction items (one to four in the table) this suggests the business game experience had a definite and positive effect on the introduction of new students to the Department. Results for game items (five to nine) are slightly less encouraging. Evidently, participation in the game caused students (by their own perception) to increase their knowledge of supermarket management but this corresponds with an overall reduction in contribution, also risk-taking. For the latter items, an explanation may be that the many interruptions during the game sessions caused by students being called away to complete their course registration meant that they were unable to perform according to their original pre-game intentions.

A comparison of post-game responses between Groups C and D indicates no statistical significance.

All subsequent analysis is centred on Group D respondents.

TABLE 2 Intercorrelations among Variables

		PRE-GAME									POST-GAME															
		Geography	Support Staff	Teaching Staff	Other Students	Motivation	Interest	Knowledge	Contribution	Risk-taking	Geography	Support Staff	Teaching Staff	Other Students	Motivation	Interest	Knowledge	Contribution	Risk-taking	Game	Team	Rank (overall)	No. of Games profitable	X, Y, MACS	Rank (last round)	Schools Visit
PRE-GAME	Geography	1	.56*	.40*	.24*	.05	.11	.23*	.04	.08	.44*	.47*	.18*	.11	-.04	.02	.21*	-.11	.02	-.15	.03	-.08	-.01	-.02	-.08	-.18*
	Support Staff		1	.62*	.34*	.05	.26*	.21*	.16*	.17*	.27*	.60*	.32*	.35*	.01	.13	.08	-.01	.02	-.18*	.05	.10	-.03	-.09	.13	-.25*
	Teaching Staff			1	.59*	.27*	.20*	.12	.27*	.11	.31*	.47*	.43*	.32*	.13	.09	.12	.12	.12	-.27*	.01	.19*	-.21*	-.23*	.22*	-.18*
	Other Students				1	.15	.26*	.16*	.28*	.21*	.19*	.39*	.36*	.54*	.18*	.16*	.12	.15	.08	-.15	.07	-.02	.08	-.03	.06	-.14
	Motivation					1	.26*	.22*	.57*	.31*	.11	-.11	.18*	.28*	.52*	.11	.08	.17*	.22*	-.17*	-.05	.18*	-.17*	-.01	.14*	-.08
	Interest						1	.33*	.21*	.26*	.05	.16*	-.01	.22*	.21*	.72*	.22*	.01	.15	-.11	.03	.08	-.09	-.11	.06	-.23*
	Knowledge							1	.19*	.21*	.00	.06	-.06	.11	.04	.22*	.58*	-.02	.02	-.27*	-.10	-.03	-.06	-.15	-.07	-.14
	Contribution								1	.46*	.14	-.05	.15	.26*	.46*	.16*	.04	.33*	.19*	-.17*	.09	.12	-.04	.05	.17*	-.16*
	Risk-taking									1	.15	.00	-.07	.22*	.24*	.17*	.11	.04	.31*	-.14	.06	.01	.12	.23*	.12	-.09
POST-GAME	Geography										1	.43*	.34*	.23*	.12	.05	.04	-.05	.10	-.19*	.22*	-.01	-.09	-.16*	.03	-.12
	Support Staff											1	.42*	.24*	-.04	.10	.06	.14	.12	-.06	.11	.03	-.02	-.10	.02	-.13
	Teaching Staff												1	.52*	.21*	-.01	.05	.13	.15	-.08	.01	.12	-.11	-.19*	.19*	-.13
	Other Students													1	.36*	.24*	.07	.12	.16*	-.16*	-.08	.01	.12	.07	.08	-.17*
	Motivation														1	.33*	.03	.33*	.22*	-.26*	-.11	.04	.02	.05	.06	.06
	Interest															1	.37*	.23*	.21*	.20*	.02	.08	.05	.04	-.11	-.10
	Knowledge																1	.09	.12	-.28*	-.27*	-.15	.04	.05	-.22*	.00
	Contribution																	1	.32*	-.06	.00	.07	.05	.20*	.09	.04
	Risk taking																		1	-.17*	.00	.19*	-.11	-.01	.17*	.00
	Game																			1	.28*	.02	.07	.24*	.06	-.05
	Team																				1	.20*	-.09	-.07	.33*	.00
	Rank (overall)																					1	-.56*	.00	.85*	-.15
	No. of Games profitable																						1	.50*	-.33*	.08
	X, Y, MACS																							1	-.01	.02
	Rank (last round)																								1	-.16*
	Schools Visit																									1

* Significant at the 5% level

Table 2 shows the Pearson product-moment correlations between all variables in the study. Significant correlations are highlighted. The correlations here form the basis of the multiple regression analyses summarised in Table 3. In this table, the dependent variables correspond with the responses to the post-game questionnaire items. The estimated regression coefficients are for a standardised regression analysis, in which all variables are first corrected for their mean and standard deviation. Where the coefficients in the equivalent unstandardised analysis have been found to be significant, this has been shown against coefficients in the standardised models. All regression equations are highly significant overall and the R square values at the foot of the table measure the proportion of the variance of each dependent variable explained by the regression relationship estimated for it. It is important to recognise that not all variables, which from Table 2 are significantly correlated with a given dependent variable, necessarily feature in the regression model for that dependent variable. (This is a feature of the kind of stepwise regression routine used for the analysis).

TABLE 3 Estimated Regression Equations

		POST-GAME								
		Geography	Support Staff	Teaching Staff	Other Students	Motivation	Interest	Knowledge	Contribution	Risk-taking
PRE-GAME	Geography	.375*	.199*	-.085				.111		
	Support Staff	-.262*	.314*	-.096	.162					
	Teaching Staff	.063	.087	.308*	-.206					
	Other Students	-.024	.053	-.103	.401*	-.044	-.103			
	Motivation		-.225*	.009	.120	.384*			-.147	-.010
	Interest		.105		-.040		.701*	-.269*		
	Knowledge					-.133	-.160	.526*		
	Contribution				-.087		-.097		.260*	-.175
	Risk-taking				.113	.023	.001			.312*
POST-GAME	Geography			.111	.042					
	Support Staff	.260*		.264*	-.100					
	Teaching Staff	.125	.292*		.419*	.009				
	Other Students	.138	-.060	.419*		.170	.081			.033
	Motivation			.015	.138		.155*		.205	.005
	Interest				.152	.331*		.431*	.086	.081
	Knowledge						.279*			
	Contribution					.127	.183*			.304*
	Risk-taking				-.004	-.020	-.017		.235*	
	Game	-.206*			-.036	-.116	-.054	-.018		-.123
	Team	.245*						-.199*		
	Rank (overall)									.176
	R^2	.394	.491	.425	.515	.460	.656	.486	.222	.258

Taking an overview of these last results, it is clear: (a) All post-game questionnaire responses depend directly and significantly on the corresponding pre-game responses; (b) Responses to induction items on the post-game questionnaire depend significantly on other induction variables, and little on game variables. (The situation is reversed for post-game responses to game items); and (c) The manner in which the business game was structured - in terms of the breakdown into game sessions and teams - had a significant effect on some post-game responses.

Additional Analysis

Some of the survey findings recorded in the last section are quite different from those anticipated. The significant fall-off in contribution is a particular disappointment but one that we have followed up with a more detailed analysis. We have also examined more closely the nominal fall in motivation as well as the apparent link between the structuring of the game and post-game outcomes.

(i) Changes in Motivation and Contribution

Of the 107 Group D respondents, 19 were more motivated after playing the business game; 29 were less so. Comparing the former students (coded 1) with the 88 students whose motivation had not increased (coded 0), a discriminant analysis (Cooley & Lohnes 1971) was performed to see if a linear combination of all variables studied - excluding those on pre- and post-game motivation - could be found which, would reliably predict which students were 'Type 1' and which 'Type 0'. A significant discriminant function was found, details for which are shown in Table 4. Parallel results, also recorded in the table, were obtained for target ('Type 1') subgroups, comprising in turn, students whose (a) motivation had dropped, (b) contribution had increased, and (c) contribution had dropped over the game period. (In each case, 'Type 0' students were taken to be those in the Group D sample who were not 'Type 1').

TABLE 4 Standardised Discriminant Functions

	Increased Motivation	Decreased Motivation	Increased Contribution	Decreased Contribution
Pre-Game Geography			-.467 (3)	
Pre Game Motivation			.700 (2)	
Pre-Game Interest	.720 (4)	-.969 (2)		
Pre-Game Knowledge	.543 (2)			
Post-Game Geography				.813 (2)
Post-Game Teaching Staff				-.430 (4)
Post-game Interest	-1.040 (3)	1.567 (1)		
Post-game Knowledge		-.523 (3)		
Post-game Risk-taking			.762 (1)	-.484 (3)
Rank (averaged over all rounds)	.477 (1)			
X, Y				-.813 (1)
Percentage of students correctly classified	72	76	67	67

(Figures in brackets indicate the order in which variables were admitted to the relevant discriminant function.)

Students predicted in an analysis as belonging to the 'Type 1' category are invariably linked with the most negative discriminant scores. Consequently, students judged as candidates for increased motivation can be seen as having low pre-game interest and knowledge and high post-game interest. They also belong to teams ranked relatively highly in the business game.

At odds with this, students chosen as most likely experiencing decreased motivation, demonstrate a level of post-game interest well below that for the sample as a whole.

Students classed as showing an increased contribution score in the game tend to have post-game ratings that are low for risk-taking but high for motivation and geography.

Those predicted will achieve a decreased contribution score, on the other hand, are 'X' type students with a post-game ratings that are low for geography but high for risk-taking and teaching staff.

(ii) Student Responses and Game Structuring

In fact, the structuring of the game had an effect on both pre- and post-game responses. (From Table 2, the Game Session variable is significantly correlated with attitudes to five of the pre-game questionnaire items).

It is believed the few major correlations involving the Team variable have their origin in the drawing of teams exclusively from the X or Y student segments.

Restricting attention therefore to the impact of aggregating students by Game Session, we arrive at the breakdown given in Table 5.

TABLE 5

		Game Session 1	2	3
a) Means				
Motivation	Pre-Game	3.92	3.74	3.53
	Post-Game	3.92	3.77	3.32
Interest	Pre-Game	2.63	2.55	2.29
	Post-Game	2.75	2.31	2.20
Knowledge	Pre-Game	2.08	1.86	1.44
	Post-Game	2.46	2.07	1.80
Rank (last round)		3.58	4.81	5.15
b) Percentages				
X students		62	19	63
Schools Visit		25	26	46

The table shows that the reduction in overall post-game motivation (Table 1) is located entirely in Session 3 of the game. In part, this may be explained by the correspondingly low level of post-game interest, with the latter possibly being traced to the comparatively high proportion of students attending Schools Visits (who because of their relative familiarity with the Department were perhaps in greater danger of becoming disaffected by the proceedings). However the average team rankings in the last round of the game suggest another possible cause, not mentioned until now: in the last round of the game, session 3 participants played conservatively as if the last round was still some way off, while their colleagues in other sessions 'went for bust' in an effort to finish well. Arguably, this made the difference. All students in sessions 1 and 2 belonged to teams that finally 'crashed' in the game. Their obvious amusement with the results for these sessions compared with the conspicuously restrained reaction by session 3 students to their own last round output.

CONCLUSION

By most of its own criteria, the study vindicates the use of simulation gaming for student induction. However, there are reservations from the work about the types of game that are most appropriate to orientation. A drawback with competitive games, as we have seen with the SUPERTRAIN sessions, is that they may encourage lower scoring students to lose interest and become demotivated - though this was not found to be a significant problem in the survey, here. At the same time, quantitative games can lead to more numerate newcomers becoming uncontributive.

A factor that cannot easily be dismissed is that of student entertainment. Irrespective of the package chosen, the game experience may be the more productive for finishing on an enjoyable note.

REFERENCES

Cooley, W. W. and Lohnes, P. R. 1971. Multivariate Data Analysis, Wiley.
Freeman, J. M. 1984. CAT for Supermarket Management, Training Officer, Vol. 20, No. 8, 229-233.
Kingsbury, R. 1974. The Realities of University Life, University Tutorial Press.

Orbach, E. 1979. Simulation games and motivation for learning: a theoretical framework, Simulation and Games, Vol. 10, No. 1, 3-40.
Pascarella, E. T., Terenzi, P. T. and Wolfle, L. M. 1986. Orientation to College and Freshman Year Persistance/Withdrawal Decisions, Journal of Higher Education, Vol. 57, No. 2, 155-175.
Patel, J. K., Kapadia, C. H. and Owen, D. B. 1976. Handbook of Statistical Distributions. Marcel Dekker Inc.
Warren, M. R. 1975. Training for results, Addison-Wesley.

SIMULATION REFERENCE

SUPERTRAIN. In Freeman, J. M. 1984.

APPENDIX

UMIST BUSINESS GAME QUESTIONNAIRE - PRE-TEST

Introduction to Management Sciences Department

Based on your experience to date, as a new UMIST student, how do you rate your present knowledge of Management Sciences Department according to the list given below. (Please circle the number that best summarises your opinion).

	Very Good				Very Bad
1. Geography of the Department (e.g. the whereabouts of toilets, library, computer room etc.)	5	4	3	2	1
2. Support staff in the Department (e.g. part-time teaching assistants, programmers, etc.)	5	4	3	2	1
3. Teaching staff in the Department (e.g. Game lecturer, Game administrator)	5	4	3	2	1
4. Other new students	5	4	3	2	1

The UMIST Business Game

Based on the briefing for the Business Game, indicate your expectations for the exercise, using the list below.

	Very Much				Very Little
5. I am motivated to participate in the Business Game	5	4	3	2	1
6. I am interested in supermarket management	5	4	3	2	1
7. Supermarket management is something I know about	5	4	3	2	1
8. I will contribute to team decision-making	5	4	3	2	1
9. For my part in the Game, I am prepared to take risks	5	4	3	2	1

If you have any further information you regard as relevant to the Business Game exercise, please write in the space below:

Name Game(Red/Blue/Green) Team(1-9)

Desirable characteristics and attributes of a business simulation

Richard Teach
Georgia Tech, USA

ABSTRACT: This paper reports the results of a three hour workshop that reviewed the basic premises of a computer generated business simulation and the efforts of that workshop to design a new hierarchical, multi-functional simulation.

The workshop participants reviewed the evaluation process for the players of simulations and for the performance of the teams. Great differences in opinion existed between Americans and Europeans on the performance evaluations and criteria.

In the simulation design phase, a game that had two levels of hierarchy, Top Management at one level and Marketing and Manufacturing at a second level, was developed. The designed simulation will require extensive negotiations between the two operational divisions, Marketing and Manufacturing and between each division and Top Management. Top Management will control the operating divisions by the setting of budgets and goals. Extensive forecasting and planning will be required of all levels.

KEYWORDS: hierarchy, negotiations, multi-functional, evaluation, simulation, design, budgeting, control.

ADDRESS: College of Management, Georgia Tech., Atlanta, Georgia, 30332-0520, USA.

INTRODUCTION

This paper reports the results of a three hour workshop. The participants of this session were given the task of considering the design of a new business simulation without constraints. That is, if one were to start from scratch, what characteristics and attributes would a 'new' computer based, business simulation have. The participants reviewed the basic premises evaluating the players of a business simulation and discussed what would be the best design attributes and the most appropriate characteristics for a new multi-functional, total enterprise simulation. The particular advantages and limitations of the use of microcomputers were considered. For simplicity, this paper will describe the newly conceived simulation as one which produces consumer goods, but in reality an industrial product could just as easily have been used.

PERFORMANCE EVALUATION CONSIDERATIONS

In the early part of the workshop, it became evident that divergent views were held by those at the workshop when it came to evaluating the participants of a business simulation and/or the results. The evaluation methods differed substantially between the Americans and the Europeans.

The Americans seemed to want to evaluate simulation teams based on 'Bottom Line' performance at the end of the simulation and determine a 'winner'. Measures of profitability such as earnings per share, market share, and return on investment were the choice of the Americans for evaluating business simulation teams. They seemed to think that an optimal solution is feasible for every decision and the major issue in participating in a business game is one of finding this optimum. This search for optima increases a team's dependence on computers to do the analysis and to find the supposed optimum decision rules. While difficult, optimum decisions can be determined but it requires a simultaneous decision process; a rather unrealistic assumption. This emphasis makes the decision process more mechanistic and less interactive among the members of each team.

Europeans, on the other hand, tended to be much less concerned with the determination of winners and seem to be more interested in the 'experience' obtained from playing the game. Europeans wanted to enhance the interaction of the team's members. They were concerned about the intrusion of computers and its affect on the team members' interactions and the quality of the decision making process. Europeans tended to see the decision making process more as a political process and less as one involved in finding the one best solution. They were interested in providing the teams with the experiences of joint decision making where decisions are frequently made in a serial fashion where there are few optimal decisions. Either consensus or power plays within the team was seen as determining the final decisions in each round of play. After a simulation or game has run its course, the Europeans spend a great deal of time debriefing the teams and pointing out, or illustrating and explaining, the decision making interactions.

Neither the Americans nor the Europeans tended to differentiate the performance evaluation within a team. In academic situations, the Americans wanted to give each member the same grade. The Europeans were quite satisfied with leaving the gaming process upgraded, each participant getting out of the experience whatever he or she carried with them.

The workshop participants agreed that some compromise between these two views could be made. While the computer could be used as an aid to the decision making process, it was by no means to be considered as the decision maker in itself. Modeling optimality was seen as impossible due to the stochastic nature of the competitors' decisions and the sequential decision making process that occurs in firms. The analysis of current operations and forecasts should make use of a microcomputer and the output of any game should be in a form to encourage, but not require, additional analysis. The reporting capabilities and comparative graphics of microcomputers were considered to be great aids in the decision making process, but the decisions had to be made as the result of some joint process involving some set or subset of each team's members.

THE STRUCTURE OF DECISION MAKING

One of the deficiencies of most of the existing business simulations was considered to be that they are structured in a way that hides or disguises the hierarchical and sequential nature of corporate decision making. Current business simulations provide the players with a set of outputs which are made up of Income Statements, Balance Sheets, a variety of other reports and the results of marketing research requests at the end (or beginning) of each period. All of this information is, miraculously, both accurate and up-to-date. Each team then faces the problem of converting the data and information in the reports to the knowledge needed to make the next round of decisions. In reality, data first arrives as estimates of what may have occurred or as forecasts of what is expected to happen in the future. As time passes the data are periodically updated and months after the fact, accurate assessments of what transpired are made available. The accounting process takes time.

Another deficiency of existing simulations is the structure of the management teams. Since total enterprise games want to create a Top Management view, all the players take on a roll as equal holders of the keys to the executive washroom. It is frequently assumed that all requests from this team of top managers are carried out in the quickest and best way possible, all in agreement with the wishes of Top Management. The truth of the matter is that decisions are carried out in a sequential order and not always as top management directs. Latitude in decision making is usually given to subordinates as long as general policies are obeyed. Budgets are established based on forecasts sent up to Top Management by division chiefs and corporate goals are set by the executive level. Marketing determines how to best utilize its budget allocation and how to follow the guidelines.

Manufacturing often determines the allocation of capital expenditures up to predetermined authorization levels and the appropriate split between research and development and quality control. Each executive understands that he or she is judged, based on performance criterion that may differ from the criterion of executives in other functions. Sub-optimization in these functions occurs because neither positions in marketing nor manufacturing are evaluated on the overall profitability of the firm. Instead, individual divisions or departments are frequently evaluated on short-term contributions to profits or as profit centers in themselves. Thus, the seeds of conflict are sown because in many ways these evaluation criterion result in a zero sum game; marketing can gain at the expense of manufacturing and vice versa. Because the process of the evaluation of executive performance, the evaluation of different departments, divisions and functions and the possibility of contribution tradeoffs; directions from Top Management are often not carried out in the manner expected. Another contributor to this alteration of directives is the fact that the forecasts on which Top Management must depend, prove to be in error and decisions must be made and plans revised at the division or department level.

THE TEAM COMPONENTS

It was deemed important that any new game has both marketing and manufacturing components, operating semi-autonomously and coordinated by the Top Management component in a hierarchical fashion. Top Management must receive forecasts and requests from both Marketing and Manufacturing and in turn produce budgets, quotas and goals for each Division. Marketing orders certain volumes of products with specific characteristics from Manufacturing, but Manufacturing has the right to question these orders when they drive up the unit costs. Also Marketing and Manufacturing need to negotiate a transfer price for the goods ordered by Marketing in order that both Manufacturing and Marketing are able to operate on a profit center basis. Whenever Marketing and Manufacturing have conflicts, negotiations to resolve these conflicts between them must take place.

Top Management Components

The management of actual firms is done in a hierarchical fashion. Those at the top set policy and budget limitations and the subordinates in lower positions attempt to carry out the policies under the imposed constraints. Generally, business simulations do not have this feature but they need to have a similar structure in order to demonstrate how the constraints, under which decisions must be made, often affect the decision making process itself. Thus, the decisions made by Top Management in a new game must include the establishment of marketing budgets, capital expenditure budgets, hiring and/or firing quotas, and the setting of sales goals. These, as well as corporate and product image objectives, need to be established by one or more team members designated as 'Top Management'. The establishment of credit lines, obtaining alternative methods of raising and investing funds, and the setting of dividend levels, all fall to the prerogative of Top Management.

Because the overall profitability and strategic direction of the firm is the responsibility of Top Management, The Top Management of each team must have the responsibility for the allocation of capital and operations budgets to each of the two major (Marketing and Manufacturing) divisions. While alterations to these budgets are possible, subordinate executives must obtain authorization prior to any deviation from the established budgets through negotiations with those playing the role of Top Management.

Manufacturing Components

The manufacturing component in this new simulation needs to include at least three identifiably different manufacturing processes, each with a different throughput and material consumption rate. Examples of these three processes might include a lathe for cutting, a milling machine for shaping and an assembly robot. (An additional possibility could be a drill press). The manufacturing process needs to consume a minimum of three raw materials, the mix of which would determine the product attributes. Examples here might include wood, plastic, and steel and possibly even stainless steel and/or aluminium. A different raw material mix would produce a different product with different attributes. The initial starting position for the firms should include two products, one of which should be identical across all firms and, the other, a product unique to each firm. The multi-machine, multi-raw material, multi-product situation reinforces the line balancing problems inherent in manufacturing.

The manufacturing component in the simulation needs to incorporate the ability to track each specific machine with its own efficiency, output and maintenance records. Newer machines should be more efficient and faster, but should also carry a higher initial or capital cost. Machines need to be able to continue to produce after their depreciation has resulted in a zero book value. Random number generators will be able to determine maintenance costs on an individual machine basis and even produce '<u>lemons</u>' or machines that never work right. The simulation needs to allow for the repair and/or the disposal of equipment that is inoperable, malfunctioning, or just no longer needed. (It may even be possible to incorporate a market for used equipment among the players and have firms be able to sell off excess equipment to the highest bidder or junked for a salvage value).

The decisions in the manufacturing component is to include routine purchase of raw materials, the purchase of capital equipment, up to budget limitations, and the leasing of warehouse facilities for all the inventory. Quality control expenditures need to affect the rejection rate of the products and increase the quality of the product as perceived by the buyer. Engineering expenditures need to increase the slope of the learning curve. Research and development (R&D) expenditures will be needed prior to bringing out a new product and R&D should enhance the quality of all the firm's products in the eyes of the buyers. In addition to these decisions, forecasts will need to be provided to Top Management regarding production capacities, utilization rates, the cost per unit manufactured, employment levels and anticipated plant expansions.

Manufacturing is to be a profit center. The direct cost of manufacturing needs to be charged to the department and its income should be the result of orders from Marketing. As mentioned earlier, the transfer price to Marketing needs to be negotiated between Marketing and Manufacturing and signed off prior to any additional decision making.

Marketing Components

Marketing is to include the traditional sales function as well as the four P's; product, pricing, promotion and physical distribution. The products in the simulation are to have identifiable and alterable characteristics that will affect their cost as well as their marketplace acceptance. In pricing the product, P-Q demand curves will have a negative slope with price elasticities that can be altered by the administrators of the simulation. The response curves for promotion will have a portion that shows declining returns for expenditures and the promotion expenditures are to directly affect the price elasticities. Thus, increases in marketing expenditures at the firm level are to result in reduced price elasticities for that firm. The elasticities of both price and promotion are to be greater for the firm than for the industry as a whole. Aspects of the physical distribution of the products, such as the carrying of inventory at the distribution points in order that the products are close to the buyers, is to be incorporated. Marketing is to bear the carrying costs of this associated inventory. The selling function is to be accomplished by a sales force whose number and direction is to be controlled by Marketing.

In the same fashion that Manufacturing is to obtain reports on the performance of each machine on the factory floor, Marketing is to receive a report on performance of each individual in the sales force. Decision to fire, train or redirect sales effort can then be made at the individual salesperson level as well as the decision to hire additional salespersons. Each action should have its associated costs.

In addition to the four P's and sales, Marketing must also determine the best market segments in which the firm is to compete. Segments of buyers will be identified by sets of demographics (or SIC's for industrial products) but there is not to be a one-to-one mapping between demographics and segments. Each buyer segment will have its own demand function. Each segment will have a 'best' product but will substitute other products for the best one, if necessary. Both the promotion and the sales force will need to be directed to the appropriate segments. Marketing is to request the production of new products, determine the product attributes and decide what, if any, new market segments are to be attacked.

Since the only major source of revenue for the firm is through the sale of products, Marketing will be required to provide Top Management with extensive sales forecasts for both the short-term (the next period) and the long-term (the next year) by product. This forecast should be in terms of both units and market share. It is from this knowledge that Marketing bases its requests for units of product to be manufactured.

Marketing is to be a profit center and, as described under manufacturing components, must negotiate with Manufacturing for a satisfactory transfer price on which it bases its profit contribution.

THE TIMING OF DECISION

The number of the decisions needed to operate the simulation being designed is large, but their timing is the most important or critical feature. In order to make this simulation perform in a realistic manner it requires that Top Management receives requests and forecasts, some as much as a year in advance. It is based on these forecasts and requests that Top Management establishes budgets and sets goals and objectives.

In general, the divisions request permission from Top Management to alter the direction of the company. For instance, capital equipment purchases, the development of new products and the removal of old products all will need approval prior to the time that they will be carried out. The amount of lead time needed for the approval by Top Management will be a function of the impact of the decision on corporate policy.

THE REPORTS

Financial and managerial reports will be produced for two periods. The report for the period of time representing the quarter just completed (T-1) will only be an early or preliminary estimate. In fact, it will be the actual data but will have administrator controlled random numbers added to each value. The only accurate information available in the report for period T-1 will be the marketing research reports and any emergency reports regarding stockouts, inventory shortages or breakdowns. The random error terms are to be controlled in a way that will result in either biased or unbiased estimates of the true values to be reported. It is anticipated that the smallest error terms will apply manufacturing costs and the largest error terms will apply to sales. The reports will be duplicated for period T-2 but the random error terms will be removed. Thus, the information will be accurate only for periods that are one or more quarters old.

THE COMPUTER INTERFACE

All of the input and output are to be available both on paper and in electronic form. The electronic form is to be produced in a format that could be imported into popular spreadhseet, database and graphic software. If the participants desired, they should be able to use microcomputers to process the forecasts and accomplish any analysis the players wanted to do.

CONCLUSION

This workshop has designed a very complex and extensive business simulation. While the author has developed algorithms for most of the simulation, it will still require a lot of time and effort to complete the design phase and to write the computer code to produce a working model.

The customer in focus

Chris Brand
Alec Keith
Maxim Consultants, UK

ABSTRACT: Customers take many shapes and forms. They might be buyers of consumer goods, social security claimants, students, pensioners, purchasers of services etc. Understanding customer needs is vital for ANY organisation's success. To satisfy these needs, the organisation must understand WHAT customers expect, and HOW it should react in particular circumstances.

The customer in focus provides a simple framework for exploring customer servicing issues and how to resolve them. The game kit allows trainers to create their own game and develop customer-related training material specific to their own organisation. Participants then explore their own customers' needs, and how to resolve them.

KEYWORDS: Customer sensitivity, customer servicing, management.

ADDRESS: Maxim Consultants Ltd, 6 Marlborough Place, Brighton BN1 1UB, UK
Tel: (0273) 672920.

INTRODUCTION

The customer in focus was originally designed for a major corporate client. The game was used to explore their specific customer servicing issues, and possible channels of response. The game has since been turned into a kit for trainers to create their own game. This article provides an introduction to the customer in focus. It describes how the game kit is tailored to a specific organisation, how to play the game, key issues raised, and potential applications.

THE GAME

The customer in focus is a simple, but powerful tool for exploring customer servicing issues. The game is ideally played in groups of four to six participants. To gain maximum benefit from the game, groups should consist of participants from a variety of different backgrounds and with differing levels of experience. This ensures a range of viewpoints and facilitates cross-fertilization of ideas.

Using packs of cards, teams generate a series of situations requiring action by their organisation. Each situation includes: a type of customer, an issue raised by that customer, and a point of contact within the organisation. Teams decide: key actions required to resolve the issue, and methods of strengthening the organisation's relationship with the client. The game is normally played over 60-90 minutes with an additional 45 minutes feedback.

DESIGNING YOUR OWN GAME

The customer in focus is a self-contained kit, from which users design their own game. The kit contains participants' instructions, organisers notes, decision forms and blank cards.

Before the game can be played, the kit must be adapted to raise customer servicing issues typically encountered by the participants' own organisation. This can be undertaken by the game organiser, prior to starting the game, or by teams themselves during a game design session. Tailoring the game entails recording typical customer types, issues raised and contact points within the organisation on the blank cards provided. Once this design stage is complete, teams are ready to play the customer in focus.

PLAYING THE GAME

Teams shuffle the packs of cards thoroughly and arrange them face down on a large desk. Each team then turns over the top card in each pack to reveal a customer, an issue and a contact point within their organisation. Together, these cards represent a unique customer servicing situation facing the organisation (e.g. a high street retailer enquires about quantity discount, by telephoning your sales department). Teams must:

a) Decide the three most important actions their organisation should take to ensure customer satisfaction, and

b) identify the key selling/reselling opportunity arising from the situation, e.g. selling new products/services, encouraging continued use of existing products/services etc.

Participants are encouraged to be <u>creative</u>. The aim is to identify how each situation would <u>ideally</u> be handled, not necessarily how it should currently be resolved within their organisation. In addition, any selling/reselling opportunities must be practical <u>and</u> relevant to the situation faced. Participants must avoid actions which are inappropriate, or likely to damage the organisation's image. Decision are entered on 'record sheets' for each problem and handed to game organisers. Used cards are then placed at the bottom of the pile and teams proceed to another customer servicing problem.

The customer in focus becomes a vehicle for raising specific training issues relevant to the teams' own organisation. By discussing such issues with colleagues, exploring alternative approaches and exchanging ideas, participants develop their awareness of customer needs/expectations and how their organisation should resolve them. These key learning points are reinforced during feedback.

FEEDBACK

Simple game structures such as the customer in focus provide a very powerful framework around which a trainer can develop specific skills. To derive maximum benefit from playing the game, it is vital the trainer conducts a structured feedback session. Whilst the precise nature of feedback depends on objectives for playing the game, it normally operates over three stages.

Stage 1

When the exercise is complete, either the trainer, or each team selects one situation encountered during the game. The situation chosen might be: a particularly sensitive issue, a controversial issue within their organisation, and an unusual problem, pertinent to their organisation's current practice. Each team then prepares a short presentation on: approaches to resolving the issue, alternative courses of action considered, possible selling/reselling opportunities, degrees of conflict within the group, and decision making processes.

Approximately 10-15 minutes is allowed for preparation.

Stage 2

A representative from each group in turn gives a presentation using a flip chart. Members of other groups are encouraged to comment on the approach taken and to identify alternatives.

Stage 3

After each team has given its presentation, main issues arising during the game are briefly summarised. These very much depend on the nature of the organisation, participants' background and experience etc. The following are some basic issues which often arise: the customer's importance to the organisation; sensitivity of the issue raised; how the issue was raised; the organisation's speed of reponse; channels of response; method of response; follow-up action required; likely success/failure of selling/reselling opportunities.

APPLICATIONS

The game was originally designed by Maxim Consultants Ltd, for a major international charge card company. The game was run with 60 supervisors and managers from the company's customer servicing departments at their annual operations conference. On this occasion the game was run on a competitive basis. Decisions were scored by the company's senior directors, according to creativity, feasibility and practicality of response. The game was an outstanding success, and generated several ideas later discussed at departmental meetings and integrated into company policy.

The game has since been turned into an off-the-shelf kit. Its simplicity and flexibility make it ideal for integration into customer service training programmes. Used this way, the game helps reinforce key customer servicing issues. In addition, the game can be tailored to raise specific servicing issues, according to the organiser's needs. More importantly, the game helps participants relate theory to practical situations encountered in their own work environment.

The customer in focus is currently used by retailers, banks, educational establishments, county councils, pharmaceutical companies, insurance companies and breweries, amongst others, and is available from the authors (cost £95 plus VAT).

Thwarting the solution — a memo writing simulation

Bill Robinson
University of Technology, Papua New Guinea

ABSTRACT: There is more to learning memo writing than linguistic considerations of register can account for. Memo writing might best be practised in the dynamic context of a simulation.

A memo is the result of a many forces in the work place, including situation, style of management and workload. The simulation described in this chapter attempts to model these elements within the context of a small company suffering under an authoritarian administration.

A full description of the activity is given, together with a note on its genesis and the ISACA '86 presentation.

KEYWORDS: Management training simulation, ESP (English for Specific Purposes), memo writing, in-tray out-tray exercise, business correspondence, ESL (English as a Second Language).

ADDRESS: Department of Language and Communication Studies, University of Technology, Private Mail Bag, Lae, Papua New Guinea.
Tel: 43-4999.

BACKGROUND TO THE SIMULATION

Genesis

This simulation was designed for Papua New Guinea (PNG) undergraduate students of electronics to practise memo writing: it forms part of two final year subjects 'Report Writing and Supervision' in one course and 'Verbal Communication' in another. As highly trained people in a developing country, these final year students can expect rapid promotion in their profession. They need some practice in memo writing in a non-threatening environment, to prepare them for the supervisory roles they will soon be playing in real life.

Many of the courses in 'Business English' for foreign learners consider that the main problem of business correspondence lies in the style and the choice of the correct expressions; i.e. they treat the problem at the level of register. (Readers who have received overseas business correspondence may occasionally have seen the results of this approach). In particular, both letter and memo writing assignments tend to have rubrics such as 'You have received the following letter/memo from Mr. X. Write a suitable reply, bearing in mind the points discussed earlier...'

In this type of exercise the learners have too little stake in the situation - who really cares about "Mr. X" and anyway this "Mr. X" has no way of responding to the learner's correspondence. In the current language teaching jargon this is non-communicative language teaching. On the other hand, in a simulation the players have a greater degree of personal involvement in the situation, and their actions do strike up responses in other, real, people: i.e. the teaching has become communicative. It seems to me that business correspondence is a prime candidate for treatment by simulation.

In this simulation the problem of register is dealt with only at the debriefing stage. In the simulation itself two other, perhaps more important, elements of the real world are being modelled: the pressure of distracting administrative work in the office and the power that can be wielded through the memo.

Aims

The simulation has the following aims:

(i) to give practice in the writing of memos

(ii) to give experience of the pressure of administrative work felt by middle managers (i.e. it is also an in-tray out-tray/delegation exercise)

(iii) to allow participants to feel something of the power and importance of the memo.

The simulation is therefore not limited to use by foreign language (FL) students, but would be suitable for any trainee manager. It just happens that the simulation was originally designed in PNG and so makes use of a certain amount of local colour in some of the pre-written memos. It should perhaps be pointed out that all formal education in PNG is conducted in English, so the students have had many years experience of learning in an English medium environment and their standard of communicative English is very high.

The Situation

The simulation takes place in a small (fictitious) company of electrical contractors in Port Moresby, Papua New Guinea. The company has six departments/ sections, each under a Head of Department/Section (HOD) and an Administration, represented by the Company Secretary (see Fig. 1).

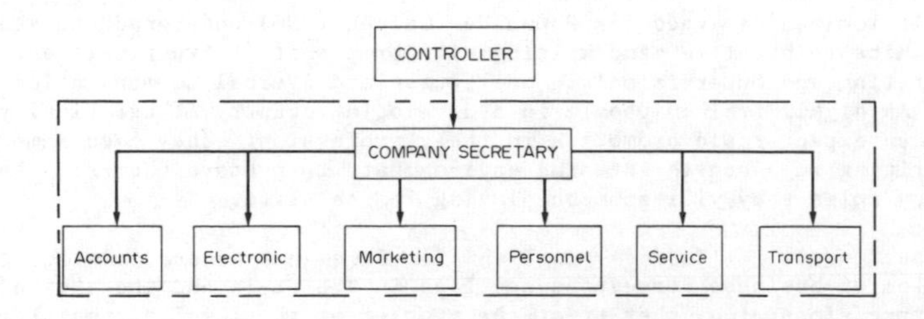

Fig. 1 Organisation of the simulation and the company structure.

The administration of this company, 'Electrical Contracts (POM) PTTY', has decided that there is too much absenteeism from the workplace, because staff are leaving the premises to buy snacks and cigarettes from local stores. The Company Secretary has therefore decided to install vending machines in each of the six departments/sections.

In order to estimate how many machines to install, and which goods to stock, the HODs are required by the Company Secretary to conduct a survey of the estimated consumption of these snack-bar articles throughout the company.

DESCRIPTION OF THE SIMULATION

The participants assume the role of the salaried staff in the six Departments/ Sections and sit at the separate Departmental desks. One other player is the Company Secretary. The only mode of communication between these Departments is through memos.

The instruction to conduct the survey is conveyed to the HODs by the initiating memo (see below, 'Materials and equipment used'). As soon as the survey is under way the Company Secretary starts to send out other (prewritten) memos on totally unrelated matters. This Secretary suffers from a degree of isolation from the rest of the company's staff and its activities. As a result this person has become very authoritarian and slightly paranoid, as is evidenced in the style and the tone of many of the subsequent memos, e.g. the Administration consistently makes a distinction between 'Salaried staff' and the 'operatives' and expects different standards of behaviour from these two groups. These Secretary's memos keep the HODs and their office staff busy and are designed to slow down the completion of the original survey by increasing the HOD's workload ('thwarting the solution'). For example, some memos give the HOD information or instructions that must be passed on to all staff in their department via notices on the Departmental noticeboard: these notices need to be drafted and posted. Some memos call for references to be written for ex-employees of the company, and others relate to various crises that have arisen and require HOD's assistance in their solution. (See Appendices for some examples of these memos).

The simulation ends when the Company Secretary informs all HODs that the suppliers of the vending machines will not be able to deliver for a further six months and calls for a meeting of HODs to discuss what should be done to prevent staff leaving the premises to buy their snacks.

Duration

This simulation has been run with non-native English speakers (i.e. FL students) in as little as one and a half hours, but for that, the initialising memo needed to have attached a pre-printed summary form, and other simplifications were necessary. A two to two-and-a-half hour gaming time is more useful. The language teaching equivalent of a debriefing would perhaps be the marking of the memos written in the simulation and the follow-up lesson. For native speakers of English a half hour debriefing would allow time for discussion of the Company Secretary's paternalistic management style and its effects on the recipients of his/her memos.

Materials and Equipment Used

A. The controller has the final MEMO which ends the simulation.

B. The Company Secretary has 3 kinds of pre-written memos:

1. the MEMO to initiate the survey (an individualised copy to each HOD).
2. those addressed to all HODs, concerning company policy.
3. those addressed to specific HODs,. concerning specific Departmental matters.

C. HODs receive the Departmental packs containing:

- a sheet of background information on the company
- a number of noticeboard blanks
- a number of memo blanks
- a file in which to file copies of memos sent
- an in-tray and an out-tray
- sheets of carbon paper
- drawing pins for the departmental notices
- a hole puncher, to enable the copy file to be kept tidy

It is very helpful to have access to immediate photocopying facilities to assist with the production of the filed copies of correspondence. However, I have only once been so privileged when using this simulation!

Number of Participants

With six departments/sections and the Company Secretary, the absolute minimum for the simulation would be seven participants: however, each department would be better served with 2 office staff, making a useful minimum of 13 participants. I have run with up to 4 staff per department, making a maximum of 25.

Special Requirements in Participants

The Company Secretary needs to read the prewritten memos and to judiciously send them to any HOD seen to be progressing toward completion of the initial survey ahead of the others (on the advice of the Controller). The Company Secretary should be capable of dreaming up other similar delaying memos.

PARTICIPATION AT ISAGA '86

17.30 on the first day of a conference on the Cote d'Azure is not guaranteed to attract many participants to an unknown gamer's offering from PNG! However, 12 participants were kind enough to attend. Most were native English speakers, but at least two were not, and came from a variety of disciplines, including foreign language teaching, management consultancy and statistics.

This was the first time I had tried this simulation with native speakers, so I was most interested to see how it would play in this context.

A post-simulation count shows that over sixty memos/notices were produced by this group, so they were certainly participating actively in the simulation.

The paranoia implicit in many of the Company Secretary's memos caused some amusement and, in the spirit of the simulation, a number of memos were passed which complained to the Company Secretary about the poor performance of the messenger, a role played by the Controller.

The participants reported that they found the simulation enjoyable to play and an interesting way of teaching memo-writing. Much of the enjoyment seemed to come from the exercise of the power of the memo: picking up the Company Secretary's pompous and authoritarian tone, participants wrote quite strong memos to each other, for example:

> "this is inconcise and unclear. Please clarify."

or back to the Company Secretary:

> "this instruction is discriminating..."
>
> "my staff are out selling, not sitting in their office, so they will not need your vending machines".

As a further distraction, and to add a little local colour to the session, I had given each Department a copy of an old PNG newspaper. These papers all contained sensational or exotic news items: "Terror Along the Highway" or "Dinner Table Carnage" (actually a story from USA!) or "World fame for Trobriands' yams", which I had thought participants might enjoy if interest had flagged in the simulation. In the debriefing one of the participants, Elizabeth Christopher, made the interesting suggestion that the newspaper might be exploited as a useful resource: players should be asked to peruse the paper and draw the attention of the relevant department to any item that might affect the company's policy, image or sales. This seems to me to be a very positive suggestion.

Participants were kind enough to request all my spare copies of the prewritten materials and agreed that the basic model of this simulation - an authoritarian administration thwarting progress towards the solution of a problem through the imposition of unrelated tasks - could be made to work in other teaching and training situations.

APPENDIX

This Appendix contains five examples of the pre-written Company Secretary's memos. Some of these memos are fairly colourful, but most of the Secretary's other memos are of the "inform staff and operatives that..." variety, or are asking for a reference for an ex-employee.

1. ARTIFACT SELLERS.
2. Precautions in the Rainy Season.
3. Managing Director's Visitors Tomorrow.
4. Injury and Compensation Claim.
5. DELAY IN DELIVERY OF VENDING MACHINES.

ELECTRICAL CONTRACTS (POM) PTTY.

Proprietor and Managing Director: J. Sikio, BSc. MA.

PO Box 99
Port Moresby.

MEMORANDUM

To: ALL HODs/SECTION HEADS FROM: COMPANY SECRETARY

ARTIFACT SELLERS

It has come to my notice that there are a large number of artifact sellers coming onto company premises during working hours to sell their handicrafts.

These people have no authorisation to enter the premises.

Please inform all your staff that any further visits by these artifact sellers are to be reported directly to you.

You will be responsible for escorting such visitors off the premises.

P. Kawage
Company Secretary

ELECTRICAL CONTRACTS (POM) PTTY.

Proprietor and Managing Director: J. Sikio, BSc. MA.

PO Box 99
Port Moresby

MEMORANDUM

TO: ALL HODS/SECTION HEADS

FROM: COMPANY SECRETARY

Precautions in the Rainy Season

The Company spent a great deal of money on the new carpet in the main entrance area to the premises. It is apparent that many people are coming straight into the building with wet shoes and dripping umbrellas, thus making the carpet very wet: there are already signs of mildew along the centre.

Please make sure your staff and operatives follow the new regulations:

1. all non-salaried staff wearing wet thongs will take these off before they walk into the building.

2. salaried staff will ensure their shoes are dried on the mat OUTSIDE the main doors before walking in.

3. NO WET UMBRELLAS are to be brought into the building. They must be left OUTSIDE the doors in the rack provided. The Company cannot accept any responsibility for lost/stolen umbrellas.

Your active co-operation in this will help the Company's image.

P. Kawage
Company Secretary

ELECTRICAL CONTRACTS (POM) PTTY.

Proprietor and Managing Director: J. Sikio, BSc. MA.

PO Box 99
Port Moresby

MEMORANDUM

TO: ALL HODs/SECTION HEADS FROM: COMPANY SECRETARY

Managing Director's Visitors Tomorrow

The Company's Managing Director, Mr. J. Sikio, will be bringing a party of visitors from the Rotary Club round the premises tomorrow.

All Departments/Sections should be prepared for this party of visitors and must ensure that they receive a good impression of the work the Company does, and of the quality of our staff and operatives. Please ensure that all staff and operatives are aware of the particular importance of punctual arrival at work, and that they do not make the hall carpet wet.

P. Kawage
Company Secretary

ELECTRICAL CONTRACTS (POM) PTTY.

Proprietor and Managing Director: J. Sikio, BSc. MA.

PO Box 99
Port Moresby

MEMORANDUM

TO: TRANSPORT SUPERVISOR FROM: COMPANY SECRETARY

Injury and Compensation Claim

I have just heard by telephone that the 15-seated Honda bus, which we need for transporting the Rotary Club visitors tomorrow, has been held up by some villagers at 10 Mile. They are holding it until we pay their compensation demand for K20,000 for the friend who was killed falling off the back of the Toyota pick-up last Monday.

You need to get hold of the community leaders and start negotiations. Draft me a letter and let me see it before it goes out.

You will also need to provide a suitable vehicle for tomorrow's visitors. I suggest you find out about hiring one - get a budget for this from the Accountant.

THIS IS URGENT.

P. Kawage
Company Secretary

ELECTRICAL CONTRACTS (POM) PTTY.

Proprietor and Managing Director: J. Sikio, BSc. MA.

PO Box 99
Port Moresby.

MEMORANDUM

TO: ALL HODs/SECTION HEADS FROM: COMPANY SECRETARY

DELAY IN DELIVERY OF VENDING MACHINES

We have just received notification from the manufacturers of the planned vending MACHINES, VENDORS LTD., that we must expect long delays in the delivery of any orders placed with them.

The Company cannot then expect early implementation of its proposal to install vending machines in every Department.

The Company has no wish to return to the old ways when operatives, and on occasion staff also, were constantly leaving their place of work to purchase snacks, cigarettes, etc. We must therefore devise an interim supply of the necessary goods, until the vending machines are installed.

Would HODs/Section Heads therefore arrange to hold URGENT DEPARTMENTAL/SECTION MEETINGS to discuss ways of making this interim supply available to our entire workforce.

The Company will be required to then hold a Meeting of HODs/Section Heads to discuss these findings so that we can implement the solution that is best in line with Company Policy.

P. Kawage
Company Secretary

Communications role playing exercise for executives

Joe Kelly
Bakr Ibrahim
Concordia University, Montreal

ABSTRACT: This simulation of an executive meeting reveals the paradigm of corporate drama that integrates the structure of roles (the actors), the process of events (clarification, evaluation, decision) and values (democratic and task). Simulations reveal that the top management meeting is more likely to be an adversary proceeding - not an open clash of wills, but still a set of opposing viewpoints trying to obtain dominance with very little concern for truth and beauty. Such a meeting will usually be conducted in a courteous, formal, almost legalistic atmosphere, but in essence is a game of getting the upper hand. If you would survive in this paradoxical environment, it is important that you learn how the people interrelate with each other, what roles they play, and what weapons they consider acceptable.

KEYWORDS: Structure, roles, communication, values, video.

ADDRESS: JK: 116 Marlin Crescent, Pointe Claire, Quebec H9S5B3, Canada.
Tel: (514) 697-7213.

The Drama of the Existential Encounter

Running a meeting is an exciting activity, but considered scientifically, it breaks down into long boring lists of 'do this' and 'do that'. When you start with the absurd person singular, the manager who has to actually run the meeting, you soon realize the need for myth, magic and meaning. Producing and directing meetings becomes an art, transforming inputs into outputs with value added. This 'two plus two equals five' aspect of meetings demands a certain savoir faire which owns much more to theatre than to science.

THE INTRODUCTION

Now the objective of this paper is to come up with simulation of a meeting which feels right to practicing managers. An increasing number of managers are realizing that a corporate meeting, when it is not a black tragedy, is often a gloriously funny comedy if seen from the right optic, posture and perspective. Behind these stories is a specific paradigm - drama. And this paradigm of

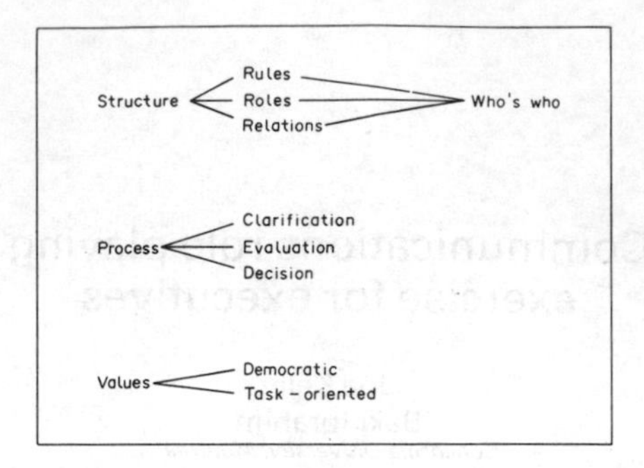

Fig. 1

drama assumes an imaginary integration of structure (the cast, dramatis personnae), the process (sequencing of events), and values (myth, magic and meaning) - see Figure. The meeting becomes the threatre of action; sometimes the theatre of the absurd and occasionally the theatre of cruelty.

This occasionally zany and crazy characteristic of meetings has driven sophisticated executives to recognize the 'I'm insane, you're insane' dimension of interpersonal life. Perhaps it is not too far fetched to argue that every organization should have printed over the doorway, 'To All Ye Who Enter These Portals - You don't have to be crazy to work here, but it certainly helps'.

THE EXERCISE

The group is exposed to the leaderless group problem. The leaderless group (or emergent group; there are leaders) situation has been used extensively as a selection device. The method consists in asking a group of candidates to discuss a controversial topic for a period of 30 minutes or more with the object of arriving at a group consensus. No official head is appointed. Observers record the attempted leadership acts and the effectiveness of the leadership behaviour of each participant. Perhaps the leaderless group might be better described as a stress group task. 'Leaderless group' is intended to indicate that no leader is initially chosen, but it might imply that none ever arises. In fact, the group structures itself in a particular way; roles emerge and people move to their appointed stations.

Description of the Method

A group of executives is divided into two groups - the participants and the observers. The participants are placed around a table in the centre of the room and are given the problem described below.

Communications Role Playing Exercise

Objective: To give participants experience of operating in senior management group.

Exercise: A top executive training and development program for people to become presidents.

You are the top management (just below Vice President leve) of a firm in the word processing machine business employing 5,000 people. You are mainly employed in strategic planning, personnel and administrative duties. You have been instructed by the Chief Executive Officer to set up a top level executive training and development programme. You are to recommend a selection procedure, and, secondly, 'the subjects', experiences and type of career guidance to be included in the programme.

The programme will last for 24 months and is expected to produce a cadre of presidents who can manage the companies that your firm is planning to take over. The CEO is insistent on having a contingency plan for executives who 'fail' the programme.

Role for Observers

Search for roles emerging in discussion groups.

1. Task specialist (look for the person who uses phrases such as 'zeroing in on the problem', 'let's bear down on the subject', 'Let's attack the problem', and so on).
2. Human relations specialist (his style is essentially placatory - uses phrases such as 'how do we feel about ...?', 'this is kinda fun, kinda interesting' (generally tries to de-tense group).
3. Eccentric roles (look for bizarre and stochastic behaviour).
4. Scapegoat roles (who is the group picking as the fall guy? Does he welcome it?)
5. Isolates (not part of the group proper).
6. Observes (scientifically detached!).

Look for the stage of Clarification, Evaluation and Decision.

Clarification (what definitions and assumptions are made? What is the problem?).

Evaluation (how do they feel about the problem in emotional terms?).

Decisions (what did they decide to do?).

The Exercise

The exercise is instrumented. Instruments are developed to enable the group to objectively define group processes and roles. The instruments include:

a) Analysis of speaking time by participants.
b) Bales Analysis of the Pattern of Interaction which defines the process of the discussion: Clarification, Evaluation and Decision.
c) Roles are allocated - task specialist, human relations specialist, eccentric, isolate observer and so on.
d) Material is taped and made available for play-back.
e) Some sessions are video-taped so that the total effect of one's participation can be gauged.
f) Check lists are provided to measure the chairman's performance.

The actual problems which are presented to the group are capable of some degree of analysis.

Training is Needs-oriented

The participants are allowed to carry on discussing this problem for approximately half an hour and then the observers input by way of feedback of their observations. There is a remarkable ;degree of similarity between the tape recording of one group and another. The Bales' analysis of the interaction pattern is particularly revealing showing how the group moves through three phases in solving the problem. These are clarification ('What is it'), evaluation ('How do we feel about it?'). Most groups find it very difficult to balance the 'work' against the 'non-work' (the de-tensing of the group). In general terms at this stage of the development the participants find it very difficult to recognize that the group has a hidden agenda.

It is usually possible for the observers to identify the task specialist, the human relations person, the various isolates, the scapegoat and the eccentric.

For the group to operate properly, there has to be polarization of opinion. And the real problem is how to control the anxiety that has been generated in these groups so that it can be mobilized effectively to help them to solve the problem.

A Moment of Agonizing Truth

The meeting usually begins by one member going straight to the solution of the problem and spelling out the subjects to be taught and the difficulties to be overcome. This is quite inappropriate at this stage as the rest of the group is still tied up figuring out the problem. Most members are in stacking orbits trying to work out their contribution; they have got to be allowed to land. Meanwhile the group zig-zags around the problem going through the phase of clarification - making the assumption they need to proceed.

Slowly the group makes the assumption to get forward, and they begin to get into the phase of evaluation. The critical issue here is usually - 'How do we feel about management training?'. Members repeat their experience of various courses.

CONFLICT IN MEETINGS

The executive who has studied group dynamics might say there is nothing excitingly new in all this. But executive meetings have not stood still. A dramatic change in tone and content has become obvious. The significant difference is the emergence of naked conflict in meetings, conflict which is so common and visible that executives have been compelled to find both a rationale and a ritual for it.

The executive as entertainer can learn much from Harold Pinter or Edward Albee on how to figure out the scenario, interpret the script, set up props, make the right entrances, and exit on cue. He must grasp the meaning of the term 'we will entertain this motion' - we will take into cosnideration, fool around with it, pummel it and see generally what it is made of.

Executives in these videos of meetings often provide good examples of old-fashioned double dealing, double binding, doubling up, and diabolical role playing.

Like a good chess player the successful top manager is a sportsman while he plays. he never squirms while he is losing, never crows when he is winning.

Just as task-oriented tough-minded executives try to out-wit and out-psych their opponents so do chess players try to get ahead of their challengers.

In the executive meeting, you sometimes see the same kind of confrontation. As the contestation goes on, the mental, physical and emotional strains go off the graph. Two executive grand masters fighting it out. One, with a poker face, when he has the game under control, there is the slightest flicker of triumph in his eyes, and his eyes linger on his enemy for a moment.

But all this goes beyond the laws of logic, for so much of what is happening in meetings goes beyond logic. For our contemporary executives like Sartre and Camus, the absurd bears the stamp of truth - for the unexpected by its very lack of probability appears to have enhanced its probability. 'What else is new?' would be the executive's response to this non-logical statement.

In this topsy-turvy world, the absurd moves into the realm of the possible. And in this world of Organisational Alice no one is guilty or responsible. When people cannot be held exclusively responsible, classical management with its managerial mandates ('I hold you totally responsible for ...') ceases to be relevant. We are into a new management thesis called existential systems theory.

One possible explanation for these behaviours is that the non-linear part of a manager's brain is involved in this process of running meetings. There seems to be a kind of ballet-dancing aspect to running meetings which requires an executive to have some sense of where his body should be, and what is the required response to a particular situation. What this amounts to is that the right hemisphere of the brain which is responsible for intuitive spatial relations, creative thinking and simultaneous understanding which is required in such skills as sculpting, dancing and body language has to be brought into play to supplement the analytical skills which an executive brings to the meeting in any case. These analytical skills of course, are concerned with language, linear logic and sense of time. Apparently managers seem to have this intuitive ability to get both the right hemisphere and the left hemisphere moving in balance to achieve what they in fact want. Apparently managers are very skillful in getting this left side going and all this means is that they have to develop their intuitive skill in a non-linear logic. And much of this, of course, is revealed in how in fact they have been able to physically intervene spontaneously in a meeting. And yet they have no idea just where it comes from. What is needed is creative chaos to get both hemispheres working in the right ways.

The Values

The two kinds of meetings' values reveal a dilemma. Democratic: 'I can say my piece'. 'You mean I have to say what I think?'. 'We have to go by the majority'. Task-oriented: 'You mean we have to finish this agenda?'. 'If we don't decide, you will - now?'. Putting both of these values together can cause a lot of tension. Public Agenda - developing a policyh for promotion based on merit versus Hidden Agenda - trying to put paid to the 'Young Turks' or sinking the 'Old Guard'.

CONCLUSION

This technique in using simulations to study meetings had enabled us to bring familiar information into fresh combination and to draw out unexpected and interesting findings. The video allows us to enlarge our understanding of meetings, allows us to review the evidence in tranquility and to work out the deeper structures of the executive narrative.

Video studies enable us to take a longer, wider view of the dramas of meetings. And what is revealed is an adventure story sustained by continuing interplay of driving energy, bold executive personalities and recurring accidents. The video allows the viewer to study the background and character of these executive meetings and to focus attention on the structure and meaning of their work. The videos are like a composition of Brueghael drawn with a delight in people and showing executives with a consuming interest in the whole business of managing and are filled with movement.

The simulation allows an extended voyage of discovery and into deeper structures that lie behind executive behaviour - to go beyond the arithmetical collection of data provided us by people like Sune Carlson and look at the calculus of behaviour; look at the differentials and integrals of behavioural change. Now we have executive behaviour with many cunning passages but the main line is clear enough.

What Simulation is Telling us About Meetings

As is well known, management is an interpretive science in search of meaning; the aim is to make meaningful, at least in this particular case 'the facts of executive meetings'. It is at once empirical, interactive and critical. This critical knowledge of executive behaviour transcends both empiricism and interpretation and it produces the emancipating effect of freeing us from false consciousness and eliminating forces beyond our control. The whole point is a critical aspect of executive behaviour to provide a guide to revolutionary action.

Methodological issues and taxonomies in simulation: An introduction

Jan H G Klabbers
State University at Utrecht, The Netherlands

The papers that are included in this section are rather diverse with respect to subject matter and form. Yet they all are concerned with similar issues. They deal with methodological problems related to the general field of gaming and simulation. In particular they have relations with the different origins of games, like for example role playing, operational games, mathematical game theory and simulation and with the way these distinct roots are bringing forward a new research tradition. Questions that are being raised touch fundamental scientific problems like: "What are the appropriate methods of defining the domain, entities and processes in the field of gaming/simulation?", "How can gaming/simulation be fruitful in investigating the relevant problems and how can it support in constructing theories?", "What logic is appropriate in this realm of scientific investigation and justification?", "Can we stick to formal methods, time-invariant solutions and rational behavior for well-defined issues, or on the contrary do we have to cope with ill-defined problems for which solutions are context dependent?"

If we accept that we are dealing with situations for which there are no clear-cut solutions, then we have to ask ourselves when we better stop looking for ever more formal/mathematical tools, and where we enter the domain of ambiguity, equivocality, meaning processing, and preservation of identity in adverse circumstances. How far can we go in pursuing academic rigour without destroying what is intrinsic to the problem? But what the intrinsic problem is depends from scientific point of view on the choice of the appropriate methods of defining the domain, entities and processes! We have to admit here that we have closed the circle, finding ourselves in a trap of own making.

The broad variety of methods and techniques in the field of gaming and simulation, and the very different assertions that are being used about what logic is appropriate, stems from the fact that the various disciplines, that are part of it, perform different scientific rituals. In this regard what is presented here shows an interesting cross-section of what gaming/simulation can offer. Each contributor has his own narrative, and some of them are moralistic.

In trying to gather up the threads of our stories, we have chosen to begin with papers that are concerned with the appropriate methods of defining the domain, entities and processes of the field. Subsequently we move to the discussion of the methods and techniques for investigating specific problems and for constructing domain specific theories. While starting with rigorous methods, gradually the scope is broadened and we notice a shift in viewpoint in terms of the type of knowledge that is being produced and disseminated. Into every tale threads are woven concerning assertions about the appropriate logic for scientific investigation and justification.

Klabbers presents a general frame of reference applicable to the whole field of gaming and simulation. The aim of it is to improve communication within the community of game and simulation designers, and between designers and users. In the taxonomy three interrelated features i.e. 're-presentation of reality', 'rules', and 'actors' are elaborated in great detail. They constitute the overall game theoretical space.

Switalski pursuing similar objectives, focusses his attention mainly on two general features that support his typology namely, the degree to which roles played by the game participants are diversified, and the nature of the interactions that are present in the game. To prove that the body of knowledge has matured sufficiently to consolidate the simulation's languages and methodology he coins the term simulogy. Switalski's typology is mainly concerned with the feature 'actors', with their divers roles and interaction zones.

In their paper Underwood and Duke discuss a conceptual framework and methodological requirements to establish sound design and application principles for policy formulation. They point out that gaming meets those requirements. They trace the history of strategic gaming and view policy gaming as the civilized variety of the traditional war gaming. While evaluating traditional approaches of formulating policies, i.e. formal modeling and expert assessment, they discuss primarily the 'rules (of conduct)', and conclude that formal models are limited because decision options are narrowed down to a small number, and they do little to assist in developing contingencies. Underwood and Duke claim that gaming is uniquely appropriate for exploring ill-structured environments.

Cecchini attempts to classify different types of conflict on the basis of Rapoport's interpretation and criticism of the Theory of Games. He distinguishes three fundamental modes of conflict: fights, games, and debates. As games have been classified systematically, fights and debates so far have not. Cecchini discusses the taxonomy of games and provides a rough outline of fights and debates. Apparently he too restricts himself to the feature of the 'rules' in the game theoretical space. All three modes of conflict are composed of matches, which in turn can become a game, a fight, or a debate or a set of these three modes in a linear or branched succession, interacting, retroactive and nested. Each match consists of a number of moves. With these concepts Cecchini develops a phenomenology of conflict, moving from the game matrix cage into play, from the abstract into the real performance.

Aurifeille and Morel discuss the assignment problem, which is one of the classical operations research problems, similar to allocation of resources, and competitive routing problems, Several actors have to share a series of goods or locations. They give their preferences simultaneously by weighing each alternative. By choosing an algorithmic approach an otpimal solution is obtained. However this approach has some disadvantages. For example actor's preferences need not be stable, but apparently will depend on the mix of preferences. As a consequence solutions tend to become unstable. To illustrate some of the weaknesses of the algorithmic approach Aurifeille and Morel have compared stochastic simulation experiments with an interactive (judgemental) approach. An interesting prospect for further research is to skip from a simultaneous to a sequential process of indicating preferences. But this would suppress equality of the players. This is exactly what Selbirak is trying to figure out.

Selbirak assumes only two players, the leader and the follower. The leader occupies the higher level of the hierarchy. He possesses some information about the follower and he has the right to make the first move. Hierarchical game theory is a rather recent branch of the classical game theory and is more susceptible to information patterns between players than the classical approach.

Was in the classical approach the analyst an interested but neutral observer, in the hierarchical theory he chooses for partizanship with the leader, and consequently becomes an actor himself. Selbirak discusses three solution concepts all related to Stackelberg equilibria. The results of the mathematical techniques are being applied to the management game EHPORT, which seems to be very promising.

All the examples of game theory mentioned above emphasize the search for optimal 'rules of conduct'. Except for the hierarchical game theory, traditional game theory assumes one single reality for all players/actors. Evidently mathematical game theory has produced an important body of knowledge. As the recent branch of the hierarchical game theory is showing however, some of the basic puzzles are inherent in the theory from its beginning. The theory of games is based upon a rather simple conception about the way humans act, and consequently it is considered unconvincing in real-life situations. As a result these types of methods have lost most of their appeal to top corporate management. The two following contributors value realism or verisimilitude very much, especially when it comes to (strategic) social problem solving in urban and land use planning.

Bottari describes three different gaming procedures that are nested in the overall game CITPLAN. CITPLAN provides a learning environment to get to grips with the complexities of the interrelationships of 're-presentation of reality', 'rules of conduct', and 'actors' with their diversity of roles, interests, expectations, beliefs, etc. The game board, which represents reality in a certain way, together with the rules bring forward the context for interpreting all sorts of information exchange i.e. communication. It is clear, that reality can be re-presented in numerous ways, and that each specific re-presentation is based on a particular view on the part of the designer, which has to be made explicit in order to prevent that theories and ideologies are concealed.

That is exactly what Law-Yone points out in his contribution. He states that the structure of a game hides an ideology. He stresses basic requirements of games and simulations such as 'simplicity', 'fairness', and 'competition' with regard to games, and 'structural isomorphism' and 'reproducibility' with respect to simulation.

In his view apparently simulation has to deal with the feature' re-presentation of reality', while gaming focusses more on the 'actors' and their 'rules of conduct'. SPACE is a generic frame game with off-springs like SHIKUM, SPACESET, SPACECIT. Its main actors are Capital, Labour, and State, all engaged in the class struggle concerning the spatial problematic under capitalist urban development. The actors (players) are being confronted with the discrepancies between 'rules' and 're-presentation of reality', each of them trying to legitimize their definition of reality.

In this section it will become clear what the impact is of the preliminary choice of the domain of investigation i.e., what type of reality will be re-represented and what rules of conduct are allowed, on the appropriate logic for investigation and justification. We hope to provide the readers with a challenging and stimulating stream of thoughts.

A user-oriented taxonomy of games and simulations

Jan H G Klabbers
State University at Utrecht, The Netherlands

ABSTRACT: In this paper a taxonomy of games and simulations is presented that is aimed at improving the quality of the consult of the potential user with the designer respectively the distributor of games and simulations. In other words it is intended to provide the user/game-operator, with a language to reach a mutual arrangement with the designer (distributor). As more and more games require micro-computer facilities and thus move into the direction of what traditionally has been the realm of computer simulation, it is considered necessary that the potential user knows what the impacts are of buying a game or simulation. In addition the taxonomy stresses the designer (distributor) to be much more explicit with regard to the kit that is being offered. In the long run we expect that the methodology of the design process will become more explicit and subject to scientific reflection. We consider such a process a necessary condition for the further development of the whole interdisciplinary field of gaming and simulation. Consequently we hope that it will lead to a better communication between scholars, old hands and new comers.

KEYWORDS: Actors, rules, (re-)presentation of reality, semantic-, syntactic-, technical-, pragmatic-, utilization-, and evaluation aspects.

ADDRESS: Faculty of Social Sciences, State University at Utrecht, Heidelberglaan 1, 3584 CS Utrecht - The Netherlands. Phone: 030-534880

INTRODUCTION

Tracing the history of gaming and simulation shows that both fields have a different background, different objectives and different contexts of use. Simulation is generally considered to belong to the realm of the so-called hard sciences, while gaming, especially the role-playing variety, is having more affinity to the social sciences. One could also state that simulation is predominantly concerned with description of general characteristics and ultimately control of reality. Gaming on the other hand is more receptive to making sense of reality and to meaning processing between human beings. Thus historically gaming and simulation belong to two different academic cultures. Yet more and more their mutual fate is being controlled by advances in research and development that force the two distinct scientific communities, that utilize either gaming or simulation, to combine efforts. As a striking example it is sufficient to note the rapidly increasing supply of user-friendly software packages for microcomputers, that help end-users (learners) in conceptualising reality in various ways. In using these packages, the distinction between gaming and simulation will become more and more blurred. Let us try to explain this.

Basically games are defined by three interrelated features:

- (re-)presentation of reality
- rules (of conduct)
- actors (roles, beliefs, norms, values, etc.).

Simulation is a widely used tool for 're-presentation of reality'. In gaming this feature is moulded in the reference system i.e. 'what's on the board'. Rules can be represented in many different ways as we know from mathematical game theory and from frame games. A rather recent way of representing rules is demonstrated by knowledge-bases which are part of expert systems. All these approaches to rules emphasize another dimension of the game-space. By combining two characteristics mentioned above i.e. 're-presentation of reality' (simulation) and 'rules' (knowledge-bases), finally the actor(s), come 'into play'. Increasingly and perhaps unintentionally gaming-type configurations are emerging in the fields of simulation, artificial intelligence, cognitive sciences, software engineering and information technology.

During the last decades gaming and simulation increasingly have drifted in the same direction and it seems that now the time is right to integrate efforts in the set up of joint research programmes, graduate and doctorate educational programmes, etc. The aim of such an endeavour should be to develop a (practitioner's science of design. Although gaming and simulation are being recognized as fruitful scientific methods of enquiry in their own right, the journals 'Simulation and Games' and 'Simulation/Games for Learning' and many other ones, not to forget the numerous books, are examples that prove our point, we have to admit that those who represent both the gaming and simulation communities have not yet developed a common language, which is a pre-requisite for the emergence of such a new discipline. As a consequence, communication between both fields leaves much space for improvement, and cross-fertilization. Maybe the potential user will eventually benefit from it.

A bottle-neck in the dissemination of games and simulations is the fact that it is very difficult to evaluate their effectiveness. Especially games are notorious in this regard. How should a potential user of a game or simulation learn about its usefulness, when the producer/designer is rather limited in his/her means to describe a game with all its variety? How can a potential user know that s/he will not buy a pig in a poke?

Both to improve communication within the communities of gaming and simulation, and between designers and users of games and simulations, we present a taxonomy that aims of facilitating their detailed description and that shows their richness, variety and complexity. It is being assembled in such a way that games and simulations can be characterised both independent of and coupled to each other. Primary goal however is to provide a common language for communication about games and simulation.

B. GLOSSARY

In order to be practical a glossary is presented. Its hierarchical structure fits the general set up of the taxonomy, which is based on a semiotic point of view that is applicable to software engineering (Bemelmans et al. 1984) and, with respect to gaming, has been proposed by Marshev and Popov (1983) as well. The main entries are presented in section B.1.

1. Main Sections of the Glossary

1. Semantic aspects. Semantic aspects deal with conceptualization of reality, and interpretation of components of the simulation/game. They deal with the conditions of use, that influence the correspondence between game/simulation and reality.

2. Syntactic aspects. Syntactic aspects define the rules or regulations via a combination of signs or symbols such as a language, sign-language, codes, etc. Syntactic aspects are defined by the format or mode of operation of a simulation/game.

3. Technical aspects. Technical aspects are considered a subset of the syntactic aspects. For various reasons, not in the least the use of special signs and symbols, it is suitable to have a special entry for these aspects to be able to describe computer-based games and simulations in more detail.

4. Programatic aspects. Pragmatic aspects define the potential impact of the method of preparation and conduct of a simulation/game. Under the specific heading of complexity, factors that influence the conduct of the actor(s)/user(s) are listed.

5. Utilization aspects. Utilization aspects are a subsect of the pragmatic aspects. In the former section emphasis is placed upon the conduct of a simulation/game session, while utilization aspects inform the potential user about the consequences of buying a simulation/game. Conversion-, adaptation-, operation-, and possibly dissemination-costs for running a simulation- or gaming-session can be indicated.

6. Evaluation aspects. Evaluation aspects permeate through all aspects mentioned above. Main criteria for evaluation are: validity (of primarily the semantic aspects), reliability (of primarily syntactic aspects), and utility (of some semantic, some pragmatic and of mainly utilization aspects). The main reason for limiting evaluation to some of all possible aspects is to be economical while at the same time to maintain a sufficient robustness.

2. Sub-sections of the Glossary

2.1.0 General information. This entry provides the title of the game/simulation, the source (producer/designer and/or distributor), purchase cost and a summary of the purpose of a session in which the game/simulation is being used i.e. what message is to be conveyed. Related terms are: scenario, and concept report.

2.1.1 Semantic aspects.

1. Scope of use. Reality is conceptualised as a (social) system. The horizontal scope refers to the subject matter or content of that particular game/simulation. For example when the 'individual' is the subject matter, the game/simulation deals with issues like learning, memory, decision making, problem solving, and other skills. Literature on gaming/simulation shows a wide variety with regard to topics that fit into this scope. Many games/simulations integrate more than one topic, like for example business games, human settlement- and socio-economic games/simulations. The vertical scope refers to the assumptions underlying the way of handling the subject matter. When dealing with pure simulations, it is assumed that no exogeneous actors are involved and that relevant processes are the result of an internal 'machine' like for example an algorithm or rule-base that drives the evolution of the respective system. In that case we are dealing with a zero-actor game/simulation. it is obvious that if the

development over time of the game/simulation depends on one or more exogenous actors, i.e. exogenous to the processes defined at the level of the horizontal scope, we are dealing with one- or multi-actor games/ simulations. These actors can focus on one single issue from the horizontal scope like health care or financing, or they may integrate several issues, as is the case in many business games/simulations. This option is indicated by 'single' respectively 'integrative (comprehensive)' function.
The environmental scope is only relevant when dealing with an open system. Events, coming from outside and beyond direct control of the system may be beneficial or threatening and they may improve or worsen internal affairs.

2. Rule base. Games/simulations are based on rules, which may be so strict that there is no space for spontaneity or surprise, i.e. a rigid rule situation. Rules may also be defined just to bound the initial situation, while leaving open the way actors handle it, i.e. a free form game.

3. Context of use. This entry deals with the functional aspects of the game/simulation. While verisimilitude is important in the educational setting, in operational use it is realism that must be stressed.

4. Target population. This population constitutes the source(s) from which the players (actors) are recruited.

5. User investment. This condition of use indicates the effort that has to be put into the game/simulation. It makes quite a difference to have a large group of students for a period of months, like in business games in graduate schools, or to have four executives for only two hours.

2.1.2 Syntactic aspects. Elaboration of these aspects is based upon the format described by Ellington et al. (1982). Under this heading primarily those aspects are listed that apply to games, whether or not assisted by a computer-simulation.

Psychomotor dexterity relates to all games in which psychomotor skills are the dominating and decisive factors for success.

Field games and table games, as scaled down field games, belong self-evidently to the syntactic aspects.

Electronic based dexterity games (arcade games) need some explanation. Meyers (1984) calls them reflex games, which require a lot of manual dexterity i.e. accuracy, speed.

Examples of home-computer based games: PAC-MAN; SPACE INVADERS; ONE-ARMED BANDITS.

Examples of TV games: BASKETBALL, SOCCER.

Cognitive and social dexterity refer to intellectual abilities and social skills that are necessary conditions for participating fruitfully in a game. This entry does not emphasize the dexterity of the players as such, but the mode of operation of the game that challenges or triggers those abilities.

Examples of manual games:
Simple manual games: role playing, cross-word puzzles, CHARADE, BAFA-BAFA.

Card games: BRIDGE.

Board games: CHESS, GO, MONOPOLY, POLICY NEGOTIATIONS, THEY SHOOT MARBLES.

Device based games: ROULETTE, RUBIK'S CUBE.

Examples of home computer games/role playing games (Meyers 1984): puzzle/ maze role playing games, adventure drama role playing games (DUNGEONS & DRAGONS), multiplayer/competitive role playing games (war games).

Examples of games based on micro-, mini-, and main-frame-computers are the numerous business games, and CHESS.

2.1.3 Technical aspects. Technical aspects are only applicable to computer based games/simulations. They form part of the syntactic aspects. As they require a particular sign-language and codes to implement games/simulations in combination with well-defined modes of operation, those aspects need close attention by the potential user/buyer. The terms that are used are the familiar jargon of programmers, and software engineers. Those who are not accustomed to this language should be rather careful, because adopting such a game/simulation may require extensive conversion of the software, which requires skill, time and money. The effort may cause quite a headache.

2.1.4 Pragmatic aspects. The conduct of the players is certainly influenced by the complexity of use. Eight complexity factors are listed that give an idea of the player's investment during the game/simulation session, and the pressure under which sessions are carried out.

2.1.5 Utilization aspects. Consequences of not considering the technical aspects carefully will become clear when taking into account the utilization aspects. Simple manual or board games are usually easily ready for use. Complex computer-based games need much investment to buy, to maintain and to run on a main-frame or mini-computer. On micro-computers that become increasingly powerful, games/simulations are much easier to run.

When dealing with a generic game/simulation that has been disseminated widely, it is assumed that users/game operators can readily exchange experiences, up-dated versions etc.

Utilization aspects are the main responsibility of the producer/designer and/or distributor. The user should better be critical in evaluating these aspects.

2.1.6 Evaluation aspects. Evaluation aspects penetrate all other ones. To prevent the user of this taxonomy from losing track, we will limit them to those aspects that are most evident. Three criteria will be applied i.e. validity, reliability, and utility.

Lederman (1984) distinguishes face validity and construct validity. Face validity or verisimilitude means 'correspondence with or recognition of real world counterparts'. Construct validity or realism means 'correspondence between concepts and the activity designed to represent those concepts: Lederman refers to process reliability (a predictable process) and product reliability (a predictable outcome). Will various groups of players (actors) engage in similar processes and produce the same kind of results? (How did it happen, what did they do, what are the implications?).

Utility is viewed as a weighing of the costs in terms of time, money, energy, (concentration) and emotional expenditures versus the outcomes and benefits.

Validity is mainly related to the semantic aspects, that is conceptualization of reality.

Reliability leans mostly on the syntactic aspects i.e. the mode of operation and the signs and symbols that are related to it. In this regard the reproducibility of processes and products depends on the technical aspects as well.

Utility for obvious reasons depends on semantic aspects (user investment), pragmatic aspects (complexity of use) and utilization aspects.

C. INTENDED USE

It is assumed that while characterising a particular game or simulation by this taxonomy both producer/designer and potential user/buyer can communicate more fruitfully. It may enhance the development of a good image or cognitive map concerning that particular game/simulation. Each entry can be supplemented with comments and references. An additional advantage may be that thorough comparison of different games/simulations will improve, resulting in a balanced cost-benefit analysis and in a more realistic choice about buying and using.

In the long run utilization of a taxonomy like the one presented here may stimulate and improve the quality of the debate between professionals concerning methodological issues and advances in the field of gaming/simulation to provide a more solid foundation for user-oriented design.

On the basis of this taxonomy a data-bank on games and simulations is being set up that will be used as a clearinghouse. As it will be implemented in a Local Area Network that is connected with world wide networks, the facility will be open for inter-national consultation.

REFERENCES

Bemelmans, Th., Van der Pool, J. and Zwanenveld, N. (eds). 1984. Polyautomatiseringszakboekje. Arnhem: PBNA.

Ellington, H., Addinall, R. and Percival, F. 1982. A Handbook of Game Design. London: Kogan Page.

Horn, R. and Cleaves, A. (eds.) 1980. The Guide to Simulation/Games for Education and Training. Fourth edition. London: Sage.

Lederman, L. C. 1984. Debriefing: A critical reexamination of the post-experience analytic process with implications for its effective use. Simulation & Games, 15:4.

Marshev, U. and Popov, A. 1983. Elements of a theory of gaming. In Stahl (1983).

Myers, D. 1984. The pattern of player-game relationships. A study of computer game players. Simulation & Games, 15:2.

Siebecke, R. 1984. Nomenklatur zur Beschreibung fuer Oekonomische Spiele. Friedrich Schiller Universitat - Informationszentrale fuer Oekonomische Spiele. Jena. DDR.

Stadsklev, R. 1979. Handbook of Simulation Gaming in Social Education. Second edition. University of Alabama: Institute of higher education research and services.

Stahl, I. (ed.) 1983. Operational Gaming: Frontiers of Operational Research and Applied Systems Analysis. Vol. 3. New York: Pergamon Press.

TAXONOMY GAMES AND SIMULATIONS

LEVEL 1 USER - ORIENTATION

0. GENERAL INFORMATION 1. NAME OF GAME/ SIMULATION: 2. PRODUCER: 3. DISTRIBUTOR: 4. PURCHASE COST: 5. PURPOSE (SCENARIO): (concept report)		DATE:			
SEMANTIC ASPECTS (content) 1. SCOPE OF WORK	1. horizontal scope 1. individual (intra-personal) 2. social/human relations 1. group communication 2. organizational comm. (incl. mass. comm.) 3. cross-cultural comm. 3. family 4. women 5. consumer 6. drugs 7. educational organization	1. administration 2. research			

continue scope of use		3. teaching	1. curriculum 2. classroom 3. school	1. kindergarten 2. elementary school 3. high school 4. college 5. polytechnic 6. university
	8. health care 9. human services 10. religion 11. economics 12. energy 13. environment 14. technology 15. military 16. settlement: urban/rural 17. law 18. politics (international relations) 19. history 20. future			
	21. business	1. marketing 2. financing 3. production 4. human resources (manpower) 5. R & D 6. logistics (transportation)		
	22. demography 23. geography 2. vertical scope 1. zero-actor (pure simulation)			

	2. one-actor (one player group)	1. single function 2. integrative (comprehensive) function			
	3. environmental scope 1. risks/opportunities 2. constraints 3. resources 4. absorption of products/services 5. irrelevant				
2. RULE-BASE	1. rigid-rule 2. free-form (frame)				
3. CONTEXT OF USE	1. education (learning) (verisimilitude)	1. concepts 2. processes 3. rules			
	2. training	1. social skills	1. team work 2. negotiation 3.		
		2. decision making 3. policy formation			
	3. research	1. experimentation 2. theory testing 3. theory construction			
	4. operational (realism)	1. management support function 2. manager support function 3. decision support function 4. planning support function			
	5. recreation				

4. TARGET POPULATION	1. family				
	2. fraternity/sorority				
	3. elementary school				
	4. high school				
	5. college/university				
	5.1. undergraduate				
	5.2. graduate				
	6. business organization	1. low management 2. middle management 3. top management			
	7. government	1. local 2. regional 3. national			
5. USER-INVESTMENT	1. group size 1. size subgroups 2. duration per session				
II. SYNTACTIC ASPECTS (format/mode of operation)					
1. PSYCHOMOTOR DEXTERITY	1. field games 1. football etc. 2. volleyball etc. 3. tennis etc. 4. dart etc. 5. golf etc. 6. gymnastics etc.				
	2. table games (scaled down field games)				
	3. electronic based	1. home computer game 2. TV game 3.			

2. COGNITIVE/SOCIAL DEXTERITY	1. manual	1. simple manual game 2. card game 3. board game 4. device based game		
	2. computer based	1. personal-computer (home computer) 2. micro-computer 3. mini-computer 4. main-frame-computer		
III. <u>TECHNICAL ASPECT</u>				
1. CONFIGURATIONS	1. hardware	1. central processing unit	1: 8 bit 2: 16 bit 3: 32 bit	
		2. internal memory	1: 64 Kb 2: 128 Kb 3: 256 Kb 4: 512 Kb 5: 1 Mb 6:	
		3 input	1. card reader 2. papertape reader 3. optical reader 4. bar code reader 5. terminal keyboard 6. magnetic tape unit 7. (floppy)disc drive 8. AD-converter	1. joystick

			2. voice	
			3. x-y tablet	
			4. mouse	
			5. track ball	
			6. touch screen	
			7. pedal	
	4. output	1. printer(s):	1. alpha-numeric	number: 1/2/
			2. graphic	number: 1/2/
		2. plotter		
		3. terminal(s)	1. screen (display)	number: 1/2/
			2. paper	number: 1/2/
		4. magnetic tape unit		
		5. papertape puncher		
		6. (floppy)disk drive		
		7. microfiches		
		8. DA-converter	1. voice	
			2. pedal	
	5. external memory	1. card		
		2. papertape		
		3. magnetic tape		
		4. (floppy)disk		
	6. local/wide area network (LAN/WAN)			
2. software	1. operating-system	1. unix		
		2. MS-DOS		
		3.		
	2. model-base	1. single program	1. basic	
			2.	

			2. modular program	1. pascal 2. fortran 3. c 4. Lisp/Prolog 5.
		3. data-base	1. hierarchical 2. relational	
		4. knowledge-base 5. method-base	1. linear programming 2. simulation 3. forecasting 4.	
		6. conversational system (dialogue components)	1. process-oriented	1. pascal 2. c 3. Lisp/Prolog 4.
			2. object-oriented	1. small talk 2.
		7. (local)network software 8. documentation		
IV. <u>PRAGMATIC ASPECTS</u>				
1. COMPLEXITY OF USE	1. operator's and player's manual (No. pages) 2. number of actors (players resp. roles) 3. number of variables and degree of interaction between domains of horizontal scope (see I.1.1.)			

	4. degree of interaction between actors (player groups)	4.1. no-interaction 4.2. exchange of information only 4.3. mutual adjustment of decisions per round 4.4. mutual adjustment of policies		
	5. number of decisions per round 6. number of decision-aids (algorithms internal to game or simulations, data-base-management, accounting sheets, etc) 7. number of rounds 8. number of reports generated (size of computer-output)			
V. UTILIZATION ASPECTS 1. CONTRACT CONDITIONS	1. conditions	1. duration of contract 2. number of test-runs 3. delivery-time		
	2.	1. initial investment 2. conversion of software 3. standard-software 4. new versions 5. training 6. assistance with 1. conversion 2. initial use		

2. MAINTENANCE RESP. PREPARATION OF USE	1. maintenance costs	1. hardware		
		2. software		
		3. personnel (expertise)	1. game-operator(s) 2. software-engineer(s) 3. programmer(s)	
	2. operational costs (sessions)	1. preparation effort	1. staff 2. equipment 3. manuals	
		2. facilities	1. building 2. computer 3. services	
		3. evaluation	1. staff 2. equipment 3. report	
		4. expendables		
3. DISTRIBUTOR (PRODUCER)	1. course(s) for game operator(s) 2. documentation 3. consultance 4. release new versions			
4. GENERIC VS SPECIFIC	1. generic	1. predecessor	1. adopted version 2. adapted version	
		2. successor(s)	1. adopted version(s) 2. adapted version(s)	
	2. specific			
5. DISSEMINATION	1. number of (off-site) users	1. local 2. national 3. global		

VI. EVALUATION ASPECTS				
1. VALIDITY	1. semantic aspects	1. scope of use	1. horizontal scope 2. vertical scope 3. environmental scope	
		2. rule base	1. rigid rule 2. free form (frame)	
		3. context of use	1. education 2. training 3. research 4. operational	
2. RELIABILITY	1. syntactic aspects	1. psychomotor dexterity 2. cognitive/social dexterity		
	2. technical aspects			
3. UTILITY	1. semantic aspects	1. user-investment	1. group size 2. size sub-groups 3. duration per session	
	2. pragmatic aspects	1. complexity of use	1. simple 2. moderate 3. complex	
	3. utilization aspects	1. contract conditions	1. conditions 2. costs	
		2. maintenance resp. preparation of use	1. maintenance costs 2. operational costs	

Gaming simulation: An attempt at a structural typology

W. Richard Switalski
University of Warsaw, Poland

ABSTRACT: This paper is concerned with typology of games used mainly in management and economics. Two general structural features are employed as criteria to construct this typology. They are: the degree to which roles played by game participants are diversified and the nature of interactions that are present in the game. Three basic game types are distinguished. They are briefly characterized in terms of four more properties.

KEYWORDS: simulogy, typology of gaming simulations, role diversity, interaction zones, simulogic models of economic systems.

ADDRESS: Dept. of Cybernetics and Operations Research, University of Warsaw, Dluga 44-50, 00-241 Warszawa, Poland: Tel. Warsaw 31 32 02.

INTRODUCTION

There is growing awareness that certain aspects of, and notions contained in, simulation have to be examined again as new experience is amassed and reflected upon. Ören (1984), among others, discusses the developments that recently occurred in the design and use of simulation models. He feels that the time has come to consolidate simulation's languages and modeling techniques. Also Crookall et al. (1987) attempt to redefine some of the central concepts related to simulation. They advocate that such notions as system, model, role-play, simulation, game etc. should be accorded new insights.

Before addressing issues to which the main part of this article is devoted we would like to suggest to simulations' designers and users a term that, in our opinion, better expresses the essence of the approach and, also, reflects on the degree of maturity that this body of model-based methodologies has acquired. We propose that the term simulogy (and consequently simulogic, simulogist etc.) is used to indicate the type of approach, the methods of conducting experiments as well as the technique of achieving aimed at objectives.

The Meanings of Simulation

There are numerous definitions of simulation. Each author proposes his own. Some differ in immaterial particulars while others reveal essential divergencies.

We may agree that there are sufficient reasons to accept the existence of many definitions and that there is always something that validates them in a particular place and time. As observations contained in this article rely on examples drawn from economics and business we need to specify, in a more detailed manner, those features of simulogic approach that seem to be the most essential. Also to avoid the emergence of peripheral problems, on the one hand, and to embrace a possibly wide fields of simulation approaches, on the other, let us agree that:

Simulogy is a body of methods, techniques and languages that aims at the construction and application of behavior specific models of systems in order to analyse reactions of a model while its internal relations as well as external conditions that replicate the modeled system's environment are allowed or forced to change. The model-based simulogic experiments are supposed to make people familiar with the operations of certain systems or yield indications regarding the probable behavior of systems constituting the prime interest of a researcher, designer or decision maker. Conclusions drawn from the results of simulogic experiments can be used to improve the operation of existing systems by introducing organizational changes studied in model terms; they can also be used to design a new system that would effectively carry out objectives and tasks that are set forth.

Gaming simulation is regarded, very rightly perhaps, as a further or deeper refinement of simulogic methodology. There is a whole diversified family of gaming simulation techniques, some of them are termed role playing experiments, business games, decision games, management games, operational games and so on.

Simulogic games should be distinguished from games in the mathematical theory sense. The latter are used to visualize situations rather than to cope with processes and their dynamics. Nevertheless, recent developments in the mathematical theory of games resulted in multistage, hierarchical games which, in some way, acquire the capacity to cope with sequences of situations. The mathematical theory of games and gaming simulations were created and developed independently. In the evolutionary process both disciplines borrowed ideas and notions from each other. Gaming simulation took some formalism from games in the mathematical sense, while the latter drew on the concept of time lapse and probably on the idea that strategies are developed and introduced into the game as it progresses.

As of now we do not have a general enough theory of gaming simulation. It is rather doubtful that such a theory is ever going to be developed but, taking into account the observed convergence of gaming simulation approaches and mathematical games, we may hope that eventually there will emerge a discipline that would be as wide and open to possible areas of applications as simulogy is and that it would be equally susceptible to algebraic presentation and the formal treatment as the mathematical theory of games is. At present the flexibility, openness and versatility of gaming simulation (these qualities are a natural outcome of gaming being rooted in simulogic approach) offer better chances to represent situations and processes more realistically.

The Notion of Interaction in Simulogic Models

It is widely accepted that gaming simulation incorporates all the essential features of simulogic approach. However, since the notion of a simulogic game rests on the idea of concurrent active involvement of many participants (individual or grouped into teams), who perform certain roles, there emerges a new quality in gaming, viz. that of interaction. Although it is intuitively felt that interaction is a sort of mutual interrelationship that happens among entities (participants, models or formulae) present in the game, it is not easy to characterize its substance.

In a recently published article Bryant and Corless (1986) list four types of interaction: "(i) within teams, (ii) between teams, (iii) between the teams and the facilitator and, (iv) between the team and the structure of the game". Role-play is the substitute of functions that are carried out by policy or decision makers as well as by those who execute these decisions in real systems. The rules that are imposed on participants' roles replicate the ramifications of the inner part of the real system. Some specific areas or the environment of the real system are reproduced by additional sets of rules or by formal (i.e. mathematical) models. Role playing and participants' behavior elliciting component of a simulogic game, given the concurrency of their activities introduce, normally, the need to communicate between and to exchange information among participants. That side of interaction stresses its informational aspect.

Simulogic gaming also creates new planes for interaction among and between aspirations, objectives, decisions and their outcomes within as well as outside the individual team. that is achieved not only through communication but, more importantly - especially when the between teams contacts are considered - it emerges thanks to the existence of interaction zones.

An interaction zone is created by this particular part of simulogic game in which a given participant (individual or team) is forced to compete or cooperate with others to achieve the imposed or the arrived at objectives. Formation of coalitions, bargaining, bidding, auctioneering or outright manifestations of conflict may occur in interaction zone. In some cases these forms of interaction are facilitated by the model or another umpiring device. What seems to be important here is that in interaction zones the informational aspect is overshadowed by the cause-effect mechanism with emphasis put on the effect side. At times the uncertain, and intension-independent, character of outcome of contacts within the interaction zone manifests itself with all lasting consequences for a participant.

The importance of interaction in both meanings (i.e. as a vehicle to facilitate communication and in its role as means to exert influence on the course of game evolution) is mentioned by Crookall *et al*. (1987). In fact, they seem to regard the presence of interaction as the sufficient condition for a model to be classified as a simulogic one.

STRUCTURAL TYPOLOGY OF SIMULOGIC GAMES

Criteria

Typologies of simulation games can be built with all attributable intricacies and sophistication. Classifications pertain modeled domains, areas of application, functional properties, model types etc. (Elgood 1984). This paper examines two most relevant features of any gaming simulation experiment: (i) number and the degree of diversification of roles played by participants, and (ii) number and nature of interaction zones.

The first feature when used as a classification criterion translates into: *If there are two or more participants (counting teams as composite participants) in the game, do the roles they enact diverge?* Answer in the negative indicates that the same singular role is played by each participant. A "yes" answer tells that there are at least two participants that follow different sets of game rules. Thus all exercises termed gaming simulations can be split into two classes: either homogeneous (H) or diversified (D) roles games. A simulation in which there is only one participant (or role) is, obviously, of H class. In such a model there exists a limited room for interaction, i.e. it takes place between the model and the participant, only. A game with many participants/teams - each enacting the same role, also falls within class H. If participants are grouped into teams then team members would have to communicate among themselves. That does not necessitate the emergence of interaction in the second sense that was discussed earlier. There are, however, games that belong to class H and that do involve interaction between teams.

When in a simulogic game or in a description of simulogic exercise we encounter interaction involving different teams or role playing participants then such a model can be termed interactive (I) or interparticipant (interteam) game. Conversely, to the non-interactive (nI) game class belongs every game in which participants/teams are effectively isolated from each other. It follows then that in an nI game the participants neither communicate between teams nor even interact indirectly, e.g. through the formal model or an umpiring device.

In fact the simulations that entail non-communicating and non-interacting teams should be regarded as infra- or sub-games. However, not only the long established tradition but also the fact that, at times, nI games become, eventually - after some experimenting, interteam interactive - justifies their inclusion in our analysis.

The two criteria (i and ii) put together make it possible to distinguish four types of simulogic games:

i \ ii	interteam interaction	
	absent (nI)	present (I)
homogeneous roles (H)	HnI	HI
diversified roles (D)	DnI	DI

In this formally arrived at typology there is a place for games that involve diversified roles but are noninteractive.

We have to examine two cases here: a) either roles of teams are diversified or b) roles of team members are diversified. In both cases, since by definition there does not exist interaction, games of type DnI are decomposable into a number of isolated games of HnI type. In case a) there will be as many HnI games as there are teams and in case b) the number of HnI games depends upon the number of team members and teams as well. Other possible situations are covered by HI and DI games.

Six attributes (including the two criteria) are used to discuss the three main types of simulogic games: (1) the domain of the original problem (system, process) that is represented in the game; (2) the form and structure of the model around which a given game is developed; (3) the number and diversification of participant roles or team positions in the game (cf i); (4) the data; (5) the number and character of interaction zones (cf ii); and (6) the game rules.

<u>A Homogeneous Roles Non-interactive Game (HnI)</u>

A. the domain - elementary operations that take place in a modeled system, e.g. allocation of tasks,scheduling of job execution; in some instances activities comprise composite processes like manufacturing, inventory control, transport, sales, etc.

B. The model - usually a simple algebraic or functional expression that links decision taken by the participant with outcome of that decision.

In most HnI games the formal models correspond to these interaction zones that cannot be covered by active encounter of game participants; cf block (2) on Fig. 1; SV - D stands for interaction between SV (scenario variables) and D (decisions taken by participants).

C. The participants - typically 2 to 8 individual participants - all play the identical role of a production manager, project supervisor or a salesman who design and carry out certain policies. As the game participants do not interact among themselves either directly or indirectly, their upper number is not limited by the model.

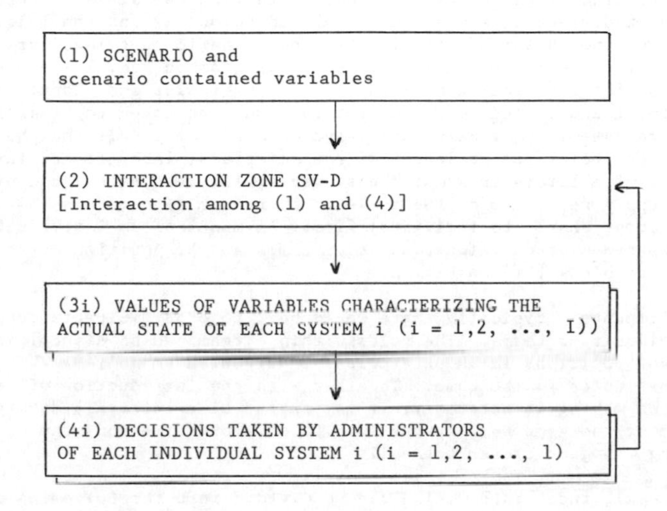

Fig. 1. Flow diagram of HnI game

D. The data - at the initial stages of the experiment, the categories and the values of all variables are identical for each participant. At subsequent stages individual state variables may vary from one participant to another, situational (environmental) variables, that are fed into the game as it progresses, are the same for all participants.

E. Neither horizontal nor vertical influences of one participant on others are encountered. The only interaction that appears in this type of experiment takes place between the participant and the model - cf block (2).

F. Games of this type do not usually have an autonomous convergence mechanism that would preclude their length. The game is played for a definite number of stages set in advance although it does not necessarily have to be made known to participants. Sterman and Meadows (1985) recently designed a game that is a good example of HnI game. Radosinski and Szczurowski (1985) give description of another, more complex, game of this type.

A Homogeneous Roles Interactive Game (HI)

A. The domain - operation of a real sytem. The systems that are modeled in the game could be manufacturing and/or retail enterprises, property developers, construction firms, etc. Some games may center on processes in which systemic content is only implicity present (i.e. the institutional framework does not play an important role in the game, e.g. the inventory control games). Games of this type can also entail processes in which no tangible goods are subject to negotiations and exchange, e.g. the legislative game that assumes that members of the house have interests in passing certain bills but cannot directly control the voting and have to trade their influences on issues that are divorced from these interests for favors granted them by others.

B. The model - it is usually a set of models that serve various purposes. There would normally be three basic models, viz. the model that simulates implementation of individual decisions (see Fig. 2, block 7; this roughly corresponds to the model described in point B of previous section), the outlet model of interaction zone in which outcomes affecting the state of supra-system as well as of individual systems are calculated (block 8) and the inlet model of interaction zone in which interplay between scenario contained variables and results of previous decisions takes place; cf diagram on Fig. 2 - block (2). SV - GR stands for interaction between scenario variables and global results (GR). The inlet and outlet models of interaction zone taken together form the model of suprasystem. Some games are based on a set of models that help to control physical relations (balances of raw materials, labor force, investment capacities), price levels (raw materials, wages, interest rates) and overall demand and supply relations on the market. There are also models that allocate production factor shares to individual firms. A model of an individual system helps to keep track of decisions, their outcomes and of position of each participant; cf block (9i) on Fig. 2.

C. The participants - typically there would be 2 to 8 participants ;that are either individuals or teams. The roles within a team can be diversified so that important positions in an enterprise, institution or policy-making body can have substitutes in the game. Together with the introduction of teams, the group decision-making is necessary; it may extend time intervals devoted to a single stage of the game but, by the same token, more dimenions can be included, thus bringing the game closer to reality.

D. The data - all input information can be divided into the following categories: (1) a subset of external state variables: these variables, identical for all participants, are either released from the scenario or are a result of past developments that took place in game; (2) a subset of internal state variables that are dependent only on a given participant's previous decisions; and (3) a subset of internal state variables that depend on decisions taken by a given participant and the resultant interaction of decisions taken by all other participants.

E. The interaction zones - at least one interaction zone in this type of experiment is usually placed on the output side; that is to say that the final outcomes of decisions or activities result from an interplay of participants' intensions. This would, in most cases, be a market on which the teams try to sell their products. There could also be interaction zones on the input side, e.g. on activities concerning hiring labor or buying raw materials or commissioning new production facilities or securing bank loans. In each of these cases the participants may attempt to obtain a certain quantity of inputs, but what they actually get would depend upon the total supply of respective items as well as upon conditions (prices) offered by all involved individual participants. The interaction zones could be segmented, that is to say that in a given sector of the zone only some teams are active.

F. The rules - they have to cover the technical side in which the stage interval is divided into subintervals devoted to data collection, bargaining, decision-making and to computation of new state variables arising from decisions and external factors. There are also rules that explain what moves are allowed. Such games could be played either during a predetermined number of stages or until a monopolistic position is attained by one participant. Thus, the game can have a built-in autonomous convergence mechanism.

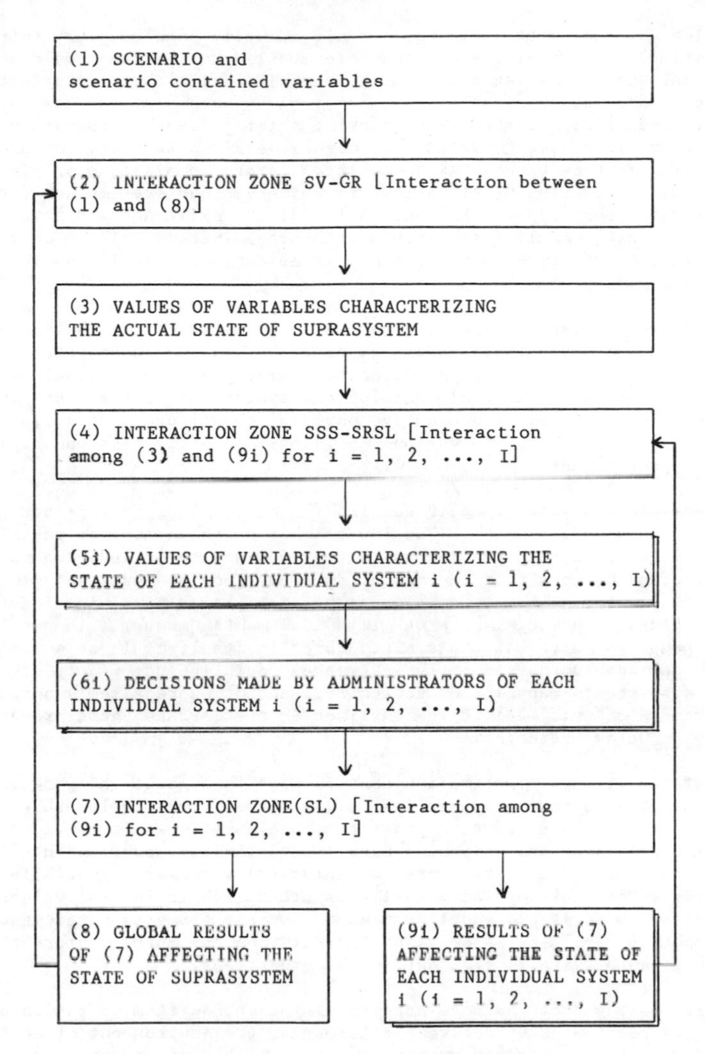

Fig. 2. Flow diagram of HI game

There are many good examples of games that fall within this category. One of the most beautiful, as far as simplicity and emotions it arises are concerned, is THE COMMONS GAME developed by Powers (Greenblat 1984).

Among others STARPOWER (Shirts 1969) is well know and widely used. A complete mathematical model together with the description of a simulation that involves dynamic interaction of multiple firms under conditions of central planning is given by Switalski (1984).

A Diversified Roles Interactive Game (DI)

A. The domain - it may embrace activities of a real system in which a hierarchical structure of its elements can be observed (vertical decomposition of a system) or in which various units of the system are mutually complementary (horizontal decomposition). The real systems that are modeled in the game could be planning or long range policy making centers together with units that are affected by activities conducted by overall system control centers. At the lower level there are industrial, commercial or banking enterprises that operate under guidelines or directives formulated by higher level units (e.g. corporation headquarters, central banks, tax authorities, state commissions of planning etc.). There could also be situations in which various systems are forced to cooperate with other systems indispensable for the existence and development of the former or are compelled to compete with other systems. There could also be games in which strictly economic content is absent, e.g. social policy making games or war games.

B. The model - similarly as in HI games, sets or networks of models are usually involved; since the processes that take place within the suprasystem are to a greater extent modeled by the participants themselves, formal models need not replicate the missing active elements of the system but, rather, are intended to imitate the responses of non-active physical constituents of a system as well as of the suprasystem to various decisions and activities undertaken by individual participants.

C. The participants - there could be either individual persons or teams positioned in charge of separate systems. The lower limit of the number of participants would rarely be less than ten, although the tendency to include too many players is counteracted by more than the proportionately growing model and game complexity and the amount of time required to carry out activities involving interaction of participants. In social or economic policy making games it is possible that a participant can represent the society which in itself could be a part, if not the whole, of the suprasystem. Thus DI games could include widely divergent roles of subsystem's managers or controllers and of suprasystem's administrators. A good example of this feature is contained in the CAPJEFOS game presented at ISAGA '86 by Cathy Greenblat.

D. The data - all input information that is used throughout the game can be divided into the following categories: (1) a subset of variables that characterizes the state of the suprasystem, one part comes from the scenario, the other constitutes the composite result of activities carried out (i.e. it contains the suprasystemic feedback element); (2) a subset of variables that characterizes the state of individual subsystems. Their initial values are calibrated so as to assure equal chances of success for each participant; and (3) a subset of variables whose values result from the mutual interaction of all activities that are carried out in the experiment.

It is worth noting that, as more aspects of the suprasystem operation and behavior are replicated by active participants, the environment of an individual team is modeled, to a greater extent, by the participants themselves rather than by the artificial, to some degree, formally preset procedures and models. One can thus say that data that characterize the suprasystem's environment come from outside the game, whereas data that describe the position of the subsystem as well as the state of its immediate environment are generated during the course of the experiment.

E. The interaction zones - there are numerous spheres and hierarchical levels at which the participants have to exchange information and interact with each other. Usually interaction zones are system- or process-specific, i.e. only some participants would normally interact in a given zone. Thus price controls imposed by price authority on certain goods would affect only those subsystems that, either utilize or produce a given commodity, whereas total demand and supply of this commodity would depend on the response of both the users as well as the suppliers. In hierarchically organized suprasystems the systems that belong to different levels would interact both ways: the rules or directives

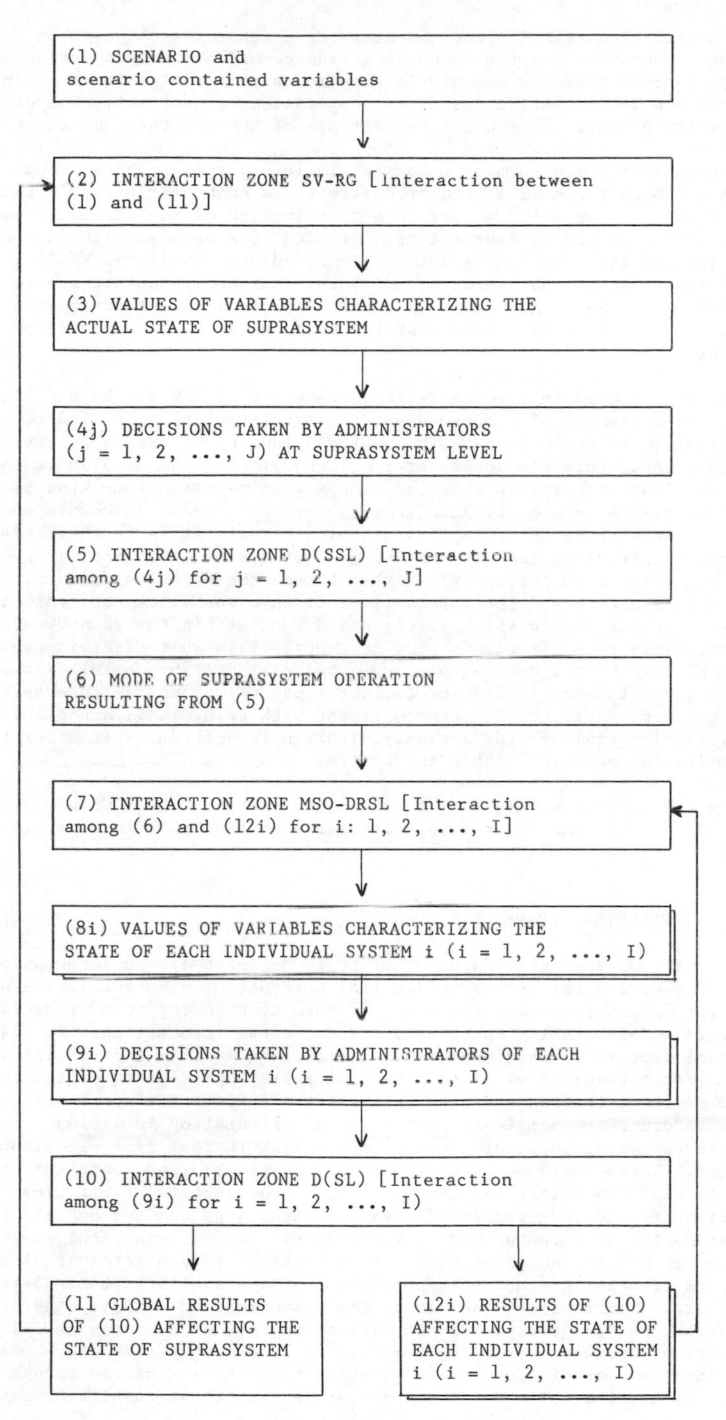

Fig. 3. Flow diagram of DI game

set by superiors would influence the behavior of subordinates which in turn may force the former to alter the rules in pursuance of an improved performance of the whole suprasystem. We may distinguish zones of positive interaction, i.e. cooperation, and of negative one, i.e. competition or conflict. Cooperation and competition zones are the natural outcome of the way the game was structured.

F. The rules - they are in many respects similar to those that are applied in HI games although there is a greater diversity as each participant's role has to be defined in a detailed manner. The termination of the experiment would be governed either by the achievement of some kind of stable equilibrium, which in turn would indicate that the experiment; ends up with positive results, or by a complete breakdown of the suprasystem. Rules that define objectives for individual participants have to be adequately harmonized, otherwise the process of convergence of results toward improved efficiency of the suprasystem may never take place.

A game that is rooted in the theoretical framework elaborated by Dantzig (1963) could serve an example of a macroeconomic gaming simulation applied to planning of a centrally planned economy. The central planner and the manufacturing enterprises constitute the suprasystem. There emerges the problem of coordinating production plans in firms so that the economy can maintain some kind of equilibrium. The central planner could command firms to turn out specific amounts of all goods. As the experience of central planning reveals it is neither effective nor it secures equilibrium. An alternative mode of system operation may require the central planner to set up prices that induce the firms to produce the demanded quantities. The game and the interactions it involves allow the central planner to arrive at prices under which supply and demand within the economy are balanced while the firms take efforts to produce exactly this what otherwise would be expected of them by the central planner. Switalski (1979) described the structure of such a game although it did not evoke the central planners' interest. They seemed, until recently, to be more confident with traditional approaches to planning or with implementing strictly algorithmic procedures in order to retain their dominating position within the system.

A more recent development in this field is the exciting CAPJEFOS game in which horizontal and vertical interactions replicate village life in a developing country.

CONCLUDING REMARKS

The realm of gaming simulation experiments is centered around informational processes, i.e. acquisition or collection, perception, structuralization, transformation and final utilization of information in the form of decisions and reports. Models help to structure and process information. To visualize the spatial aspects of participants' activities as well as the spatial spread of suprasystrem together with its component parts, boards that depict locations, networks of communication and transport links and flows between locations are needed. In order to facilitate participants' orientation in various items, dummies to represent physical elements that exist in real life situations are introduced. These would be surrogate money and pieces that substitute for capital assets, raw materials, commodities, infrastructural facilities, etc. Different forms and ledgers serve to record situations and the overall state of the experiment. In computerized versions of simulation games, the gaming equipment can be less numerous since it is possible to use terminal displays; hard copies of maps, graphs and reports can be made available to individual participants. Computer also serves as the communication facility that links all players. But computerization imposes stringent demands on size, diversity and quality of software servicing a game. On-line computerized versions can speed-up the process of conductging experiments. It is achieved thanks to the automatic calculation of results of each stage. As these results have to contain the effects of interaction, the functions and formulae are sometimes complex enough to justify the use of a computer (Coote et al., 1985).

We should be aware, however, that in studying the dynamic properties of any real system in which human factor exerts influence, it is not computer speed which contributes most to the success of modeling effort. More important is an adequate model representation of the nature of information processes carried out by people. The circulation of information, perceptions of its form and meaning, accompanied by inherently vague and otherwise difficult to ellicit preferences together with the use of information - particularly in group decision-making - constitute the most essential aspect of real systems' operations that are to be replicated in a gaming experiment.

The game is the means of activating informational processes in a controlled setup. Feedback controlled processes are omnipresent in the operation of all, or nearly all, real systems. By designing a game with a simulogic model proper built into it we actually make up for the link that is missing in other modeling techniques. It allows us to simulate the activation of matter and energy, of resources, skills and know-how in the pursuit of goals. In model terms, as opposed to physical reality - in addition to information processing - we need simulated environment which provides the closure of the feedback loop.

The typology that is presented on these pages certainly is not the most important among all other typologies that could be conceived. It allows, however, to see clear line in the evolution of simulogic models. In our opinion the trend in this evolution is clearly advancing toward games that are of DI class. These games are better equipped to cope with the complexity of real life situations and thus render higher quality of reality's representation in model terms. It was somehow unexpected, that our approach to the study of structural differentiation of simulogic models resulted in a more or less streamlined sequence of three types of models. This ordering (HnI, HI, DI) indicates the directions in which efforts should be made if one wanted to construct a perfect simulogic model. Starting from a single man-model simulation we can build up a framework to accommodate multiple parallel simulation (HnI) that enables to compare performances of many participants. By introducing the interaction zones into HnI models one can further make the performance of one participant dependent on actions of others. In these cases the model builder does not have to guess what would be the behavior of other participants - the probability that replacing the man-model interaction by participant-participant interaction yields better results than any formal model representation of the impaired behavior is quite high. Finally, the diversification of roles within a gaming simulation context implies that immediate environment of the role-playing participant is replicated and, moreover, that interaction between a participant and its close environment need not follow rigid ramifications contained in a mathematical model. The course of events is mutually dependent upon the roles and personalities that are active in the game.

It is worth to note that rules, roles, models and paraphernalia (equipment) together with the situational scenario of the simulogic game constitute the reference metamodel of the game while the implementation of the scenario and actual playing of the game (i.e. the execution of rules, enactement of roles, activation of models) in a given setting with a given group of participants is the realization of the reference metamodel. Unlike models that belong to other modeling paradigms (econometrics, mathematical programming, operations research, industrial dynamics) the gaming simulation models do not always yield the same results even when the same reference metamodel is activated. The fact that they generate greater diversity (even increasing when we follow the sequential ordering of game types) suggests that they offer a chance to counteract the diversity inherent to real processes and systems. That, in turn, implies some kind of superiority of simulogy, in general, over rigid experiments with other classes of models.

REFERENCES

Bryant, N. and Corless. 1986. The management of management games. Simulation/Games for Learning, 16:3.
Coote, A., Crookall, D. and Saunders, D. 1985. Some human and machine aspects of computerized simulations. In van Ments and Hearnden (1985).
Crookall, D., Oxford, R. and Saunders, D. 1987. Towards a reconceptualization of simulation: From representation to reality. Simulation/Games for Leaming, 17:4.
Dantzig, G. B. 1963. Linear Programming and Extensions. Princeton: Princeton University Press.
Elgood, C. 1984. Handbook of Management Games. Aldershot: Gower.
Greenblat, C. S. et al. 1983. Research version of THE COMMONS GAME. Mimeo.
Ören, T. I. 1984. Model-based activities: a paradigm shift. In Ören et al. (1984).
Ören, T. I. et al. 1984. Simulation and Model-Based Methodologies: An Integrative View. Berlin-Heidelberg-New York-Tokyo: Springer-Verlag.
Radosinski, E. and Szczurowski, L. 1985. Computer simulation applied to education in a firm's finance. Simulation & Games, 16:4.
Sterman, J. D. and Meadows, D. 1985. STRATEGEM-2: a micro-computer simulation game of the Kondratiev cycle. Simulation & Games, 16:2.
Shirts, R. G. 1969. Simile II.
Switalski, W. R. 1979. An interactive simulation model of two-level planning system. Oeconomica Polona, 2.
Switalski, W. R. 1984. A competitive market model with the delivered price base. Oeconomica Polona, 4.
van Ments, M. and Hearnden (eds.). 1985. Effective Use of Games & Simulation. Leicestershire: SAGSET.

Decisions at the top: Gaming as an aid to formulating policy options

Steven E. Underwood
International Institute of Applied Systems Analysis, Austria

Richard D. Duke
University of Michigan, USA

ABSTRACT: The field of policy gaming - the application of gaming to strategy formulation in a non-military setting - needs a conceptual framework on which to develop sound design and application principles. In this paper we address this need by delineating some methodological requirements for a policy formulation technique and suggest some characteristics of gaming that meet these requirements. We trace the history of strategic gaming from World War II to the present, describing how the institutional environment has evolved during that period. Top level strategic problems in the 1980s are characterized by high levels of complexity, conflict, and risk; the effective management of strategic issues requiring managers to formulate policy with incomplete and qualitative information, ambiguous organizational goals, and little ability to predict outcomes. The traditional approaches of formulating policy - formal modeling and expert assessment - are viewed as limited in this context. Policy gaming is proposed as an alternative approach that is more appropriate than the traditional approaches for exploring the structure and impacts of policy options in a complex and turbulent policy environment.

KEYWORDS: Policy gaming, strategy gaming, business, decision making, strategic management, public policy, ill-structured problems, conflict, communication, military gaming.

ADDRESS: SEU: Program in Urban, Technological, and Environmental Planning; University of Michigan; 218 Carver Building; 506 East Liberty St; Ann Arbor, MI 48104; USA; Work: (313) 763-7205; Home: (313) 453-1771.

Gaming as an aid in formulating policy and strategy is not new. Strategy gaming[1] originated with early exercises designed for exploring and practicing strategy in preparing for war. One can imagine the concept of gaming originating in ancient times when a military officer sketched a battle plan in the dirt to assess and communicate alternative strategies for conducting battle. A simple "what if" game would emerge as that sketch was modified to explore alternative plans and their impacts. Strategic was gaming has evolved and endured through the centuries; the war games that are in existence today have a distinctive "high tech" character and use exotic technology to enhance realism. The basic purpose of gaming has not changed much; they are still used for exploring and communicating strategic options in a simulated, low risk context.

Policy gaming, which is the civilian cousin to the war gaming, was introduced into the public and private sectors after World War II (WW II). Early civilian gaming applications addressed training and educational needs rather than policy. Still in an early stage of development, policy gaming is promising to emerge as an effective approach to formulating organizational strategy.

A. REQUIREMENTS OF POLICY FORMULATION

The environment and requirements for formulating public and private sector strategy have evolved radically in the last forty years. The post WW II period saw the emergence of a need of a new approach to explore strategies in an evolving institutional environment. Economic, political, and social demands on organizations increased rapidly in the decades after World War II. In this section we will describe the related evolution of strategy making in public, private, and international contexts, with a special focus on the use of gaming for exploring policy options.

The years immediately following WW II were characterized by rapid growth and relative prosperity in the United States. A similar wave of economic growth was experienced a few years later in Western Europe. In this setting, organizations, public and private, could set relatively optimistic growth targets and expect to meet them through effective budgeting and financial control. Because this environment was so plentiful and supportive of organizational growth, competition and strategy were minor concerns of top management. Decision making at the top echelons of an organization was directed toward rapid growth and based primarily on efficiency criterion.

The organization was viewed as an independent and closed system that could be buffered from outside influences. Scientific management, and a reductionist approach to organizational control were useful for promoting efficiency in this context, as were the analytical techniques of operations research and management science, which were exciting and burgeoning fields of study during this period. Methods of strategy were neglected. Gaming, a method which had once flourished as an approach to formulating strategy, was viewed strictly as a device for management training and executive development, a perspective that would predominate for at least the next thirty years.

In the early 1950s, short term budgeting and financial control activities of organizations were, of necessity, augmented with formal long-range planning techniques. Long-range planning emerged as an organization-wide effort to define goals, programs, and budgets over longer time horizons. Top level organizational managers used trends and projections to set objectives, in an expanding economy, and in a world less dominated by escalating change and rapid fluctuation they were free to assume that the rate of growth from previous years would continue into the future. Systems analysis, trend analysis, and other quantitative and reductionist techniques became the predominant approach to corporate planning and forecasting in a relatively stable and predictable environment.

In the 1960s, the growth began to level off, and the attention of top managers turned from efficient operations to marketing and diversification in order to cope with increasing competition. Managers in the private sector were forced to consider more carefully the needs of the consumer and the activities of the competition. Thinking about the future was formalized into a _strategic planning_ process which, for the first time, addressed forces external to the organization, assessing industry trends and market opportunities.

During the 1970s and 1980s, as the institutional environment became even more complex, turbulent, and intrusive, it became increasingly difficult for organizations to plan for external events. _Strategic management_ emerged as an integrative, organization wide, approach for assessing opportunities and threats in the environment, exploring strategic options, and specifying and implementing contingencies. Traditional planning methods were recognized as limited in this environment, so new methods were sought to synthesize information from many perspective and sources, and cope with the complexity and ambiguity that now permeated the decision environment. Strategy gaming has resurfaced in this climate as an effective approach to aiding top management in exploring policy options.

Organizations in the public sector experienced these demands from the institutional environment much earlier than their private sector counterparts. The difficulties of managing competing public demands and complex social problems have been a painful reality for public officials since the mid 1960s. Policy gaming used in an educational or training context played an influential role in preparing officials for the complexities of managing public issues. An example in this arena is the huge urban policy game METRO/APEX which was begun in 1964, put into initial operation in 1967, and subsequently modified and used in many cities and Universities in North America and Western Europe. Moreover, in recent years, there has been increasing pressure placed on government organizations to become more responsive to fiscal concerns. In a sense, the problems of public and private organizations appear to be converging as they are both wedged between external influences of the public and internal demands of fiscal responsibilities.

At a global level, issues requiring international exchange and cooperation have remained complex, while technological influences have made them even more urgent. War gaming has continued to be used for exploring strategies for issues involving direct conflict. Moreover, in recent years gaming has been adapted to explore international issues requiring shared interests and cooperation (Underwood and Toth, 1987).

There are two predominant methodological approaches to making strategic decisions for organizations; both of these generally support an intuitive and informal process of discussion, assessment, and taking responsibility. The first approach involves the application of formal quantitative models to decompose and analyze the problem prior to selection of an option; finance, budgeting, and trend analysis models predominate in this approach. The second approach involves the consultation and assessment of experts acting either as individuals or groups.

Both the formal "model" and "expert" approaches have limitations. Formal models tend to reflect a narrow perspective, based on a single set of underlying assumptions. As these models become increasingly large and sophisticated their flexibility is limited. The vast commitment of time and resources required by this approach affords little opportunity to experiment with different perspectives and configurations. Considering the high degree of uncertainty and complexity in strategic problems, such models have clear limitations. Similarly, the practice of eliciting expert judgments, while being capable of bringing together a variety of perspectives, also has significant limitations. The approach is static; it cannot deal effectively with uncertainty; and the political stakes of the experts involved in the process are only rarely considered in an explicit manner. It is in this environment that "strategic" gaming has re-emerged as a new approach to formulating and exploring policy options.

Gaming has been successfully applied in several recent cases involving situations characterized by:

(1) ill-structured and consequential problems,

(2) cooperation and input from many parties, and

(3) the need for a holistic overview for judicious action.

B. POLICY PROBLEMS AND METHODS OF FORMULATION

As modern organizations become more vulnerable to forces outside their control, the problems that management faces become less well-defined and more difficult. It is the task of the top management to grapple with sets of interrelated problems. If one problem is addressed in isolation from the others, the actions taken may actually exacerbate other problems. Solutions in this context must provide an answer to what should be done, rather than merely how things should be done; strategic problems must be carefully distinguished from operational problems.

Well structured operational problems are typically based on a closed systems perspective of the organization where there is a high level of internal control over key variables. In contrast, less structured strategic problems are viewed as open systems subject to less internal control (see below):

TABLE 1 Comparing Operational and Strategic Problems

Problem Attributes	Type of Problem	
	operational	strategic
(1) ability to predict	high	low
(2) complexity	low	high
(3) conflict	low	high
(4) risk	low	high
(5) clarity of goals	clear	ambiguous
(6) information	quantitative	qualitative

Operational problems can usually be appropriately addressed with quantitative technqiues such as operations research, trend analysis, budgeting, etc; a more holistic and intuitive approach is needed for addressing the ill-structured problems characteristic of strategy making. Formal models are generally appropriate and effective in resolving operational problems, but they have been less successful when applied to strategic problems. Similarly, expert assessment is more effective at addressing objective phenomena than at exploring human responses to various strategy options.

The remaining sections of this paper consider six characteristics of strategic problems and evaluate the appropriateness of the three methods (formal models, expert assessment and gaming) for exploring policy options.

1. Ability to Predict

There is always uncertainty about the outcomes of decisions. However, the level of uncertainty increases in relation to the level of complexity when exogenous forces influencing the outcome. This makes management decision-making more difficult because it becomes less clear which option will improve and which will aggravate a given situation.

In the years immediately following World War II, when there were fewer exogenous institutional forces impinging on an organization and the rate of change was considerably slower, it was possible to review past decisions to gain insight into a current problem. Historical trends could be used to provide the basis for the planning of future operations. This was especially effective where numerous cases permitted a probability function to be constructed and the level of uncertainty could be determined through statistical inference. However, present policy situations with this level of mathematical definition are uncommon; it is more common to encounter a decision environment subject to discontinuities, surprises, and catastrophes. The statistical approach to uncertainty has not proven effective in this kind of situation.

Another element of strategic problems that contributes to the difficulty of prediction is the extent of human involvement. While operational systems are typically mechanistic and involve a small amount of human input, strategic decisions are highly influenced by human actions that are more likely to result in unpredictable outcomes. For example, in planning the Alaska pipeline, it was probably easier to predict the volume of flow than it was to predict the final location of the pipeline.

Expert assessment of impacts is superior to the formal modeling approach in situations involving high levels of uncertainty. Through experience, knowledge, and intuition, the expert can formulate policy options and assess impacts in highly complex and uncertain environments.

Similarly, gaming has the potential for superior results in an environment of high uncertainty. Gaming provides an opportunity for multiple experts to interact in exploring the design and impacts of policy options. Moreover, it provides contextual detail to stimulate thought and discussion among the participants, and a framework for exploration that addresses the issues of timing and future events in an explicit manner.

2. Complexity

It is relatively easy to make a decision when there are few well-defined decision options that impact directly on quantifiable variables for which decision criteria have already been established. However, as the number of criteria, options, intermediary variables, and relationships between these variables increases, the complexity of the situation also increases, making it more difficult for any one person to grasp the entire picture and make a fully informed decision. Decisions at the top are more likely to resemble a complex web of interactions rather than a simple decision tree. If one variable of a complex problem is altered, there is potential for movement in many related variables.

In a highly complex situation the reductionist approach of formal analysis is likely to over-simplify the problem. This may result in misrepresentation of the problem environment and in a related decrease in validity. An expert, on the other hand, may comprehend one aspect of the situation but have little knowledge about another. The use of multiple experts, whose collective knowledge covers the problem space in a more comprehensive manner, can improve upon this situation. The gaming approach not only synthesizes the opinions of experts, it also provides a framework for real time comparison and evaluation of opinions. By including stakeholders and experts in the expaloration, variables that are missed in the initial scenarios will be represented in the collective mind of the participants. If a significant variable is overlooked, it is likely to surface in discussion. Moreover, by playing through the interaction of variables in the game, the participants get a more holistic, contextual, and concrete perspective on the situation.

3. Conflict

Decisions may be difficult when there is only one decision maker; when there are multiple decision makers (or groups, coalitions, etc.), the difficulty is likely to increase as a function of the both level of interdependence and conflict that exists among them. If the individuals involved in the decision are identical in their values and objectives, they can be considered monolithic and the collective decision is similar to an individual decision. If the individuals involved in the decision have conflicting objectives, then resolution of conflict becomes an integral part of resolving the problem. When this situation exists, the decision makers need an approach to:

(1) elicit and clarify objectives,

(2) clarify differences in beliefs and values, and

(3) identify and explore options for possible agreement.

Formal models can be used for analyzing conflict and negotiation, but they have limitations with respect to identifying and exploring new options. The expert approach is usually based in the consensus mode taking great care to minimize conflict. Gaming, however, originated from the need to explore the dynamics of conflict; to develop strategies under extreme conditions of conflict - war. Exploring solutions to institutional conflict is a natural extension of the gaming approach. Through the play of a strategy game, the underlying objectives of the parties crystalize, surface and become more explicit; points of conflict can be identified; and potential options for resolution can be developed and tested. The potential of a game for assisting individuals in viewing the situation from other perspectives through playing other roles is a major benefit of this approach.

4. Risk

Managers must formulate strategies that will prove effective in terms of the overall mission of the firm. If they succeed, the firm succeeds; if they fail, they may risk the very survival of the organization. Management may elect incrementalism, delayed action, and/or conservative policy to reduce risk, but many decisions still involve a high level of risk. Decision aids cannot reduce the stakes involved in making these decisions, but they can assist in reducing uncertainty by collecting and focussing pertinent information. Managers need a problem solving approach that helps them collect and use information while considering the range of potential options and related outcomes from many perspectives.

Formal models are limited in this respect because they usually involve narrowing the decision options to a small number which are then decomposed and analyzed in terms of one aggregate decision criterion. They can usually formalize the level of uncertainty by assigning a probability to it, but they seldom actually reduce the level of uncertainty by adding new information. Moreover, formal approaches do little to assist in developing contingencies.

Expert analysis may reduce the level of uncertainty by bringing new information to bear on the problem, and by introducing new options, however, the expert approach has a limited capability for integrating knowledge and exploring potential impact. Gaming, however, can be applied to decision problems of this nature to help management explore options, contingencies, and impacts from a variety of perspectives. Each game participant brings their unique knowledge and perspective to the exercise. As issues develop in the game, this knowledge is introduced as it is needed. Through the interaction of the participants the perspectives, knowledge, and information is synthesized and applied to the problem in a systematic, yet intuitive manner.

5. Clarity of Goals

The foregoing discussion assumes that the mission, goals, and objectives of the organization are known, accepted, and internally consistent throughout the decision process. This is not always the case. It is not uncommon to find uncertainty, ambiguity, and a lack of clarity regarding the mission and goals of the organization. This points to another limitation of formal modeling in the context of strategy formulation: different assumptions regarding the mission are seldom substituted in normative decision models, because of the time and expense; descriptive models do not address the question of objectives at all. When formal models are applied, it is most common to assume one set of quantifiable objectives and decision criteria. If part of the problem is that the objectives are in question, then this approach will usually be misdirected.

The expert approach also tends to overlook the exploration of objectives. The experts will typically assume the initial or explicitly stated objectives of the decision maker; the hidden agendas of the experts are seldom made explicit or challenged in a systematic manner. If goals are ambiguous they generally remain that way.

In gaming, the mission, goals, and objectives are explored as part of the overall situation. The gaming approach to policy formulation assists in making the participant's goals and objectives be explicit; the mission of the organization is not isolated from the rest of the situation. Through gaming, modifications of the mission can be explored and tested like other components of the overall problem. By accelerating the simulated passage of time, games require the participants to act and to interpret their actions in terms of outcomes. It is though the retrospective assessment of actions in debriefing that objectives are clarified and perhaps formulated for the first time. Through assessment and judgement of the gamed behavior the participants can formulate and test new objectives. Entirely new objectives can emerge out of the juxtaposition of behaviors, rationales, and participant responses.

6. Information

In theory, all problem situations can be quantified; in practice some problems are more amenable to quantification than others. There are a number of possible reasons to use qualitative data rather than quantitative data. A sample of possible reasons are listed below:

1) contextual and holistic presentation of data is more meaningful and preferred,

2) quantified data is not available and is expensive to collect,

3) data cannot be easily defined by exhaustive and mutually exclusive attributes,

4) inferential statistics are not appropriate (e.g., the sample is not large enough, etc.), and

5) the relationships between the variables are ambiguous, unknown or cannot be accurately represented mathematically.

This is not meant to be an exhaustive list, but rather to represent various occasions where qualitative data may be preferred to quantitative data. if the data that is central to the problem happens to be qualitative, or the quantitative data required for the use of a formal model is not easily available, then the formal modeling approach may not be appropriate for the support of policy formulation. In these situations, the expert approach is usually better at accounting for both the quantitative and qualitative aspects of a problem. However, there is a need to integrate quantitative data of formal models, qualitative data, and the knowledge of the expert in support of strategy formulation. Gaming can be used as a process for synthesizing, presenting, and assessing information from a variety of sources; subjective and qualitative information as well as quantitative data can be employed.

Gaming has major advantages over formal models and the expert assessment techniques. We are not suggesting that these approaches to assisting policy formulation be replaced by gaming; rather they can be effectively supplemented through the gaming approach.

In conclusion, the evolving nature of strategic problems in both the public and private sectors indicate the need for a new approach to problem exploration and assessment that combines the natural advantages of the expert approach with procedural structure that:

1) synthesizes information and perspectives of stakeholders and experts,

2) provides contextual detail to stimulate thought and discussion among stakeholders and experts,

3) provides a systematic framework for addressing the timing of strategies and events,

4) presents the complex relationships among objective and subjective variables,

5) clarifies objectives; differences in beliefs and values; points of potential conflict; and possible options for agreement among key stakeholders,

6) targets essential information from multiple sources,

7) treats goals as part of the problem and explores their nature through action and retrospective assessment, and

8) supports the use of both quantitative and qualitative information.

In our recent experiences in addressing strategic level problems in both the public and private sectors, we have found strategy gaming to be compatible with this set of specifications. Moreover, gaming appears to be uniquely appropriate for exploring strategic options in an ill-structured environment characterized by unpredictability, complexity, conflict, high levels of risk, and ambiguity.

REFERENCES

Duke, Richard D. 1974. Gaming: The future's language. Beverly-Hills: Sage.
Underwood, Steven E. and Toth, Ferenc. 1987. Improving the policy/science interface: A teleconferencing exercise for managing long-term, large-scale issues. IIASA Working Paper. Vienna, Austria: International Institute of Applied Systems Analysis.

NOTES

1. The terms policy gaming and strategy gaming are used interchangeably throughout this paper. Although policy making generally refers to the public sector decision process, and strategic planning refers to the private sector decision process, the problems addressed by each of these approaches are similar in terms of overall structure. Moreover, the practice of gaming in these contexts is quite similar.

Threat and negotiation in gaming and simulation

Arnaldo Cecchini
University of Architecture, Venice, Italy

ABSTRACT: It is an attempt of classifying the different types of conflict using Rapoport interpretation and criticism of the Theory of Games. Particular emphasis will be given to the role of threat and negotiation as communicative and simulation strategies within the various types of conflicts and also to the opportunity, illustrated by examples, to apply to these conflicts not so much the techniques of game but those of play where simulation and communication play an essential role.

KEYWORDS: game, fight, debate, threat, negotiation, communication.

ADDRESS: Cannaregio 2669, 30121 Venezia, Italy. Phone: 041/720264.

INTRODUCTION

In these pages I would like to focus our attention on the phenomenological aspects of conflict, but without analysing the causes which have given rise to it or trying to find out which actions, once removed, would prevent its existence.

Above all we shall see how to handle a situation of extreme conflict, i.e. a crisis, and in so doing we might discover that intervening in such situations and taking exceptional measures often constitutes an even locally ineffective strategy.

One of the mistakes often made in crisis analysis is to forget that phases of extreme local conflict do not always imply a worsening of the global conflict. A succession, even violent, of assaults, of military attacks or of verbal aggression, might not be the prelude to an escalation of conflict but an instrument of pressure in order to maximize negotiation effectiveness. We simulate war, we threaten opponents with indiscriminate use of violence, we put into motion the entire gameboard in order to negotiate with greater strength, credibility and pressure. Although it may seem cynical, it is wrong to forget that a bomb in a hotel might even be a message asking for an armistice, an agreement or peace. In these contests another common mistake has often been the application of the badly sharped dichotomic rasor of the question "cui prodest?".

Another point worthy of attention is the frequent operation of exchange which takes place when an external threat is enlarged, expanded or even made up in order to control, suppress or eliminate an internal conflict, or in order to build up an effective structured threat against internal rivals. Another mistake is to think that there could exist a satisfactory model for every type of conflict, derived from the set of simulation techniques known as Game Theory. I shall deal later and in detail with Jon Von Neumann's concept of a game. For the moment let us say that in a game the objective of each player is to maximize rationally defined profits, or as Rapoport (1960) says "to outwit the opponent". So, although games tend to assume these characteristics, a real-world situation very often does not. And in fact one of the objectives of rational conflict management could be to introduce this characteristic into the situation. From time immemorial, especially in international conflicts, it is by no means uncommon that at least one of the party's main objectives is to harm the opponent who is regarded as different, and as absolute evil (this is a fight, "to harm the opponent"). Obviously, in such a situation the techniques necessary to win are different. Even where, as in some religions or political ideologies, the objective of conflict is to persuade the opponent of the indisputable truth of their opinions (this is a debate, "to convince the opponent") the distribution of the techniques must be different. Once again the hypothesis of an equilibrium obtained as the necessary result of rational moves is weakened.

On the other hand, the final aim of a protagonist could be the conversion of the other. To obtain this result use is made not only of sermons and theological summae (debate) but also of regulated competitions (game) and/or the radical intervention of the Grand Inquisitor to estirpate heresy (fight). On the other hand the final aim could be the opponent destruction. yet we cannot refuse to admit that indiscriminate fights and terrorism (fight) might not be used in addition to economic propaganda (debate) and/or unions and political battles respectful of the institution rules (game).

It is worth noting that quite often the real opponent in a conflict is not the one officially announced. The fight between the English National Union of Mineworkers (NUM) and the National Coal Board (NCB) during the long strike which ended in 1985, was only superficially between these two organizations. In fact the closing down of a certain number of pits was the apparent object of the conflict. In reality it was a fight between the long-term policies of the Conservative party (i.e the reduction in power of Trade Unions) and those of the radical wing in the Trade Unions movement (i.e. the downfall of Margaret Thatcher).

A game was being simulated in order to carry out a fight. And even if we use a game mode the conflict had a pay-off matrix quite different from the reasonable one of labour conflict, and consequently it had a different optimal strategy, that is a different point of equilibrium. (1) We assign to the NUM three possible strategies: Negotiate (N), Conflict (C) and Total Conflict (T), corresponding respectively to an institutional negotiation without fight, to a classic union fight and to an all-or-nothing battle. (2) We admit that these are possible strategies for a radical left (we shall refer to this as AS, i.e. incarnated in the person of Arthur Scargill). (3) We assign to the NCB the following three strategies: Dismiss (D), Come to an agreement (A), Do not Dismiss (S), corresponding respectively to the closing down of all the pits, to a compromise, and to the renouncing of closing down pits. (4) We admit, finally, that all these strategies are those that are open to the conservative Government (we shall refer to them as MT, i.e. Margaret Thatcher incarnate).

We shall see that, for various reasons, some elements of the pay-off matrix are impossible, e.g. the Negotiate/Do not Dismiss (NS) couple must be discarded since this situation cannot create a real conflict. The order of preferences for the NUM will be NS, CS, NA, TS, TA, CE, TD, ND, while the order of preferences of the NCB will be: ND, CD, CA, NA, TD, TA, CS, NS, TS. The point of equilibrium will oscillate between CA and NA, that is in the compromise area. If we assume CA as improbable, then NA will be the most probable solution; here the degree of agreement will obviously depend on the balance of power during the fight; this is true for the "simulated" game NUM vs NCB.

In the "real" fight Arthur Scargill (AS) vs Margaret Thatcher (MT), the order of preferences for AS will be: OS, TA, CS, TD, CA, CD, NS, NA, ND, while for MT it will be: TD, CD, ND, CA, NA, TA, TS, CS, NS. In this game the solution will oscillate between TS and TD, depending on the balance of power. The truth is that AS and MT are not really playing a game but fighting, and that the game is only a fictitious element, a tactical phase of a wider competition, a competition which itself might be the basis of further games or fights or debates.

Let us try to simplify the discussion by taking the example of just two components: (1) There exists only three fundamental modes of conflict which define the global strategic aim of a player: fights, games, debates. (2) While games have been systematically classified, fights and debates have not. Figures 1 to 3 attempt to explain the taxonomy of games and to provide a rough outline of the other two forms of conflict. (3) Each mode of conflict can be made up of a finite number of matches (i.e. an "atomic" element, with its own objectives, which we shall call short-term strategy) which in turn can constitute fights, games or debates. Each match can in turn become a game, a fight, a debate or a set of the three in a linear or branched succession, interacting, retroactive and nested. The phases of a match will be called moves, the objectives of the moves will be defined as tactical (Fig. 4). (4) It is important to know if we are playing a game which is a move in a match; a match which is a game of a mode which is itself a game, or if we are playing a game which is the move of a match, a match which is a fight in a mode which, on the contrary, is a debate and so on.

Sometimes the systematic defeat of a player, no matter how good he might be, can depend on a misunderstanding of the conflict mode. Reagan and Thatcher have upset the classic scheme of a bi-partisan policy which forced both parties to "occupy the centre" not only because Reagan and Thatcher have perceived a deep tendency in their societies, but especially because they have been able to give substance to it by transforming the game into a fight. it is true that we can fight against a club using a sword, but not if we believe that the opponent's club is a sword!

Now at last we are able to examine in detail the various types of conflict, and then to understand the moment when negotiation comes into play and the purpose and effect for which we can use threat. First of all we cannot ignore the close relationship between negotiation and bargaining. Bargaining is the basic technique, at least at the outset, of every kind of transaction (and this should be remembered by those who regard trade as a game).

The concept of negotiation, however, has taken on a special meaning in the field of international relations where it is identified with the concept of "diplomacy". In "internal relations", negotiation exists wherever "a social system chooses the principle of individual freedom and, at the same time, imposes the respect of authority which cannot however codify the infinite variety of relations, exchanges, transactions, rights and duties between citizens and between them and the holders of power" (Bellenger, 1984). Consequently, when a difference of opinions cannot be composed by a higher authority or by a fight which eliminates or defeats or weakens the opponent considerably, when it is impossible to find a satisfactory point of equilibrium using the unbiased (quantitative) evaluation of a pay-off matrix, then the technique to be used is negotiation.

Contrary to what has been maintained (Anzieu, 1971) negotiation is not a typical human behaviour, consequently it is not necessarily a cultural or verbal act. Nonetheless, the act of negotiating depends on believing that a difference of

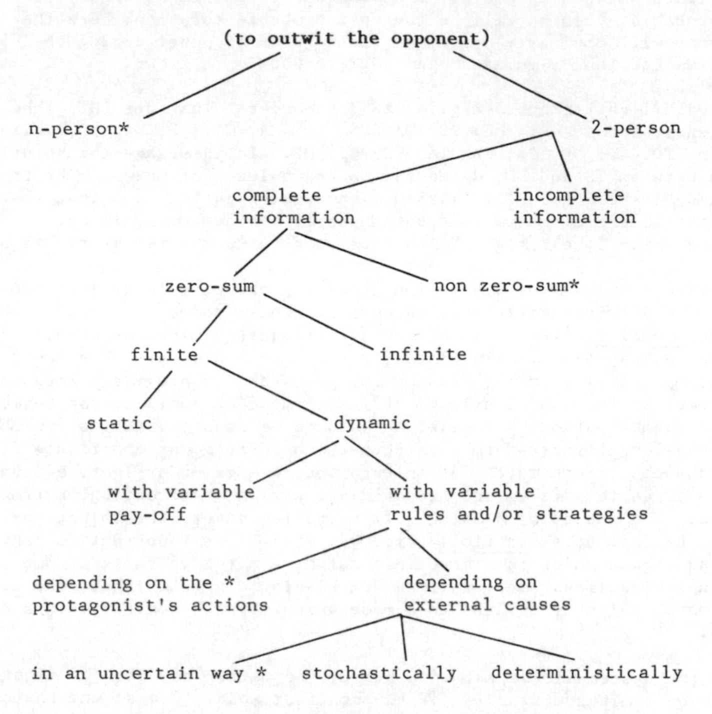

games followed by * admit and/or require negotiation

Fig. 1

GAME KEYWORDS

GAME: a game is the totality of the rules which describe it (Von Neumann - Morgenstern, 1944).

RULES: absolute commands (among which there are the beginning and the end of the game).

STRATEGY: a plan describing which moves must be made in every possible situation, yet Von Moltke (1982-1912) wrote that: "strategy is a set of expedients (...), war, like every other art, cannot be learned rationally".

RATIONAL BEHAVIOUR: "an economic subject behaves rationally if he acts so as to obtain the **maximum** utilities or profits" (Morgenstern, 1969).

OPTIMUM STRATEGY: the strategy suggested by a rational behaviour.

Fig. 1a

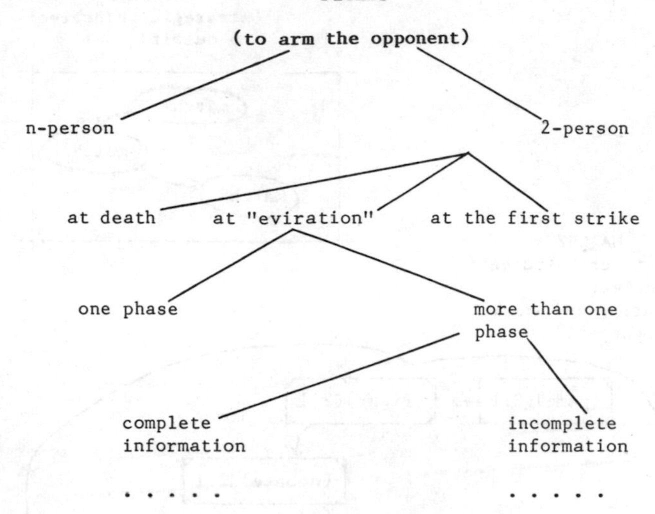

In general fights do not admit rules. In some type of fights there might be a pay-off matrix.

Fig. 2

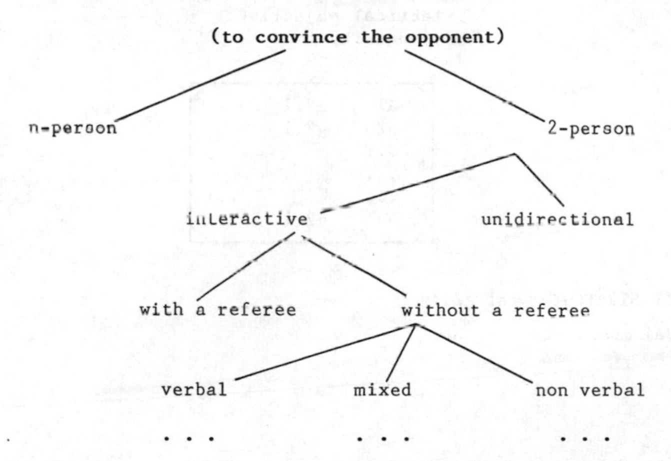

In debates rules are admitted, although not necessarily; there might be negotiation and threat.

Fig. 3

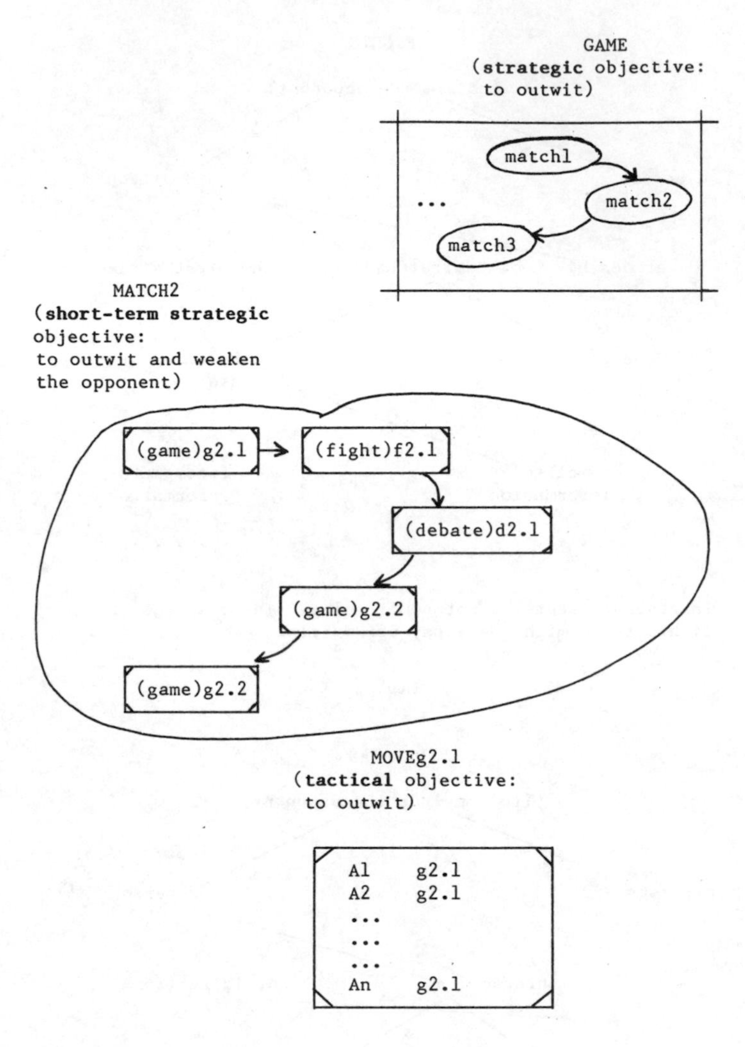

Fig. 4

opinion can be reduced by a debate: it is an information transfer during an agreed-upon (bon gré, mal gré) meeting between two individuals in order to benefit one of them. The objective of negotiation is an agreement (a compromise, concessions, "innovations"). Negotiation assumes that all the main characters have an aim and a margin of manoeuvre and that there is a balance of power.

Negotiation techniques represent a particular and relevant aspect of communication skills, continuously subject to thorough analysis, especially in commerce. We shall not deal with them extensively: let us just mention the win-win negotiation, the principled negotiation, the agreement/authority negotiation technique, the soviet style negotiation (Fisher and Ury, 1981; Cohen, 1980; Blaquère, 1983; Bellenger, 1984; Howard, 1976; Shelling, 1970).

Now we shall examine carefully the point in which, within the different modes of conflict, a negotiation can arise (see also figures 1-3). Tactically it will be used in a fight either as a plotting technique or because we are forced by the balance of power. On the other hand it may be used as a structural aspect of a debate, although not all the time. In fact debates often put principles to the fore, a choice which excludes negotiation since principle do not allow this.

Finally negotiation will be an essential component of various games, e.g. in two-person games. First, non zero-sum games, where negotiation can prevent the occurrence of the disastrous outcomes of an equilibrium, based on "rational" choices, such as in the prisoner's dilemma. Secondly those games where the pay-off matrix is dynamic and there is either complete uncertainty about its changes or such changes depend on the players' behaviour, e.g. the irresistable ascent of the dollar in the well-known game auction dollar. Finally games where the possible strategies can vary, varying the rules as a consequence. In n-person games (n>2) negotiation is almost always a component which is necessary to establish compensations which allow realizations of coalitions among the players (Howard, 1976).

Clearly, as soon as a game models a real-world conflict situation, even if only tenuously, we cannot avoid resorting to the use of negotiation. Consequently, the idea of using the Theory of Games as a General Theory of Conflicts is also limited in its application, which is the inevitable occurrence of negotiation, a phase which, of course, can be modelled and codified but is basically creative. This does not mean that it should be considered as a poetic or miracolous process. As a "negotiator must combine the quickness of an excellent fencer with the sensitiveness of an artist" (Neirenberg, 1970) he strongly resembles the character of Don Isidoro Parodi, created by Borges, who is engaged in the solving of mysterious cases in his cell and whose "cold specultive mind confirms the genial artist's intuition" (Borges, 1983). Don Isidoro belongs to the category of the great detectives whose technique is based on the ability of making shrewd inferences, in a vertiginous crescendo which springs from the single audacious abduction (Eco and Sebeock, 1983) to serendipity (Cerulli, 1975), a fragile and gigantic castle, its precarious nature built upon by the great "bloodhounds", from the founder August Dupin (Edgar Allan Poe), through Sherlock Holmes (Arthur Conan Doyle) and Nero Wolfe (Rex Stout), to the "mastiff" William of Baskerville (Umberto Eco).

But what about threat? At what point of a negotiation or of a conflict does it arise? And for what reason? What are the possible outcomes? The Italian Penal Code defines a threat as "the offence committed by a person who provokes in others the fear of unjust reprisal evoking future harm, the occurrence of which depends on the agent's will". According to this definition, if in a game type conflict a player states his intention of adopting a strategy allowed by the game rules, (e.g. in a labour conflict the strategy of "strike") this is not a real threat, although this action may appear fearful for the opponent. Threat obviously is not an action, but it is neither a "potential" action; it is the conditioning of other people's behaviour. The announcement of a future action is made for the purposes of avoiding action. Again this is part of a communicative strategy as well as negotiation; it is a different mode of the same strategy.

In terms of the Theory of Games a threat couple (Nash, 1951), is the point determined by the two strategies A1c and B1c, which are not in a Pareto equilibrium. A threat strategy A* (Blaquière, 1983) is the strategy of the first player A, known to his adversary B, whose purpose is to limit the number of wins. Quantifying this limit, we can define the threat value of player A on player B as mAB = -V2(A*,mB') where B' is the defensive strategy player B will adopt if he accepts the limit established by strategy A*. However, B can respond with an attack strategy B", a fact which will determine a risk for A, i.e. the risk of having to limit his wins because of B's attacks. Then the risk value for player A as regards his threat strategy will be rAA = -V1(A*,B").

Yet since these are non zero-sum games the choice of using or not using a threat strategy for A and of reacting with an attack of defence strategy will in general be matter for negotiation in a communication phase. In order to be effective a threat must (Lover, 1982): (1) be credible and/or perceived as credible, (2) be based on accurate information, (3) allow information and intention to be retained (i.e. dissimulation), (4) be based on a certain balance of powre, (5) allow the protagonist to conceal the "quickness" of his own actions, (6) be carried out with time to spare, and (7) not be based on skates that are too important.

Let us analyse some of these points, in particular No.4. An action with a potentially suicidal outcome actually has the effect of increasing, in a desperate but not ineffective way, the force of an opponent, thus re-establishing an equilibrium, sometimes in an exceptional way, in the balance of power. This happens because the risk value of the attacker was considered as very low, which raised the comparative way of his threat strategy. This leads us to No.7 which introduces a (further) element of subjectivity into the conflict. Some games, apparently of the non zero-sum variety, does not properly follow this pattern for the following reason. The opponents assign a thoroughly different value to the winning of a certain sum and/or to its loss.

In order to establish in which kind of conflict situation threat appears, I shall point out, as I have done above, that it can be present in a number of situations, the first being negotiation. It can be found (1) in fights, but in this case it can be a simulation of a threat (e.g., a provocation to force the opponent to fight). (2) The third situation is a debate. Let us remember that the weapons of a verbal debate are included in the set of codified rules referred to as Rethoric and that among them two of the most effective, although "unfair", techniques are logical fallacies and entymemes. Some of the false arguments seems to be real threats (one of them is called ad baculum), e.g. "if you don't believe what I say..."; "you'll make your lover suffer, if..."; "remember that Aristotle said so"; "you'll go to hell, if...".

Pragmatics has a prevailing informative function, but only in the pious intention of enlightened logicians. In the collection of "dirty" techniques used in debates we can in fact find personal attacks, slander, procès d'intention, bad faith, intimidation, ironicising, derision, corruption, defamation, scandal, beating about the bush, blackmail, double bind (Bateson, 1986), etc., until we arrive at torture and physical elimination tout court. Yet our analysis cannot be satisfactory if we forget to point out that threat sometimes can be seen as a real move, not only as a tool to impose negotiation.

In the game perverse philanthropist the threat for the players is to fail to find an agreement. Let me remind you of the game. A rich man possessing a great, but undetermined, amount of money offers $1000 to players A and B for the first minute of play if they agree on the way of dividing the money between themselves. If they do not he will take the money from the stakes and put another $1000 in the game. The game will presumably last for a long time and in any case it will be decided by the philanthropist. Apparently things are easy: let's go fifty-fifty.

But let's see what happens if A adopts a threat strategy and says: "the only way of sharing I accept is $900 to me and $100 to you", adding confidentially,

"You see, outside there is an enemy of the philanthropist who is offering me $850 for every game that I cause to fail, so the only way I have of winning is for you to accept my proposal. You'll see that this is the best way for you because I earn $50 a minute and you $100". In this case B is in a tight spot; in fact he might believe that A is bluffing and then in turn bluff himself, but he will always fear that A is not bluffing. If we remain in the abstract game, the threat of A has no solution. If it has been carried on well (remember the 7 points) it puts B in a difficult position, with almost no way out. A waits quietly, sticking to his offer, avoiding any discussion or argument and rejecting any pleading and enticements. Is he really bluffing? Time passes and A stays put, saying he is earning $900 each minute. His bluff, if any, is really an <u>investment</u>; if he bluffs he loses money now to earn more later. Time passes. He cannot be punched because rules do not permit it. The time pressure leaves him cold; if he does not bluff he will lose money, but a great deal less than B, both in total ($50 against $100) and in percentage (5.5% against 100%). Time passes.

This game is less rare than might be imagined. It is a model of, and a metaphor for, many real-world conflicts; it is present in the deep nature of law and in questions of equal rights. If all of us have the same rights, why should someone else have more? Perhaps because he presumes that he descends from the gods, or that he deserves more from society, or that he has other means?

This situation is very interesting since it leads us, at last, into the heart of the problem, which is that of the term <u>game</u>. The only choices that B can exercise to get out of the game matrix cage is to step from the game into <u>play</u>, from the abstract matrices to a real <u>performance</u>. Before choosing a game strategy, even before a communicative strategy, he will need to have an informative and forecasting strategy; he will play not against a rational and abstract opponent but against a real person. He will need to know who he is playing against; he will provoke the other to sound out the latter's credibility, his interests and faults, the importance he attributes to money, who he knows and how he has behaved in other situations. He will also try to know who the philanthropist is, if he has got any enemies, of what type, why he invented the game, etc.

Doubtlessly there are games without play, but when in a game negotiation and threat become possible or necessary then play inexorably appears. Consequently we enter the field of <u>Played Simulation</u> which enables us to analyse and forecast before choosing the appropriate strategy.

REFERENCES

Anzieu, D. 1971. <u>La psychologie de la négociation entre les groupes</u>. Louvin: Université de Louvain.

Bateson, G. 1972. <u>Steps to Ecology of Mind</u>. London: Chandler Pu. Co.

Bellenger, L. 1984. <u>La Négotiation</u>. Paris: Presses Universitaire de France.

Blaquiére, A. 1983. Introduction a la Theorie Mathematique des Conflits in <u>Etudes Polemologiques</u>, 27 Avril.

Borges, J. L. and Casares, B. 1983. <u>Sei problemi per Don Isidoro Parodi</u>. Roma: Editori Riuniti.

Cohen, H. A. 1980. <u>You can Negotiate Everything</u>. New York: Bentham Pu. Co.

Cecchini, A. 1980. Simulazione, Giochi, Giochi di Simulazione in <u>Enciclopedia della Pianificazione Territoriale e dell'Urbanistica</u>, Vol. vii, Milano: Franco Angeli.

Cerulli, E. (ed). 1975. <u>Peregrinaggio dei Tre Giovani Figlioli del Re di Serendippo per opera di M. Cristoforo Armeno della Persiana nell'Italiana Lingua Trasportato</u>. Pisa: Accademia dei Lincei.

Eco U. and Sebeock, T. A. 1983. <u>Il Segno dei Tre</u>. Milano: Bompiani.

Fisher, R. and Ury, W. 1981. <u>Getting to yes. New York</u>: Houghton Mifflin.

Howard, N. 1976. A dynamic Theory of Games. Bruxelles: Centre d'Etudes de Recherche Operationelle.
Lover, M. 1982. The Crime Game. London: Robertson.
Morgenstern, O. 1963. Spieltheorie und Wirstschaftswissenschaft. Wien: R. Oldenburg.
Nash, J. P. 1951. Non-Cooperative Games in Annals of Mathematics. Vol. 54, No. 2.
Nierenberg, G. 1970. L'art de persuader et de bien négocier. Paris: Tchou.
Rapoport, A. 1960. Fights, Games and Debates. Ann Arbor: University of Michigan Press.
Schelling, T. C. 1970. The Strategy of Conflict. Howard: Howard University.
Von Moltke, H. K. B. 1892/1912. Militarische Werke. Berlin.
Von Neumann, J. and Morgenstern, O. 1944. Theory of Games and Economic Behaviour. Princeton: Princeton University Press.

An interactive solution to the assignment problem

Jacques-Marie Aurifeille
Jean-Michel Morel
Université Paris-Dauphine, France

ABSTRACT: In the assignment problem, a series of goods, places, etc. has to be shared out among several participants. Each of them has to indicate his preferences by giving to each good a weight. The optimal solution is then obtained by a classical operations research algorithm. This paper proposes to dynamize this usual process, replacing it by a series of simulations which allow each participant to get informed on the preferences of the others and to adjust his own choices. This leads to an interactive assignment procedure which has been experimented in real problems and simulated at ISAGA '86. A detailed account of the ISAGA session is given.

KEYWORDS: assignment, interactivity, simulation, strategy, game, heuristic, system, control.

ADDRESS: CEREMADE, Université Paris-Dauphine, Place Du M[l] de Lattre de Tassigny, 75775 PARIS Cedex 16 France. Tel. 4258 5846.

A. IMPORTANCE OF SIMULATION AND COMMUNICATION IN THE ASSIGNMENT PROBLEM

1. What is the Assignment Problem?

A nonzero sum game between n players who must share a gathering of places. Each player can only obtain one of these places and naturally tries to get the one that he prefers. For example: a group of foreign missions to repart between experts, periods of professional probation between students, professional transfers in large organizations, flexible timetables (in banks, supermarkets...)

Until now, the game has only been faced through static aspects: A list of possible choices is presented to the participants. Each participant has to indicate his preferences, giving them a "weight" expressing their level of interest. For example, in the case of foreign missions: Mister E1 asks for Rome (weight seven), Athens (four), Brussels (one) and Mister E2 asks for Rome (five), Paris (five), Geneva (two). This defines a matrix of choices:

Towns:	Athens	Brussels	Geneva	Paris	Rome
Player E1:	4	1	-	-	7
Player E2:	-	-	2	5	5

The matrix of choices can be treated on computer according to some operations research algorithm. This algorithm has two goals: First it maximizes the size of the solution, i.e. the number of participants effectively assigned. Then, it maximizes the value of the solution, i.e. the sum of the weights of the selected couples. With the matrix given above, we easily find an optimal assignment of size two and value 7+5=12.

Next chapter will show that even a static process such as this one permits the existence of individual or coalized strategies. Thus one has to deal with a game, but a static one with a single step, dedicated consequently almost exclusively to the expression, collect and treatment of datas.

2. Inadequacy of the Noninteractive Approaches

We shall point out three phenomena: The heterogeneity, the instability and the underoptimality of static approaches.

a. Heterogeneity of the problem. According to the static approaches, the optimality of a solution depends only on its size and value. These approaches rely on the concept of utility, derived from microeconomy, where people are supposed to express stable preferences with identical criteria. This is not verified in the case of assignment problems, because of the nonlinearity of the preferences. It is obvious in the following example that weight 2 has not the same meaning for players E1 and E2:

Weights given by E: 7,2,2,2,2.
Weights given by E2: 5,4,3,2,1.

b. Instability. If a player knew that one of the proposed places is much on demand, he might give up the struggle and transfer the corresponding weight on more accessible places. Then, it is not the solution which is unstable, but the problem itself. Informations on the structure of the game should then improve the quality of the obtained solution.

c. Underoptimality and heretical strategies. To inform the players is a priori an advantage. However, it introduces new difficulties, particularly by giving to the participants larger possibilities of elaborating noncooperative strategies. Let us consider for instance a problem where each participant has to repart 10 points on three choices out of a list of 10 towns. If no information is available about the preferences of other players, the best strategy for every one of them is to tell the truth. Let us suppose for E1 that truth telling would give: Brussels 6, Amsterdam 3, Venice 1. if he learns that Paris and Rome are much on demand, he may follow a strategy of this kind: Brussels 8, Paris 1, Rome 1. Indeed, he knows that the algorithm maximizes first the number of assigned people. By lying on his preferences he "forces" the algorithm to assign him to his first choice. Coalitions of participants can be formed on the same principle. If E1, E2, E3 respectively prefer Brussels, Amsterdam and Venice, they may form a core of preferences:

E1: Brussels (8), Amsterdam (1), Venice (1)
E2: Amsterdam (8), Venice (1), Brussels (1)
E3: Venice (8), Brussels (1), Amsterdam (1).

All these strategies can only lower the optimality of the obtained solution. Each player appears then to have a potential power on the game. This is not taken into account by Demange (82) who demonstrates the strategy proofness in the assignment market game only when the size constraint is neglected.

Another disadvantage of information is that it risks to exacerbate personal rivalties. If E1 and E2 prefer Rome, it is a good thing to inform them on the difficulty of getting it provided that no name of participant is mentioned.

To sum up, finding an efficient assignment game requires answers to the following questions:

- Which information policy for the players?
- How to control potential powers?

B. PRINCIPLES AND RULES OF THE SIMULATION

1. Principles

The above discussion leads us to defining two types of structure variables for characterizing the problem. We shall use information variables which will enable the players to direct their choices, and control variables permitting to measure potential powers.

As information variables we shall choose:

$$C(s) = \sum_{e \text{ in } E(s)} p(e,s), \quad \text{for each place } s.$$

E(s) denotes the set of all players demanding the place s, and p(e,s) denotes the weight given by player e to the place s. C(s) may be interpreted as the share of place s, being a realistic measure of the difficulty to get s.

As control variables we define the potential power of a player e with:

$$X(e) = (1/\text{Card } S(e)) \sum_{s \text{ in } S(e)} p(e,s)C(s).$$

S(e) is the set of all the places demanded by e. X(e) measures the difficulty to assign e.

A more detailed discussion of these definitions is available in Aurifeille (84). An axiomatical definition of C(s) and A(e) might be given and lead to the above formulas. However, such an axiomatic would suppose that the assignment problem is well defined before any information process. In fact, it is the use of C(s) and X(e) which gives way to an optimization and stabilization of the problem.

2. Rules of the Assignment Game

We reproduce here the rules of the game IRENA (Interactivity Regulation, New Approach) presented at ISAGA '86:

- In this game we intend to assign 12 candidates to one week long stays in 12 different towns. There can only be one person assigned to one town. At the first step, you simply have to indicate your prefered town. If everybody can be assigned, the game will be over. If not, you will have to repart 50 points between the towns that you prefer, taking care that your level of exaction should not exceed a norm fixed by the organizators. Here is how to calculate your level of exaction. Each town has a price, which is simply the number of points given to it by the candidates. If, for instance, you put 40 points on Venice, the price of which is 80, and 10 points on Amsterdam (price 60), your exaction level will be:

(40*80 + 10*60)/2 = 1900

Another example: 5 points on Oslo (price 30), 10 points on Brussels (price 40) and 35 points on Paris (price 50) result on a level of exaction equal to

(5*30 + 10*40 + 35*50)/3 = 767

Please indicate your first wishes on a sheet!

At the end of each step the prices of the towns were recalculated and communicated to the players. A new limit was fixed for the exaction level. As long as the algorithm had not succeeded to assign everybody, players were asked to redistribute their 50 points. Before presenting and commenting the outcomes of the ISAGA session, we are now going to discuss the algorithmic problems.

3. A Heuristic Algorithm for the Assignment Problem

By heuristic algorithm we mean a process of calculation which gives a satisfactory solution neglecting too sharp combinatorial aspects of the problem. This reduces considerably the time of computation by comparison with the exact algorithms of Busacker and Gowen (1961), Edmonds and Karp (1972), Tabourier (1972), Balinski (1986). This rapidity is the key to our multistep simulation.

A second advantage of the algorithm that we present comes from its transparency, for it is based on the state variables C(s) and X(e).

a. The heuristic algorithm. The idea is to transform the initial assignment problem (P_o), defined by the weights $p_o(e,s)$, into a series of problems (P_n) defined by $p_n(e,s)$ so that:

(1) $p_n(e,s)$ tends to 0 or 1 for any (e,s).

(2) $\sum_{e \text{ in } E(s)} p_n(e,s)$ and $\sum_{s \text{ in } S(e)} p_n(e,s)$ tend to 1 for any e and any s.

Conditions (1) and (2) express that Problem (P_n) converges to a solution of (P_o), e being assigned to s if and only if $p_n(e,s)$ tends to 1. We now define iteratively the $p_n(e,s)$: We first consider, instead of the $p_n(e,s)$ the relative weights $p_n(e,s)/C_n(s)$. Then one applies to these relative weights a function g verifying:

(i) g is convex C^1.

(ii) $g'(0)=1$

(iii) $g(1)=+\infty$

A good choice is $g(x)=x/1-x$. The interest of g is, through (iii), to increase the relative weights which are close to 1 without modifying notably the small weights (condition (ii)). Our transformation is actually:

$$p_n(e,s) \rightarrow g(p_n(e,s)/C_n(e,s)) = q_n(e,s).$$

Finally, one has to restore the equality between players by setting:

$$p_{n+1}(e,s) = q_n \,/\, \sum_{s \text{ in } S(e)} q_n(e,s), \quad \text{which implies:}$$

$$\sum_{s \text{ in } S(e)} p_{n+1}(e,s)=1 \text{ for each player } e.$$

b. Explanation of the heuristic. To replace $p_n(e,s)$ by $p_n(e,s)/C_n(s)$ means that the new weight of the couple (e,s) takes into account the other demands concerning s. These demands are themselves considered after their relative weights. Thus each step explorates a wider range of interactions. The choice of g is not arbitrary. The more g increases, the more quickly the heuristic converges, but with an accentuated risk of underoptimality in the obtained solution. The rate of increase of g measures the rapidity with which one assignment decision is taken.

C. EXPERIMENTS

Two kinds of experiments have been organized to test the validity of our process.

1. Stochastic Simulations of the Heuristic

Approximately 300 experiments of this kind have been carried out concerning problems of all sizes from 10 to 150 choices by player and varied difficulties. In all cases the heuristic assigned the same number of players as the exact algorithm and gave a solution, the value of which exceeded 99% of the optimal value. The gain on computation time went from 1 to 40 times less than with the best exact algorithms. For instance a 50 players problem needed 9 minutes instead of 40 on an Apple-IIe. For 150 players and 10 choices, it passed from 18 hours to 24 minutes.

2. Interactive Assignments

The game has been employed to solve two real problems, once in a bank (Credit Agricole) to assign canvassers to geographic sectors, and once in the ENSPTT (National School for Advanced Studies in Communications) to assign students to professional probations. It was also simulated at ISAGA '86.

In each case, the players have quickly assimilated the mechanism of the process. In general their commentaries and behaviours concerned:

- The possibility of using informations to run noncooperative strategies. Particularly noticeable was the core which formed eight student of the ENSPTT by permuting the eight same choices. This coalition vanished when the norm of exaction fixed by the organizators lowered.
- Many players, particularly at ISAGA '86 pointed out the interest of the process to develop selfconsciousness of their choice criteria.
- The development of an informal negotiation process between players.

In all these cases, the process needed only three steps to assign all the participants. It was striking to compare the serene and cooperative atmosphere which prevailed with the conflicts which characterized the anterior procedures.

3. Detailed Analysis of the ISAGA Simulation

a. General observations. This experiment posed two major difficulties: The lack of effective stakes and the strong implication of players in the field of game theory which induced them into testing preferably heretic strategies. We now describe the progress of the game. To make the outcomes clear, we denote the towns by single letters from A to L. The weights given by the players to the towns at each step are given below:

	Step 1	Step 2: X(e)<2500	Step 3: X(e)<1000
Player E1	(F:50)	(F:49 E:1)	(F:39 E:9 C:1 B:1)
Player E2	(D:50)	(D:48 E:1 C:1)	(D:38 E:10 C:1 B:2)
Player E3	(D:50)	(D:25 H:25)	(H:39 C:9 B:2)
Player E4	(G:50)	(G:49 C:1)	(G:40 E:9 B:1)
Player E5	(D:50)	(D:20 F:20 1:20)	(1:48 C:1 B:1)
Player E6	(1:50)	(1:50)	(1:45 A:4 B:1)
Player E7	(F:50)	(F:30 L:20)	(F:30 A:10 J:10)
Player E8	(L:50)	(L:5)	(L:40 E:9 A:1)
Player E9	(K:50)	(K:49 H:1)	(K:11 H:10 1:19 L:10)
Player E10	(G:50)	(G:25 K:25)	(G:25 J:25)
Player E11	(H:50)	(H:50)	(H:20 L:20 1:20)
Player E12	(K:50)	(K:49 B:1)	(K:20 A:30)

We now summarize the evolution of the shares C(s) of the towns at each step:

Towns:	A	B	C	D	E	F	G	H	I	J	K	L
Step 1:	0	0	0	150	0	100	100	50	50	0	100	50
Step 2:	0	1	2	93	2	99	74	76	60	0	123	70
Step 3:	45	7	12	39	37	69	65	69	122	35	31	70

b. Evolution of the weights and of the information. Two phenomena arise in the preceding data. An additional information on the preferences of the players is obtained when X(E) is lowered. For example, Player E2, without changing the order of his choices, has enlarged progressively their range and so "admitted" places that he formerly hushed up. Player E9, on the contrary, reversed the order of his preferences when he learned that his first choice K was too risky (C(K)=123). This learning phenomenon also appears in the matrix of the shares C(s). The lower ones (towns A,B,C,E,I,J.L) increased while some of the upper ones decreased (towns D,F,G). A few intermediate cases where subject to oscillations (towns H,K). Our previous experiments showed that these oscillations depend mainly on the sensitivity of small sized problems to the rythm of reduction of the exaction norm. The faster this rythm, the more numerous the oscillations. The stabilization of the game is illustrated by the quick diminution of the variance of the C(s).

c. Prospectives. The simulation at ISAGA '86 put forth two interesting suggestions. Instead of asking the participants to use only positive weights, it might be possible to accept also negative weights indicating their main rejections. The coherence between wishes and rejections would then be kept by maintaining the same global weight of fifty points, calculated in absolute value. Another idea was to replace the simultaneous seizure of preferences by an alternate one. The first suggestion has the disadvantage of complicating the procedure; the second one would suppress equality of the players in front of the dominant coalition (in the sense of Von Neumann and Morgenstern) which would progressively arise.

REFERENCES

Aurifeille, J. M. 1984. Processus interactif d'affectation. Analyse des sytèmes. Université de Poitiers, Thèse.
Balinski, M. L. 1986. A competitive (dual) simplex method for the assignment problem. Mathematical Programming 34, p.125-141.
Busacker, R. and Gowen, P. 1961. A procedure for determining a family of minimal cost network flow patterns. Operations research technical report 15. John Hopkins University.
Demange, G. 1982. Strategyproofness in the Assignment market game. Ecole Polytechnique. Report no. A2480682.
Edmonds, J. and Karp, R. 1972. Theoretical improvements in algorithmic efficiency for network flow problems. Journal of American Computers Mathematics. V 19-2, p.248-264.
Tabourier, Y. 1972. Un algorithme pour le problème d'affection. RAIRO V3, p.3-16.

Some solution concepts from hierarchical game theory and their applications in management game EXPORT

Tadeusz Selbirak
Management Organization and Development Institute, Warsaw, Poland

ABSTRACT: The paper presents some solution concepts from hierarchical game theory and discusses their applications in concrete management game.

Considered simple hierarchical game model assumes only two players - the leader, who occupies the higher level of hierarchy and makes the first move, and the follower who acts knowing leader's choice. According to the structure of strategies available to leader, three solution concepts: Stackelberg, reversed Stackelberg and more complicated double reversed Stackelberg equilibria are discussed and compared.

In the next part of the paper concrete management game EXPORT developed in Management Organization and Development Institute, Warsaw, and used for training export managers is shortly reviewed. Participants of this game play the roles of managing directors of firms producing the same kind of good and selling it under competition on several foreign markets. For the new hierarchical version of the game, which is currently being prepared, the possibilities of applying introduced solution concepts are discussed.

KEYWORDS: Game theory, hierarchical games, equilibrium concepts, management games, simulations in training.

ADDRESS: Warsaw 22, P.O. Box 233, Poland; Management Organization and Development Institute, 02-067 Warsaw, Wawelska 56, Poland.
Tel. 251281-245.

INTRODUCTION

The paper presents some solution concepts from hierarchical game theory - a new branch of game theory which has rapidly developed during the last fifteen years, and discusses the possibilities of applying them in a concrete management game.

After pointing out the historical background, the simple hierarchical game model is described. It assumes only two players and one of them is specially distinguished. This player called leader occupies the higher level of the hierarchy; possesses some information about the second player and the right to make the first move in the game by declaring his strategy before the other's action. Such possibility enables the leader to induce to some extent desired behaviour of his partner. According to the structure of strategies available to the leader, three solution concepts: Stackelberg, reversed Stackelberg and more complicated double reversed Stackelberg equilibria are discussed and compared. Then, generalizations of these concepts for a special class of many person hierarchical games are indicated.

In the next part of the paper, the management game EXPORT, developed in Management Organization and Development Institute, Warsaw and used for training export managers is shortly reviewed. Participants of this game play the roles of managing directors of the firms producing the same kind of good and selling it under competition on several foreign markets. New hierarchical version of this game, which is currently being prepared assumes the existence of additional player who plays the role of some decision centre coordinating export activities of the firms. For such hierarchical game the possibilities of applying some solution concepts introduced in theory are discussed.

HIERARCHICAL GAME THEORY AND ITS PLACE IN CLASSICAL THEORY OF GAMES

From its origin in systematically developed form (Von Neumann and Morgenstern 1944), the theory of games has been very often criticized and attacked. Among the most crucial items in classical game theory were:

- Its strictly normative character: The main purpose of the theory was to give definite prescriptions for players to follow to achieve some specially defined game solution. Because the assumptions underlying the solution concepts were often very stringent and artificial, in such cases theory would not provide a good tool for prediction of really expected outcomes in conflicts.

- Lack of reality in modelling players' behaviour: This was due to ignoring individual perceptions. In most games players were assumed to see the same game and their information was complete. Such assumption for example underlies the Nash equilibrium concept for noncooperative games.

- Most often assumed antagonistic, two-player case: This was common in noncooperative games becasue only two-person zero-sum games appeared generally solvable. In other cases, which better reflect real-life conflicts, there is no generally accepted solution concept, which provides existence and uniqueness of solution for every game.

The criticism of classical game theory stimulated the attempts for developing new more real-life approaches to modelling conflict situations, where participants would have different subjective information about the game and non-purely antagonistic interests.

In the late sixties series of interesting papers by Howard (1966, 1971) appeared introducing elements of asymmetric information patterns and bargaining considerations into noncooperative games. Those games with new proposed elements were referred to as metagames. Players at the stage of choosing strategies were assumed to look ahead and anticipate possible reactions of their opponents upon possessed information about the game. It seems, that Howard's metagames set the origins of hierarchical game theory, though in the following years each theory developed independently in different directions.

Metagame theory, which rapidly evolved including some behavioral science concepts towards more general hypergame theory (Bennett and Huxham 1982, Hipel and Fraser 1984) became next a very promising methodology in the "soft" operations research field. It helps in better understanding of situations in complex conflicts as well as provides some formal techniques for prediction possible courses of action. However it does not propose any unique solution for a conflict.

Hierarchical game theory, on the other hand, developed into a more formal mathematical discipline including, particularly in its dynamic versions, some concepts from control theory and hierarchical systems approach. From the beginning of seventies many fundamental papers and monographs in this new born area have been published independently in both anglo-american and soviet literature (e.g. Germeyer 1976, Basar and Olsder 1982).

Hierarchical game theory is restricted to games with asymmetric information patterns between the players and asymmetric sets of actions allowed for them. It assumes that one of the players called leader, who possesses some additional information about others as well as the whole game, is in a position to announce his strategy ahead of time and enforce it on the other players. Decision problems described by these games, which incorporate such hierarchy in decision making, are also known as coordination problems since position of the leader enables him to coordinate the actions of the other decision makers. Hierarchical game theory provides mathematical techniques for assessing players' positions in concrete game in terms of best values of payoff functions, which they can guarantee themselves via applying respective strategies. According to such measures it also proposes "optimal" strategies for players.

From the above it is easily seen, that though this theory has still a normative character, it includes many factors that were not considered in classical game theory:

- information patterns are carefully studied; in modelling player's behaviour it is assumed, that each one utilizes his own subjective model of the game,
- interests of the players may be non-antagonistic,
- players do not act simultaneously; the order of their actions together with information-flows between them are meaningful elements of the game analysis.

It is also worth noting, that the role of the analyst is somewhat different here. As in the classical game theory he was almost always in the position of interested bystander, in hierarchical game theory the analyst works for a specific player, most often the leader. Hence, the analysis is always performed from the concrete player's point of view.

BASIC SOLUTION CONCEPTS IN HIERARCHICAL GAME THEORY

For the purposes of brevity and clarity in exposition of basic concepts considered in hierarchical game theory we shall assume the game model consisting of two players only. Let the first player be the leader. Then we will refer to the second as to the follower. We will assume, that the interests of the players are to maximize their profits expressed by payoff functions which depend upon two decision variables: one controlled by leader and the second by follower. Thus, we can formally note the decision structure of the game as

$$\{Q_o, Q, U, C\}$$

where

$Q_o : C \times U \to R$ is payoff of the leader

$Q : C \times U \to R$ is payoff of the follower

U is the set of decisions controlled by the leader

C is the set of decisions controlled by the follower

R is the set of real numbers.

Notice, that such notation on the first sight looks like a normal form representation of a game. However, the difference is that here sets U and C generally are not the sets of strategies. Another difference comes from the fact, that the leader always chooses his strategy first and the follower subsequently makes his choice, whereas in normal form all strategies are assumed to be selected simultaneously. The decision structure describes only formal relations between values of payoff functions and pairs of decisions. Specification of information flows between players and their behaviour is comprised in forms of available strategies and additional rules of the game. By strategies of the players we will generally understand some decisions rules, which are functions depending on additional information which will be possessed by players. Declaring his strategy each player specifies his behaviour in any situation that may appear. Since the final result of a game must be a choice of concrete decisions by all players, we will assume, that after each player selects his strategy, all decisions will be strictly determined, i.e. announced strategies are obligatory - players must act according to them.

We set the following rules of the game: (1) The leader knows exactly the model of the follower, i.e. function Q and decision set C. (2) The leader knows, that the follower acts rationally, i.e. knowing leader's strategy follower maximizes his payoff according to it. (3) Given his knowledge, the leader acts first by choosing his strategy and then communicating it to the follower. (4) The leader tends to choose such a strategy, that the guaranteed result has maximum value.

Upon the above rules, the basic solution concepts are defined according to the patterns of information flows between players. These are specified by the sets of available strategies. Of course, different patterns of information flows formally define different games, even when the decision structure remains the same.

Stackelberg Equilibrium - Game S^1

Strategy sets are the following

$$\mu = U, \ \ell = C$$

i.e. there are no additional information flows other than assumed by the rules of the game I - IV. Strategies of the players are their decisions.

For clarity, let us assume that every optimization problem considered here has a solution (in general case we may consider suboptimal solutions which always exist). When the leader selects strategy $\bar{u}$ and communicates it to the follower, the other acting rationally will choose any strategy c that maximizes $Q(\cdot,\bar{u})$. The leader according to the rule II may then count on any response c such that

$$c \in \operatorname{Arg}\max_{c \in C} Q(c,\bar{u}) = R(\bar{u})$$

and consequently, strategy $\bar{u}$ guarantees for him the result

$$r(\bar{u}) = \min_{c \in R(\bar{u})} Q_o(c,\bar{u}) \ .$$

Hence, the maximum guaranteed result for leader is

$$r(S^1) = \max_{u \in U} r(u) .$$

Any strategy $\bar{u}$ such that $r(\bar{u}) = r(S^1)$ is called Stackelberg solution (for leader) and any pair of strategies $(\bar{c},\bar{u})$ such that $r(\bar{u}) = r(S^1)$ and $\bar{c} \in R(\bar{u})$ is called Stackelberg equilibrium.

Reversed Stackelberg Equilibrium - Game S^2

Strategy sets are the following

$$\mu = \{\tilde{u} : C \to U\}, \ \ell = C,$$

i.e. leader will know follower's decision c and may declare his strategy as some function $\tilde{u} : C \to U$. Leader's strategy set is then the set of all functions depending on follower's decisions and follower's strategies are his decisions.

The follower, knowing leader's strategy, may choose any decision c that maximizes $Q(\cdot,\tilde{u}(\cdot))$. We have

$$R(\tilde{u}) = \operatorname{Arg} \max_{c \in C} Q(c,\tilde{u}(c))$$

and

$$r(S^2) = \max_{\tilde{u} \in \mu} \min_{c \in R(\tilde{u})} Q_o(c,\tilde{u}(c)).$$

Any strategy $\bar{u}$ such that $r(\tilde{u}) = r(S^2)$ is called reversed Stackelberg solution (for leader) and any pair of strategies $(c,\tilde{u})$, where $r(\tilde{u}) = r(S^2)$ and $c \in R(\tilde{u})$ is called reversed Stackelberg equilibrium.

It is easily seen, that game S^2 may be recognized as special modification of game S^1. Namely, let us change the roles of the players in the game S^1. Follower becomes then a leader and he will know the other's decision before making his own. But as a leader he has the right to make his move first and consequently he may announce his strategy as a function depending on the decision of the new follower. Game S^2 is then built from "reversion" of a game S^1 and execution of the right of new leader.

Double Reversed Stackelberg Equilibrium - Game S^3

Strategy sets are the following

$$\mu = \{\tilde{\tilde{u}} : \tilde{C} \to U\}, \quad \ell = \{\tilde{c} : U \to C\},$$

where $\tilde{C} = \{\tilde{c} : U \to C\}$, i.e. leader will know follower's strategy $\tilde{c}$ and may declare his strategy as a decision rule $\tilde{\tilde{u}} : \tilde{C} \to U$.

Double reversed Stackelberg solution (for leader) and double reversed Stackelberg equilibrium are defined upon the similar scheme as in previous games.

We can continue the reversion procedure and obtain the sequence of more complicated hierarchical games S^i, where $i > 3$, for which we may define respective solutions and equilibria. It appears however (Germeyer 1976), that only S^1, S^2 and S^3 games are meaningful, because all the games S^{2n} and S^{2n+1} for $n \geq 2$ may be reduced to games S^2 and S^3 respectively, with the same decision structure. Another important result says, that for the games with the same decision structure the best for leader is always the game S^2, i.e. the following inequalities always hold for the maximum guaranteed results for leader

$$r(S^1) \le r(S^3) \le r(S^2) .$$

Of course, the leader may not have the possibility to play an S^2 game, e.g. such case takes place when follower's decisions will not be known for him.

Presented basic solution concepts provide a framework which can be generalized upon more complicated games. One of the most often considered generalizations is the class of N+1 person hierarchical games with so-called fan-shaped structure. Such games involve one leader and N followers who act independently. Payoff functions of the players have the following forms

$$Q_o : C_1 \times U_1 \times \ldots \times C_N \times U_N \to R$$

$$Q_i : C_i \times U_i \to R, \; i=1,\ldots,N,$$

where $U=U_1 \times \ldots \times U_N$. In such a structure only vertical connections (leader - i-th follower) between the players exist. Followers are not connected, because the domain of their payoff functions are mutually separable.

For games with a fan-shaped hierarchical structure, concepts of Stackelberg solutions may be easily transmitted. Let us notice however, that independent followers cannot be formally aggregated into one player (with the exception of S^1 game) as may seem at first view. The problem is that leader's strategies communicated to different followers must be functions with disjoint domains.

It may be shown, that only S^1, S^2 and S^3 games are meaningful for the theory and the same inequalities hold for maximum guaranteed results as in two-person games (Selbirak 1984a).

For N+1 person games as a general case, considerable numerical difficulties arise while computing Stackelberg solutions and maximum guaranteed results for the leader. The number of optimization problems which must be solved increases exponentially with the number of followers and each such problem has a very sophisticated max-min nature. Even though in general games S^i practically cannot be solved, the special class of them may be distinguished for which computational problems can be radically reduced.

These games are called regular and Stackelberg solutions for them have very simple forms. Detailed discussion of regularity in hierarchical games may be found in (Selbirak 1984b, 1984c).

Of course, in the field of hierarchical game theory many other hierarchical structures are also considered. Let us only mention some directions, which are now intensively studied:

- games with incomplete information available for leader,

- games with many levels of hierarchy,

- games with interconnected followers,

- games with dynamics in decision process, e.g. multistage games and their mutual combinations.

Hierarchical games are often called games with fixed order of moves, because this feature is essential in mathematical model of hierarchy assumed here: the leader acts first and then knowing his action acts the follower. Formally, we can describe every real conflict with fixed sequence of participants' moves in terms of hierarchical game theory, even though such formal treatment of real-life games may not always reflect the position of the leader whom we usually would like to see as a coordinator of others' actions. Notice, for example, that playing chess may be considered as playing hierarchical game with leader represented by whites. Of course, the whole chess party appears to us rather not as a game with heirarchy because of its complexity - practically, two players have the same "power" at the beginning of the game. However, the analysis of the certain position on board and its possible consequences after several moves and countermoves may be well performed in terms of multistage hierarchical game S^1 with player moving first as a leader. In such a case leader's ability to induce certain desired behaviour of the follower is clearly visible.

MANAGEMENT GAME EXPORT AND IMPLEMENTATION OF DESCRIBED SOLUTION CONCEPTS

In this section we shall briefly present the management game EXPORT and discuss the possibilities of applying the described solution concepts in the new hierarchical version of this game which is currently being prepared.

EXPORT game was designed and developed in Management Organization and Development Institute, Warsaw in 1984 by Romuald Wolk (1985) with the aim of training managerial skills of directors and executive managers in the area of export activity. The computer code for EXPORT consists of about 5000 Fortan statements and is prepared to run on IBM370/148 under CMS mode. It is the latest game from the long series of business games designed by the same author during the last decade, which always reflected the actual conditions in which Polish economy operated.

Participants in the EXPORT game represent managing directors of firms producing the same kind of good and selling it under competition on several markets: one home and three foreign. Simulation spreads over three years of firm activity with one period representing three months. At the beginning of the game participants select some criteria (like: cumulated net profit, cumulated export income, value of sales, etc.) according to which effectiveness of their activity will be measured and construct one scalar performance function, which has the form of weighted sum of previously chosen criteria. During each period each firm controls up to 20 decision variables determining: investment policy, production, marketing, prices, quality of product, allocation of products, employment and other areas of firm activity. Additionally the firms may contact each other and perform some transactions, e.g. one firm may resell some units of its own products to another. After each period, the participants obtain an actual report of their situation prepared by computer. It consists of large amount of data and participants must acquire some skill in reading it properly and choosing only significant information. The whole game is controlled by the game organizer who may variate some model parameters (up to 30) influencing the course of the game and simulating special events, e.g. sale crisis. Export activity of the firm is specially preferred in the game via a system of tax reductions. It is also a necessity, because only export can provide the flow of hard currency which is needed in a production process (special raw materials must be imported).

While the EXPORT game is oriented only towards training managerial staff and propagating an export-oriented policy, the new hierarchical version of it, which is currently being prepared is not planned for educational purposes only. It is also expected to provide a tool for testing in laboratory conditions the effectiveness of some economic mechanisms. The appropriateness of such a project seems to be specially justified at present time in Poland, when parametric coordination methods are being introduced into national economics.

The hierarchical version of EXPORT game assumes the existence of an additional player on the higher level of the hierarchy which represents some decision centre having certain instruments for influencing firms towards greater export activity. The list of instruments may include, for example, tax reductions, preferences in allocations of limited raw materials, additional assignments of energy, if there are restrictions, and special stimulating funds for workers. There is essential difference between this additional player and players representing firms because their activities cannot be measured by the same criteria. Another point is, that there is no need for the centre to simulate activity of any real organization, however it may be comparable with the government, which issues decrees, acts of legislation and statutory orders creating economic mechanisms. The centre may not even be perceived by the other players as a player in the game, because its role is very similar to the game organizer's one. Such a game organizer however, in this case is no longer a passive observer, but has some interests associated with the research purposes established before the given game session. They may be (1) testing efficacy of certain mechanisms on real-life decision makers, (2) comparing several given mechanisms, and (3) searching for the most effective mechanism. Hence, the proposed game may perform two parallel functions - training managerial skills of participants and at the same time making simulations of economics with real decision makers for research purposes.

Hierarchical game theory with its methodology and techniques seems to be a very promising tool for analysing economic mechanisms by the structure presented here. Different centre's instruments induce mechanism, which may be adequately modelled by N+1 person hierarchical games S^1, S^2 and S^3. A policy of tax reductions for example, when defined by the centre as a percentage of value of export sale, induces a mechanism which can be modelled by hierarchical game S^1. A policy of stimulating funds for workers induce a mechanism described by game S^2. Game S^3 may be a good model for the situation, where the centre declares strategies of additional supplies of energy acording to projects of its utilization submitted by firms.

The formal techniques from hierarchical game theory presented above are specially useful for assessing the centre's possibilities of influencing firms under a chosen type of economic mechanism. However, as remarked on a session, assessment of "power" of a given mechanism when applied by the centre, is performed according to the guaranteed result principle, where centre making its calculations is assumed to anticipate the most disadvantageous (from its viewpoint) reactions of the firms. Consequently, the real result induced by a given mechanism sometimes may be better, but never worse, than the one calculated upon hierarchical game theory.

Since the hierarchical game model assumes a special framework of moves and communications between the players, problem of controlling them during the game session becomes very important. Each period of the hierarchical game EXPORT therefore consists of two separate decision making stages. At the first stage, the player who represents decision centre basing upon the actual information about the state of the game creates economic mechanism as a set of parameters and economic rules. This set having a form of decrees the acts of legislation is next presented to the other participants who represent firms, as a part of their report informing about the actual situation. Of course, announced rules are obligatory during the given period; if some of them are valid longer, organizer of the game informs about it all the players. At the second stage, players representing firms are making their decisions under market competition and established economic rules.

REFERENCES

Basar, T. and Olsder, G. J. 1982. Dynamic Noncooperative Game Theory. London: Academic Press.

Bennett, P. G. and Huxham, C. S. 1982. Hypergames and What They Do: A "Soft OR" Approach. Journal of Operations Research Society 33, 1, 41-50.

Germeyer, Yu B. 1976. Games with Non-antagonistic Intersts. (in Russian), Moscow: Nauka.
Hipel, K. W. and Fraser, N. M. 1984. Conflict Analysis: Models and Resolutions. Amsterdam, New York: North-Holland.
Howard, N. 1966. Theory of Meta-games. General Systems 11, 187-200.
----- 1971. Paradoxes of Rationality. Theory of Metagames and Political Behaviour. Massachusets, London: MIT Press.
Selbirak, T. 1984a. Strategies Maximizing Guaranteed Result for the Centre in Hierarchical Games with Complete and Incomplete Information About Subsystems. (in Polish), Doctoral Dissertation, Warsaw: Technical University of Warsaw.
----- 1984b. Regularity Conditions for G^2 Hierarchical Game. Archiwum Automatyki i Telemechaniki 29, 1-2, 197-210.
----- 1984c. On the Regularity in Hierarchical Games. Proceedings of the Fourth Polish-GDR Conference on Nonconventional Optimization Problems, Warsaw, May 21-25. Polish Academy of Sciences, Systems Research Institute.
Von Neumann, J. and Morgenstern, O. 1944. Theory of Games and Economic Behavior. New York: Princeton.
Wolk, R. 1985. Management Game EXPORT: User's Instruction. (in Polish), Warsaw: Management Organization and Development Institute.

CITPLAN: using gaming simulation procedures for educational purposes in land use planning teaching

Alberto Bottari
Politecnico di Torino, Italia

ABSTRACT: CITPLAN is a complex gaming activity, particularly designed for and used in a course on land use planning, where it has been associated with more traditional teaching aids. The peculiar role of gaming is assumed to be that of allowing students to have a more global and dynamic comprehension of the design process, whose conceptual and strategic components are not easily understandable by the only means of the more usual tools of communication. The paper gives an account of a session devoted to explain and discuss a first extensive experiment in using CITPLAN at the School of Architecture in Torino.

KEYWORDS: Land use planning, gaming simulation, higher education, planning design and practice, school of architecture, Italy.

ADDRESS: Corso de Nicola 22, 10129 Torino, Italia. Tel: 011 591760/5212459/5566456/5566454.

The aim of the paper is to give some general ideas of CITPLAN, an integrated system of gaming simulation procedures which has been developed and used, in association with lectures and case-studies, in a course particularly focussed on land use planning, for students of the School of Architecture of the Politecnico di Torino. The use of CITPLAN (CITy PLANning) has been considered relevant for stressing the strategic and conflictual components of technical and political decision-making in a plan design process, and for providing an artificial environment where the players can actively experience a cognitive process as decision-makers.

A. AN INTRODUCTION TO CITPLAN

CITPLAN includes GULP, SIMPLADS, MINURBO, three different procedures, each one referring to different learning situations about land use development, planning and control, in a process of urbanization of a sub-regional area characterized by the growth of a single big city. GULP (Gaming the Use of Land Parcels) allows three teams of players to develop a physical environment, buying and selling parcels of land, financing economic activities, evaluating their investments, so getting some insights on the significance of locational choices and of their relationships with functional and economic factors of development, referring both to the local and to the external market. Intentionally GULP's gaming rules stress the role of the private initiative and of a relative absence of public control and government on land uses development, in such a way that a final structure of the urbanized area is going to be determined that is a highly unbalanced one, in terms of social demand for housing, services, transports, etc. if compared with the available facilities. The analysis and evaluation of this demand and the emerging consciousness of the need of a more community oriented kind of development are at the same time the conclusive step, and the initial one, respectively for GULP and the second phase of CITPLAN, that is SIMPLADS (SIMple PLAn Design Simulation). Players must now work out a specific and detailed layout of a land use plan, especially focussing their attention on the examination of the attained levels of services equipment accessibility, looking at locations and at the demand of the overall settlement system, and comparing it all with some given standard values and thresholds. They will deal with some design strategic variables, such as housing densities, locational choices for residential and public activities, road and facilities network layouts, and have to interact with the Local Administration, which is supposed to commit the plan and must check both intermediate and final proposals.

The final steps of SIMPLADS involves all teams in discussing and evaluating the plan proposals. The third phase of CITPLAN, that is MINURBO (MINi URBan Operations), begins with the choice and approval of a specific plan, so assuming that since now land development will take place in a controlled environment and that a new set of rules is introduced (in the game) for the economic and political behaviours of the developers (the players). Moreover an implementation program must be drawn, approved and followed, and it is intended for allowing a more balanced use of physical and financial resources for the community development.

Long and short range planning choices and the eventual changements in the political, economic and social conditions will also include the possibility of changing or modifying a plan, and this may also be considered as a good starting point for debriefing the game.

In the next section of the paper a more extensive description of how CITPLAN phases work is provided. Then some notes (section C) are devoted to elucidate the nature of relationships, resources and of some communications aspects in the game; finally a short discussion will examine some critical contributions to the evaluation of this gaming and educational experience, taking into account what has emerged during the ISAGA Conference in Toulon.

B. PROCEDURES AND MATERIALS IN CITPLAN

CITPLAN is played by three or five teams, with a maximum of fifteen/twenty participants; it does not require any special computing devices, except a pocket calculator and a game-board which facilitate accounting and the perception and evaluation of the changements in land uses.

B.I. Building an Urban Environment

The first phase of CITPLAN, that is GULP, describes the development of a city in the last fifties of the nineteenth century up to the first decades of the present one, with a growing urban area evolving from the initial conditions of a mixed rural-industrial economy to a final and highly industrialized one. This means for the players to face a sequence of changes in the economic, political and technological conditions in the local area and in the 'non-simulated' world, which implies to pass from one scenario to a new one, from a set of game rules to a new and more complex one.

Therefore GULP includes one basic decision-making routine, which is well known to participants from the beginning of the game, and another one which introduces from time to time a new scenario in an exogenous and not expected way. According to the first routine, players have to (a) read carefully the rules for the running scenario, (b) buy and sell parcels of land and (c) develop land (with an open possibility for interacting with the other teams and the Game Director, but in duè times), (d) and finally fill a balance sheet where they take into account how the locational choices and the new developments and activities have affected their investments. Residential, commercial and industrial developments are allowed, and locational components heavily influence the functional and economic results of decisions, according to some basic relationships which are largely indebted to CLUG (The Community Land Use Game), a game developed by Feldt (1972).

The sequence of scenarios may be summarized as follows: (A) developing rural land uses, and exchanging goods within a local market and with an existing small town (nine rounds); (B) developing industries mainly by means of locally available energy sources (hydraulic motive-power) (three rounds); (C) urbanizing the area, that is developing residential land uses of different quality level and commercial activities, as a consequence of the population growth induced by industries (three rounds); (D) locally reflecting some heavy economic consequences of an international political and economic crisis, affecting customs duties on export/import, and of some changements in city boundaries (three rounds); (E) evaluating and accounting the effects of the crisis also from the point of view of a lack of an effecient transportation network for facilitating commercial relationships with the 'outside world' and reaching new markets, and of heavy limitations in locational opportunities deriving from the energy sources which are locally available for industries (one round); (F) developing a new growth phase for the urban area: new technologies and policies in using energy sources (electricity) and developing transportations (the railroad) offer many new opportunities and advantages for locating and developing economic activities, with relevant consequences on land uses and on the overall urbanization process (the population increases rapidly, and housing and services needs increase too, and so on ...) (ten rounds); (G) estimating the overall amount of capitals that each team has reached, and revising and discussing economic strategies followed along the game.

The physical layout of the analogical model which represents the rural and urban land uses, the parcels of land, the land estate and the activities on it, consists in a plastic board where LEGO blocks must be fastened, each one representing the minimum amount of a specific activity that participants are allowed to develop on each parcel, that is a single element of a rectangular grid superimposed on the board. GULP includes in the debriefing (cf point G in the sequence) also a critical revision of its theoretical and historical background.

B.2. Designing a Land Use Plan for an Expanding City

In SIMPLADS participants have now to leave their previous roles and game activities as landowners, contractors, industry managers, etc. since there are two roles only: the professional planner and the local authority. The second is usually played for this phase by the teacher, who will check plans according to some general aims and assumptions that the local authority has given to the planners, such as to provide a more balanced allocation of land resources between public needs and private interests; moreover he will object to a plan proposal with regard to possible impacts on the system of land values and on the local expectations (e.g., in real estate, industry, finance, social sectors), and will stress the costs for implementing the plan according to programming and budgeting in the public sector.

The planner has been committed to design a plan of land uses for an expanded urban environment, where he must take care first of all of a great amount of an unsatisfied social demand and then of an eventual further expansion. He will produce drawings with the layouts of land uses, of services locations and of roads and facilities networks, and all the papers needed for explaining his choices relating also to some basic demographic and socio-economic analyses he is asked to make. A continuous redesign of plan proposals is needed before the approval from the authority, and this is the interactive and strategic 'core' of SIMPLADS.

Some slight but significant changes are introduced in the layout of the analogical model of the game, such as the adoption of a square reference grid with the possibility of subdividing each cell, so providing an opportunity for different housing and services locations and arrangements, spatial relationships and densities, with some peculiar solutions for the roads network and the transportation system. These different available mixes of private and public land uses will allow to design a plan paying also some attention to a first organization of the urban space and form.

B.3. Implementing the Plan

MINURBO is dedicated to select one planning solution among all the proposals which have been technically approved in SIMPLADS, and to its implementation by means of a unique programming action interesting both the public and the private land development, until a plan needs to be revised or even dismissed. The decision-making basic routine is similar to the one of GULP, but the private investments have some more constraints due to the admitted land uses and the public services location in the plan, to programs which have to be submitted to a public approval, and to taxes which are imposed on building activities. In MINURBO the players must assume the roles of developers and of some other groups having a political or social relevance for the implementation process. A minimum of five roles is needed: three teams of developers, like in GULP, the Local Authority (played by participants), and the representative of the Community (three players at least, each one for one third approximately of the population, with a reference to a specific geographic sector in the urban area).

The Local Authority collects taxes, makes his own budget, provides facilities and services, and is periodically reelected (or not) depending on the results of his urban policies. The Community is continuously monitoring the local environment (checking the urban quality in terms of the unsatisfied demand for housing, services, utilities, of the level of costs for transportation, etc.), and supports a majority in case of voting for a revision of the plan or the dismissal and election of the Local Authority.

The main procedure includes the plan selection and approval in the first three rounds, then within the fifth a first program must be accomplished and its implementation will start. At the nineth round a new cycle will begin for the design of a new program. While the Local Authority can ask for a public revision of the plan whenever it is necessary, the other roles and teams may put it into discussion starting from the fourteenth round only. Voting is always needed, and based on a majority rule.

MINURBO usually does not come to an end until this revision has been fully discussed.

C. ON RESOURCES, RELATIONSHIPS AND COMMUNICATIONS IN CITPLAN

Gaming the three phases of CITPLAN involves a different mix of 'game resources' and of 'personal resources' according to each procedure. An initial amount of money and land parcels is given to the teams, together with a set of rules and forms explaining procedures, roles and the accounting system. Both the rules and the accounting procedures will grow in complexity whenever the scenario changes in GULP, while in MINURBO the financial program of the teams must take into account the costs of developing both land and some connected public facilities. Besides each player is involved in the game with his 'personal resources', that is a complex amount of capabilities he must be able to activate at the level of a tactical and strategical behaviour, ranging from economic decision-making and bargaining ability which is typical in GULP to a more complex and sophisticated policy-oriented attitude, which is needed in MINURBO where much more attention must be dedicated to some social and political components for the design and the implementation of a planning and programming process. Finally SIMPLADS is more centered around the aim of developing the technical capacities of an individual participant to deal with analysis and design problems relating to land use planning. Therefore while MINURBO and GULP ask a participant to be also able to integrate his capabilities according to a team behaviour, SIMPLADS needs a more direct and explicit personal responsibility in decision-making, that is a more 'riskful' involvement, and a more structured and controlled cognitive process. R lationships among teams and with the Game Director are partially structured, since the rules define when (in what step of the routine) and how (by means of written or oral communications, in a due time, etc.) they will take place, both in the first and in the third phase of CITPLAN, while in SIMPLADS interactions between planners and local authority are very freely arranged at an interpersonal level with the only constraint of putting an end to the procedure in a relatively short time, in such a way that MINURBO may start.

Within each team usually some kind of tasks distribution takes place (e.g. updating a balance sheet and evaluating costs and the amount of the available financial resources, or contacting the other teams for exchanging resources or looking for some political and economic agreements). The game practice and the rules do not necessarily require a random composition of the teams: they are very often some groups of students who are used to study and work together since the first years of their curricula, and a leadership will usually not become easily evident unless it is preexistent to the game.

CITPLAN is characterized by a communication system which shows some 'formal' and 'informal' features. Gestures, mimicry, using symbols and intentional recurrencies in behaving, pertain to a 'spontaneous' communication flow that a player may activate recurring to the channels he prefers, both during GULP or MINURBO. Besides CITPLAN must have a certain fixed structure for exchanging and 'formalizing' messages which is indispensable for running the simulation in due times and sequences, and for assuring the correct input for each round of the game. These formal channels and constraints are activated by the system of the rules and of the game-board with its associated pieces, since the last one serves -as Meadows (1986) points out-... "as an aid in communication, since it helps for accounting, visualizing the connections between land uses, evaluating the overall level of the urban development".

D. SOME CONTRIBUTIONS FOR A DISCUSSION

Among the issues raised during the instructional experience of CITPLAN and by the discussion and some personal contacts with game designer which followed the demonstration workshop at the ISAGA Conference in Toulon, it seems useful and interesting to start from a solicitation which comes from Greenblat (1986) to pay more attention to gaming from the point of view of the design process and not only from the educational or pedagogical one, since this allows to investigate a larger range of communicational situations. It would be interesting to explore what happens in terms of communication flows and of an effective cognitive environment whenever it is a team of researchers and designers which is working for building a game model, or is using the game design process as a tool for exploring theories and models in social and economic sciences (Feldt 1966, Coleman 1969, Bottari 1978). Another problems which has been raised is that the design prcess involves modelling and concealing some asepcts of the real world for practical, theoretical or ideological reasons; consequently it is important to give users (teachers, players, researchers etc.) the opportunity for having a deeper insight on the basic assumptions of the design process. In CITPLAN, for instance, some written schemes and communications have been prepared to elucidate the main theoretical assumptions on the urbanization development and the land use control system (which refer to some classic locational models of the economic activities and to some recommended procedures of regional acts), as well as some historical events of a city case-study assumed as a reference for some physical and political developments (that is the case of Torino). The need of 'transparency' in the design process and of the game itself as a model of the real world seems to find an agreement within researchers in the field of the social sciences (Dupuy 1973) and gamers. Apart from a not risky and limited utilization of the computer for calculations in SPACE (Law-Yone 1986), the problem of a correct use of this tool as a simulation device seems to have been successfully faced in STRATAGEM-I, a complex and highly computerized game, where ..."the model is well documented, so that users of the game can easily incorporate revisions or extensions that will tailor the game for their specific needs or regions..." (Meadows 1986). The second part of this statement introduces another issue, that is the 'adaptability' of a game to different contexts of use, which has been discussed both examining the design of a game as a 'frame' to feed with local data (Feldt 1971), and considering the different types of users or the high number of participants to some specific sessions. The possibility of being used by laymen, for instance, has been faced in CITPLAN by building in it the sequence of scenarios with a growing degree of complexity, in such a way that the participants are introduced step by step and with a new set of rules at each time up to the final state into the game model. While the rigidity of GULP comes from its prevailing nature of simulation model, SIMPLADS and MINURBO are more adjustable to different contexts, having a prevailing character of interactive procedures. Some example of adaptable game models have been cited, as CLUG (Feldt 1972), the IMPASSE trilogy (ESL 1973), UNTODES I (Duke 1981), which have been designed for fitting with different regional contexts and problems in planning and decision-making situations, while QUICK CLUG (Duke 1975) may be a reference as an example of a very simplified procedure easy to be used in large classes.

Finally CITPLAN allows to stress another relevant but scarcely explored issue, that is the use of written questionnaires particularly oriented and structured for debriefing the game in order to get some significant informations about how much this instructional game is suitable for both the course and the personal aims or expectations of the students. A significant statistical correlation between the levels of satisfaction in the participants and their previous amount of theoretical and professional knowledge and practice has been put into evidence by some first rather simple measurements, and its interest seems to suggest to carry on this exploration.

REFERENCES

Barbieri, L. 1986. La giocosimulazione nell'insegnamento dell'urbani stica. Torino: Facolta di Architettura del Politecnico di Torino.

Becker, H. A. and Goudappel, H. M. (eds). 1972. Developments in Simulation and Gaming. Utrecht: Boom Meppel.

Bottari, A. 1978. GAMES & CO: giochi e giocosimulazioni nella pianifi cazione urbanistica e progettazione. torino: CELID.

Bottari, A. 1980. Giocosimulazione Urbana e Teoria: il caso di MINURBO. Torino: Istituto di Programmazione Territoriale e Progettazioni, Politecnico di Torini.

Bottari, A. 1986. CITPLAN: Gaming Simulation Procedures for Educational Purposes in Town Planning Practice. Torino: Dipartimento Interateneo Territorio, Politecnico e Universita di Torino.

Coleman, J. S. 1969. Games as vehicles for social theory. American Behavioural Scientist, XII:4.

Duke, R. D. 1975. QUICK CLUG. Ann Arbor: School of Natural Resources, University of Michigan.
Duke, R. D. and Greenblat, C. S. (eds). 1975. Gaming Simulation Rationale, Design and Applications. New York: John Wiley & Sons.
Duke, R. D. 1981. United Nations tourist development simulation - UNTODES I. In Loukissas (1983).
Dupuy, G. 1973. L'ideologie des jeux urbains. Espaces et Societes, 4:8.
ESL 1973. Regional Planning for Monterey Bay - A Trilogy of Issue Oriented Games for Citizen Use. Ann Arbor: Environmental Simulation Laboratory, School of Natural Resources, University of Michigan.
Feldt, A. G. 1966. Potential relationships between economic models and heuristic gaming devices. In Duke and Greenblat (1975).
Feldt, A. G. 1971. Developments in the fields of (simulation and) gaming in the United States. In Becker and Goudappel (1972).
Feldt, A. G. 1972. CLUG: Community Land Use Game. New York: The Free Press.
Greenblat, C. S. 1986. Communicating about simulation design: it's not only (sic) pedagogy. ISAGA 17th Annual International Conference.
Law-Yone, H. 1986. Simulating SPACE (State Planning and Capital Expansion). ISAGA 17th Annual International Conference.
Loukissas, P. J. 1983. Gaming simulation as a communications tool in local planning. Ekistics, 50:302.
Meadows, D. 1986. STRATAGEM-I: an operational game in economic decision-making. ISAGA 17th Annual International Conference.

SIMULATION REFERENCES

CITPLAN. Bottari, A. 1986. Cf Bottari (1986).
CLUG. Feldt, A. G. 1972. In Feldt (1972).
GULP. Barbieri, L. and Bottari, A. 1986. Politecnico di Torino. Cf Barbieri (1986), Bottari (1986).
IMPASSE. ESL 1973. In ESL (1973).
MINURBO. Bottari, A. 1980. In Bottari (1980).
QUICK CLUG. Duke, R. D. 1975. In Duke (1975).
SIMPLADS. Bottari, A. 1986. Cf Bottari (1986).
SPACE. Law-Yone, H. Undated. Cf Law-Yone (1986).
STRATAGEM-I. Meadows, D. Undated. Cf Meadows (1986).
UNTODES, I. Duke, R. D. 1981. Cf Loukissas (1983).

The production of SPACE: Experiments with a Marxist game

Hubert Law-Yone
Israel Institute of Technology

ABSTRACT: This article describes the development of a family of interrelated gaming simulations for the purposes of teaching and research into urban and regional problems. Some of the drawbacks of existing urban games are described, in particular their inadequacy from a critical, neomarxist point of view. The basic dilemma in devising socialist games is analysed and the rationale for the development of SPACE-State Planning and Capital Expansion is described. Offshoots bearing the same structure yet focussing on different levels of the urban problematic, i.e. neighbourhood, small settlement, city and region, are briefly described.

KEYWORDS: neo-marxist, urban and regional planning, frame-game, gaming/simulation dilemma, structure, SPACE, SPACESET, SPACECIT, SPACEREG, SHIKUM.

ADDRESS: Faculty of Architecture and Town Planning, Israel Institute of Technology, Technion, Technion City, Haifa 32000, Israel.

INTRODUCTION

This paper reports on the development of a system of gaming simulations intended for teaching and research into urban and regional dynamics and planning. The system is based on recent studies offering a neo-marxist interpretation of the urban problematic (Scott, Harvey, Castells). The seed game is SPACE (State Planning and Capital Expansion) which was presented at the 1984 ISAGA Conference (Law-Yone).

The initial impetus for the development of the game grew out of dissatisfaction with some fundamental flaws characterising existing urban games. The main criticism of these games was that while they were ostensibly theory-free i.e. descriptive in nature, they tended to reinforce a status-quo, free market oriented ideology. Furthermore, most of the existing games simulated only the trivial among the contradictions inherent in urban spatial dynamics. We attempted therefore to devise a gaming simulation frame game incorporating recent theory and research which might be more intellectually stimulating and fruitful.

A CRITIQUE OF URBAN GAMES

The structure of a game hides an ideology. Players are supposed to represent "groups" in society, which are often defined with no theoretical explanation. They are posited however, to obey the same rules and are subject to the same penalties. It could be hence surmised that what guides this approach is a crude Weberian functionalism. Or, if it is claimed in some games that these roles represent powerful decision-making elites, then theoretical explanation can be found in approaches similar to managerialism and corporatism.

The range of decision options and outcomes i.e. the rules of the game, also reflect the way the major moving forces for change in society are perceived and presented. Reducing the payoffs to money units (justified by the need for simplification) often refer to a crude economism. Pure price competition, freely negotiated wage and trade contracts, the mandated and enforced honouring of agreements - all of these present an ideal picture of a free market. Politics, if it enters the picture at all, is sporadic, whimsical and theatrical.

In short, most games were found to be "selling" an insidious picture of reality under the cover of the "fun" of playing. What was disturbing was the suspicion that students playing these games might conceivably be led from the enjoyment of game playing to the justification of the social system which the game is supposed to represent.

Basic Requirements of Games and Simulation

There is an inherent tension between the requirements of a good game and the essential features of a good simulation. Among the basic requirements of a game are:

Simplicity. The need for an easily comprehended structure containing only the minimal number of parts. The assumption is that a game with easily learned rules is more conducive to play.

Fairness. There should be no built-in advantage for any player. Success in playing should depend only on chance or skill. It is reasonable to assume that no rational player will want to continue playing an obviously "loaded" game. A game can of course give differentially favourable results now and then but it must be fair in the long run.

Competition. The provision of a framework wherein negotiation, manipulation, outwitting, bluffing are permissible strategies to get ahead. These are part of the skills which permit players to bend outcomes to the favourable. This element also raises games above simple randomly determined outcomes and introduces inter-personal behaviour as a factor.

On the other hand, as distinct from a game, a simulation needs to have the following qualities:

Structural isomosphism. The essential form and behaviour of the game/simulation must find its parallel in reality, with a reasonable degree of accuracy. This is necessary if we wish to gain insight into the real world from the workings of the simulation.

Reproducibility. For testing of hypotheses and verification or falsification, there must be some stability and equifinality. Otherwise we would not be able to make any scientifically valid statements concerning the real world.

Now, there may be some trade-offs necessary between a good game and a good simulation. One might sacrifice some aspects of structure in order to have simplicity, or give up reproducibility in order to introduce an element of chance. It can be observed that most games involve some compromise between these various requirements. However, it is entirely plausible that these tensions could become basic incompatibilities when the gaming simulation is predicated on socialistic principles.

The Dilemma of Socialist Games

When we examine the nature of socialist games, i.e. games that involve a radical critique of the capitalist mode of production as social order, we find that there are serious philosophical as well as methodological problems. First of all, making a game simple can make it too simple and hence false. The complexity of reality is part of prevailing ideology. Trying to simplify may in fact be the wrong way to reflect the mystifying nature of reality.

Secondly, it may be impossible to make a game based on fairness and equality that is supposed to represent a world that is far from being fair and equal. On the other hand, making the game reflect faithfully the structural inequalities would tend to make it "loaded" or biased and hence not conducive to play.

Thirdly, games based on competition as a driving force could tend to rule out cooperation and planning as viable strategies. It has been suggested (Lundgren and Loar) that competition is fundamental to both successful games as well as to Capitalism. Is a socialist game then a contradiction in terms? Ollman (1983) took this to be a challenge and provided his own answer to it in the form of a board game. Unfortunately, this solution is entertaining but does not address the simulation aspects adequately. The apparently conflicting claims of a playable game and the simulation of reality from a marxist conception represented for us then the basic dilemma and challenge.

SPACE and its Derivatives

SPACE-State Planning and Capital Expansion was the frame game that finally took shape. Its main features are as follows: The main players are Capital, Labour and the State. These constitute the basic structure-Capital and Labour being engaged in class struggle and the State both intervening in this struggle as well as carrying out differential policies toward the antagonists. The class struggle consists of wage negotiations, Labour striving for a higher standard of living and Capital trying to maximize profits. The State attempts to mitigate the sharpness of the conflict by mediating wage negotiations, provision of services to Labour, ensuring profits to Capital and creating the conditions for reproduction of Labour. The game is played in rounds and each player's achievements are displayed. The set of rules expressed in the formal quantitative exchange relationships between the parties give rise to periodic crises or contradictions. The game is open ended in that these rules can be changed as part of the game.

The subsequent development of related games has shown that the above structure of SPACE can easily be transferred to different levels and instances. Thus so far we have developed four games focussing on the neighbourhood, small settlement, city and regional levels. In brief, these games are:

SHIKUM - A neighbourhood level game where the State is fractured in Local and Central Government and Capital is external to the area. Labour struggles for better services which include housing quality. The State declares the area a rehabilitation area and the idea of the game is to relate the problems of the neighbourhood to the systemic contradictions in society.

SPACESET - Labour is represented by a Development Town, (these are towns set up in development areas mainly populated by new immigrants). Capital is represented by a veteran settlement i.e. kibbutz. The wage struggle relates to two major sectors-agriculture and industry. State intervention in terms of subsidies, incentives to industry and ideological control is high.

SPACECIT - This a gaming-simulation exercise at the urban scale. The focus is on the market for the use of land. Labour tries to acquire location for housing that will maximise environmental quality and minimise accessibility costs. Capital locates production units so as to maximise profits which are affected by access to labour and distribution and retail services. The State locates public and commercial services according to policy. It also plans and regulates the use of land and thus affects land prices.

The game is based on the diffusion of impacts as each decision is made and acquired land is used. The calculations involved in assessing these impacts as well as keeping accounts of each group is by a simple microcomputer program which helps to smoothen the running of the game.

SPACEREG - This is a large scale exercise in which several teams each playing SPACESET and SPACECIT interrelate in order to simulate the dynamics of a metropolitan area with a rural hinterland. (Theoretically, it is possible to also include a team or teams playing SHIKUM. We have not attempted this formidable project). The State is divided into Local and Central Governments and both Labour as well as Capital can be broken down into further factions e.g. housing classes, industrial and finance capital etc.

Operation and Resources

Each of these off-shoots of SPACE are designed to be played in a medium sized classroom of 12 to 24 participants. However, if two or more of these games at different levels are to be played simultaneously a larger room or inter-connected classrooms would be necessary. The resources are kept to a minimum; A large blackboard, moveable tables and chairs, a stand for hanging prepared charts, coloured cards representing transaction items, paper and pencils are standard. The introduction of microcomputers at the SPACECIT level is done unobtrusively. They are meant to be data banks and decision making aids and nothing more.

Each game has a briefing period when the rules and theory behind the game is presented, several rounds of play and a debriefing period. The minimum time required for each game is two hours, the maximum can extend to several days. Communication between players and groups is not structured. In fact, how information comes to be subsequently channelled, constrained and manipulated is an important element to be examined within the debriefing process when the resemblance of the game to reality is discussed and analysed.

Evaluation

The evaluation of games is a thorny question. Like puddings, they should be assessed in their playing (eating). The games described above have been played over several semesters. Students are required to prepare reports on the discrepancies between game rules and structure vis-a-vis reality for their term papers. Based on these reports the games have been modified and tested. In this sense it can be said that they are undergoing continual evaluation and up-dating. In several ISAGA Conferences, various levels of the SPACE game have been played and received valuable comments which have led to considerable improvements. Many participants commented favourably on the property of these games of quickly and clearly presenting the major issues of a very complex problem with economical means, while retaining the intriguing qualities of a game.

SUMMARY

What we have is a basic framegame SPACE which can be transformed, while keeping its basic structure, into various levels of the spatial problematic under capitalist urban development. The games can be played individually or in conjunction, each instance reproducing the contradictions at that level. The structure of these games has been devised so as to provide some measure of solution to the intrinsic game/simulation dilemmas described above. The players or decision makers have thus been chosen so that they are pared down to the bare minimum while retaining theoretical justification. The rules governing the interactions are neither based on pure competition nor on assumptions of fairness or equal power. The built-in contradictions provide the unexpected challenges and motivations for the playing. These games have been developed and tested at the Faculty of Architecture at the Technion. Preliminary results indicate that they are both interesting and provocative, hence may hold considerable promise as teaching and research tools.

BIBLIOGRAPHY

Castells, M. 1977. The Urban Question: A Marxist Approach. Edward Arnold.
Clark, G. and Dear, M. 1981. The State in Capitalism and the Capitalist State. (In Dear and Scott. eds. 1981).
Dear, M. and Scott, A. J. (eds) 1981. Urbanisation and Urban Planning in Capitalist Society. Methuen.
Harvey, D. 1973. Social Justice and the City. Edward Arnold.
Law-Yone, H. 1984. SPACE-A Marxist Urban Game. Presented at the ISAGA Conference at Elsinore, Denmark.
Lundgren, T. D. and Loar, R. M. 1976. CLUG - The spirit of Capitalism and the Success of a Simulation. SIMULATION AND GAMES, Vol. 9, No. 2.
Ollman, B. 1983. In Search of Critical Games. MONTHLY REVIEW, September.
Scott, A. J. and Roweis, S. T. 1977. Urban Planning in Theory and Practice: A Reappraisal. ENVIRONMENT AND PLANNING. 9:1097-119.

Analyse transculturelle comparative d'un système interactif d'aide à la décision (S.I.A.D.) dans un jeu de simulation

Patrick Beauchesne, Stéphane Huchet
Collège de Rosemont, Canada

Philippe Dumas, Martine Hardy
Université de Toulon, France

ABSTRACT: Les jeux de simulation de gestion sont couramment utilisés dans les établissements universitaires du premier cycle en France aussi bien qu'au Québec.

Afin de mettre les étudiants en condition de travailler comme dans les entreprises, il leur est fournie un Système d'Aide à la Décision (SIAD) pour jouer le jeu. L'article presente les réactions des étudiants telle qu'elles ressortent d'enquêtes menées en France et au Québec. Les résultats indiquent que les simulations et les SIAD sont vécus de facon identique des deux côtés de l'Atlantique.

KEYWORDS: Simulation, business, game decision, support system, electronic spreadsheet.

DESCRIPTION DU PROJET

Le présent projet débuta en 1984 et se fit dans le cadre des échanges franco-québécois. Dans un premier temps, les deux parties concernées (Toulon et Rosemont) firent un inventaire des jeux de simulation existants utilisés dans les différents I.U.T. de France et CEGEP du Québec. Un rapport fut remis suite à ce premier travail. Il ressortait que les jeux de simulation utilisés dans les deux pays ne comportaient pas de differences significatives quand elles n'étaient pas tout simplement les mêmes. Par contre, elles ne paraissaient pas complètes dans le sens où aucun des jeux de simulation répertoriés n'intégraient un système interactif d'aide à la décision.

En effet, le monde des affaires utilise des simulateurs depuis déjà plusieurs années, en se servant des logiciels commerciaux de type tableur, (Multiplan, Lotus 123, Symphony). Le but est de permettre à l'entreprise de simuler les effets d'une ou de plusieurs stratégies opérationnelles et évidemment, de choisir celle qui lui apparait la plus rentable. L'intérêt d'un tel système est donc évident même si l'univers client est limité. (par exemple: les stratégies concurrentes ne sont pas prises en considération). Il fut donc décidé que chaque partie devait trouver ou créer un jeu de simulation ainsi qu'un S.I.A.D. Les deux produits devaient aussi être complementaires en ce sens qu'ils devaient etre d'un niveau de difficulté différent pour permettre de couvrir tout l'ensemble du programme des techniques de commercialisation des I.U.T. francais et des techniques administratives des CEGEP québécois.

L'I.U.T. de Toulon arrêta son choix sur le jeu de simulation "FIMARK" et sur le S.I.A.D. "S.O.S. Fimark"; le Collège de Rosemont, sur le jeu de simulation "CERES" et sur le S.I.A.D. "SIMULE". On y planifia par la même occasion un questionnaire commun afin de comparer les résultats de l'utilisation d'une telle technique pédagogique. Une différence cependant est à noter: Toulon administra son questionnaire en trois vagues successive afin de permettre d'étudier l'évolution des opinions dans le temps alors que Rosemont l'administra d'un seul bloc.

Les Resultats

Les résultats de l'analyse des réponses à ces questionnaires sont regroupés dans les tableaux ci-apres. En général, il ressort que les jeux de simulations sont des moyens excellents d'acquérir des connaissances (87 et 93%). "Fimark" est beaucoup plus rattaché au marketing (81%) et à la gestion (70%) alors que "CERES" est plus un jeu de simulation où les fonctions production (72%) et gestion (77%) semblent les plus importantes.

Les notions de comptabilité sont moins importantes (52% et 44%) alors que l'informatique y joue un rôle marginal (7 et 17%).

Si les étudiants francais voient dans Fimark beaucoup plus un devoir (32%), les étudiants québécois percoivent "CERES" plutôt comme un jeu (56%). A noter que 46% des étudiants francais situent "Fimark" entre les deux (jeu ou devoir).

En ce qui a trait au temps consacré aux différentes composantes décisionnelles du jeu, on note des différences marquées, ce qui est normal compte tenu des niveaux de difficultés différents des jeux utilisés. On note dans CERES que 87% des étudiants consacrent plus d'une heure aux calculs prévisionnels (44% plus de deux heures) comparativement à 53% pour FIMARK et aucun plus de deux heures. Par contre, dans les domaines comme la stratégie, la finance et les hypothèses de marché, les résultats sont similaires.

Les dépenses occasionnées par la recherche d'informations marquent aussi quelques différences. Si les pourcentages d'étudiants considérant cette dépense comme chère et non justifiés sont à peu près semblables, il faut noter que 32% des etudiants québécois n'ont rien dépensé à ce chapitre comparativement à 0% pour les étudiants francais.

Au niveau du besoin de structuration à l'intérieur des équipes, on dégage deux pôles distincts, soit aucun besoin (44 et 40%) ou un besoin très fort (22 et 34%) et ce, aussi bien au Québec qu'en France. Enfin, la majorité des étudiants ont considéré l'ambiance intragroupe comme étant excellente (90 et 84%) et l'ambulance intergroupe comme étant bonne (44 et 64%).

En terminant, une analyse des tableaux croisés faite au Québec seulement a permis de noter que l'utilisation d'un S.I.A.D. a permis de diminuer sensiblement le temps consacré à l'élaboration des stratégies, les calculs financiers et les hypothèses de marché. par contre aucune différence n'est notée pour ce qui a trait aux calculs prévisionnels ce qui semble assez contradictoire.

CONCLUSION

Bien que les cultures informatique et gestionnaire soient différentes des deux côtés de l'Atlantique, on note sur des populations d'étudiants âgés d'une vingtaine d'années des réactions étonnemment voisines par rapport aux simulations d'entreprise. L'échantillon ni les données disponibles ne permettent pas actuellement une interprêtation de ces résultats. On peut en conclure néanmoins que les simulations sont intégrées à nos cultures et vécues de facon apparemment identique en France et au Québec.

TABLEAU 1: Resultats Comparatifs France Québec des Evaluations des Etudiants

CRITERE			France %	Quebec %
1) Acquisition des connaissances			87	93
2) Rattaché à	comptabilité		52	44
	marketing		81	50
	gestion		70	77
	informatique		9	17
	production		-	72
3) Perception				
	jeu		21	56
	devoir		32	44
	entre les deux		46	
4) Temps consacré				
	calcul prévisionnels	> 1 h	53	87
		> 2 h	0	44
	stratégie	> 1 h	76	67
	finance	> 1 h	84	75
	marché	> 1 h	84	81
5) Depenses - Etudes de marché				
	aucun montant		0	32
	cher		64	61
	non justifié		33	66
6) Structuration des groupes				
	aucun besoin		44	40
	très fort		22	34
7) Ambiance				
	intra-groupe		90	84
	inter-groupe		44	64

TABLEAU 2 Temps Passe Pour les Divers Elements de la Prise de Decision Avec ou SIAD (Quebec seulement)

Elements de decision	Utilisateur SIAD %	Non utilisateur SIAD %
1) Calculs prévisionnels		
> une heure	84	86
< une heure	0	0
2) Stratégie		
> une heure	33	66
< une heure	67	34
3) Calculs financiers		
> une heure	16	50
< une heure	84	50
4) Hypotheses de marché		
> une heure	17	40
< une heure	83	60